RATTLESNAKE VENOMS
Their Actions and Treatment

edited by ANTHONY T. TU

Department of Biochemistry
Colorado State University
Fort Collins, Colorado

MARCEL DEKKER, INC.   New York • Basel

Library of Congress Cataloging in Publication Data
Main entry under title:

Rattlesnake venoms, their actions and treatment.

  Includes indexes.
  1. Snake venom—Toxicology. 2. Snake venom—
Physiological effect. 3. Rattlesnakes. I. Tu,
Anthony T., [date]
RA1242.R35R37   615.9'42   82-1376
ISBN 0-8247-1691-4   AACR2

COPYRIGHT © 1982 by MARCEL DEKKER, INC. ALL RIGHTS RESERVED

Neither this book nor any part may be reproduced or transmitted in any form or by any means, electronic or mechanical, including photocopying, microfilming, and recording, or by any information storage and retrieval system, without permission in writing from the publisher.

MARCEL DEKKER, INC.
270 Madison Avenue, New York, New York 10016

Current printing (last digit):
10 9 8 7 6 5 4 3 2 1

PRINTED IN THE UNITED STATES OF AMERICA

VOID

Library of
Davidson College

RATTLESNAKE VENOMS

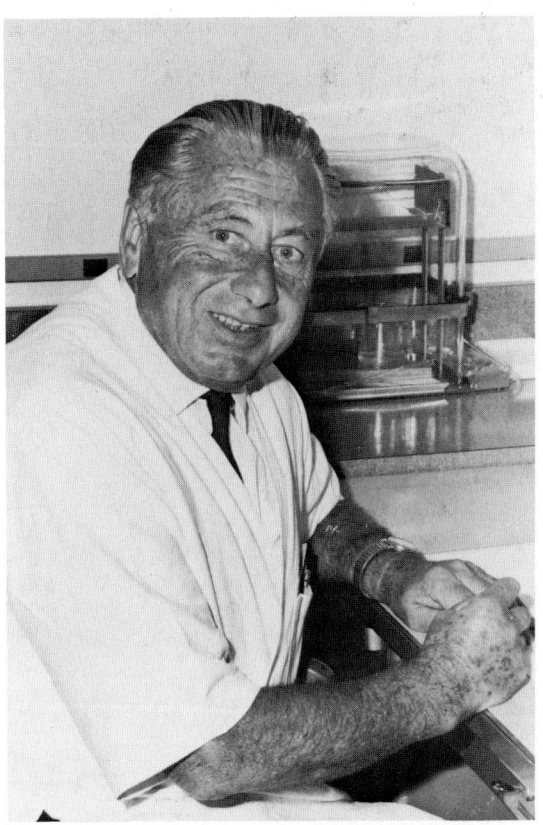
This book is dedicated to Dr. Karl H. Slotta, who pioneered modern research in the field of snake venoms by first using a fractionated component.

Foreword

In 1935, the government of the State of São Paulo, Brazil offered me the directorship of the newly established Chemical Institute in its world-famous snake and venom Instituto Butantan. Strangely enough, I was not asked to work on snake venoms, but rather to continue my research of sex hormones and also to enter a new field: the chemistry of coffee. At that time, Brazil was laboring under a surplus of coffee and was therefore looking for ways to use it industrially. However, my prime interest was to do research on snake venom, also in view of the abundant raw material available in Butantan, which received hundreds of snakes daily, shipped from all parts of Brazil. I started this research finally in early 1937 and came to the conclusion that rattlesnake venom might be a protein. In the fall of 1937 my brother-in-law, Heinz Fraenkel-Conrat, arrived in São Paulo for a short family visit, fresh from the Bergman laboratory in New York and thus an expert in proteins. Luckily, at that time I had received an outside grant toward a salary for an assistant for a year, so I suggested to Heinz that he stay in São Paulo and tackle the venom research with me. In 1938 we crystallized what we called "crotoxin" and determined its molecular structure. In a way, this discovery may have provided the impetus for world-wide research on animal venoms, the founding of the International Society on Toxinology in 1962 and its publication *Toxicon*, and all the great achievements up to the present time: research on snake venom enzymes, neurotoxins, cardiotoxins, phospholipase A, L-amino-acid oxidase, and hemolysis and blood coagulation and the search for the pharmacologic effects of all animal venoms.

The purpose of *Rattlesnake Venoms: Their Actions and Treatment* is to provide a survey of such important scientific research. Its publication definitely fills a void, and I am sure it will be widely read and thoroughly enjoyed.

Karl H. Slotta

Preface

Rattlesnakes are indigenous to the New World and are distributed throughout North, Central, and South America, possessing a unique morphological characteristic: a rattle at the end of the tail. The venoms of rattlesnakes, as compared with those of cobras, kraits, and sea snakes, are not well characterized. However, this situation is changing with the isolation of more venom components and the intensive investigation of their chemical and pharmacological properties. Since much of the information about rattlesnake venoms is widely scattered in different journals, this book attempts to assemble a great deal of this information as well as to review the various properties and actions of rattlesnake venoms systematically.

The degree of snakebite poisoning depends on many factors. Among them, toxicity and the amount of venom injected are most important. James L. Glenn and Richard C. Straight summarize these factors in Chapter 1. They also discuss and illustrate the major species and subspecies of rattlesnakes. Barbara J. Hawgood conducted rattlesnake venom research in Brazil and is presently a world-renowned physiologist on snake venom actions. In Chapter 2 she discusses the physiological and pharmacological effects of rattlesnake venoms. Charlotte L. Ownby is active in studying the pathological effects of rattlesnake venoms, particularly at the ultrastructural level, and discusses this in Chapter 3. Robert A. Hendon has been very active in research on crotoxin, and Allan L. Bieber is working on Mojave toxin. In Chapter 4 they cover presynaptic toxins from rattlesnake venoms. The chemistry of rattlesnake venom is discussed in Chapter 5 by Anthony T. Tu.

Snake treatment is an important yet variable and controversial medical practice. For the clinical aspects, two physicians experienced in the medical management of snakebite envenomation in the United States, Robert E. Arnold and Dr. Thomas Graham Glass, Jr., were invited to describe their experiences (Chapters 6 and 7). Although their views occasionally do not agree, it is important to hear opinions from both sides of this controversial issue.

I especially would like to thank Dr. Karl H. Slotta for the Foreword. As everyone in the venom research field knows, Dr. Slotta, together with Heinz Fraenkel-Conrat, started rattlesnake venom research using isolated components rather than crude venoms.

I hope this book will be a useful and interesting source of information for biochemists, toxicologists, herpetologists, physicians, and anyone else interested in rattlesnake venoms.

Anthony T. Tu

Contributors

Robert E. Arnold, M.D. Professor of Surgery, University of Louisville School of Medicine, Louisville, Kentucky

Allan L. Bieber, Ph.D. Professor, Department of Chemistry, Arizona State University, Tempe, Arizona

Thomas Graham Glass, Jr., M.D., F.A.C.S. Clinical Professor of Surgery, Department of General Surgery, The University of Texas Health Science Center, San Antonio, Texas

James L. Glenn Research Serpentologist, Venom Research Laboratory, Veterans Administration Medical Center; and Consultant Curator of Reptiles, Hogle Zoological Gardens, Salt Lake City, Utah

Barbara J. Hawgood, Ph.D. Lecturer, Department of Physiology, Queen Elizabeth College, University of London, London, England

Robert A. Hendon, Ph.D. Research Associate, Department of Biochemistry, Colorado State University, Fort Collins, Colorado

Charlotte L. Ownby, Ph.D. Associate Professor and Director, Electron Microscopy Laboratory, Department of Physiological Sciences, College of Veterinary Medicine, Oklahoma State University, Stillwater, Oklahoma

Richard C. Straight, Ph.D. Director, Venom Research Laboratory, Veterans Administration Medical Center, Salt Lake City, Utah

Anthony T. Tu, Ph.D. Professor, Department of Biochemistry, Colorado State University, Fort Collins, Colorado

Contents

Foreword *Karl H. Slotta*		iii
Preface		v
Contributors		vii

Part I Action of Venoms

1	The Rattlesnakes and Their Venom Yield and Lethal Toxicity *James L. Glenn and Richard C. Straight*	3
2	Physiological and Pharmacological Effects of Rattlesnake Venoms *Barbara J. Hawgood*	121
3	Pathology of Rattlesnake Envenomation *Charlotte L. Ownby*	163
4	Presynaptic Toxins from Rattlesnake Venoms *Robert A. Hendon and Allan L. Bieber*	211
5	Chemistry of Rattlesnake Venoms *Anthony T. Tu*	247

Part II Clinical Aspects

6	Treatment of Rattlesnake Bites *Robert E. Arnold*	315
7	Management of the Western Diamondback Rattlesnake Bite *Thomas Graham Glass, Jr.*	339
Appendix: Common Names of Rattlesnakes		361
Author Index		365
Subject Index		379

RATTLESNAKE VENOMS

part I
Action of Venoms

1
The Rattlesnakes and Their Venom Yield and Lethal Toxicity

JAMES L. GLENN AND RICHARD C. STRAIGHT
Veteran's Administration Medical Center, Salt Lake City, Utah

The Rattlesnakes 3
Introduction • Species and distribution

Venom Yield and Lethal Toxicity 57
Introduction • Venom Apparatus • Factors Affecting Venom Yield • Factors Affecting Venom Lethal Toxicity • Venom Yield and Lethal Toxicity by Species/Subspecies • Comparative Lethal Capacity of Rattlesnakes

References 111

THE RATTLESNAKES

Introduction

Rattlesnakes, genera *Sistrurus* and *Crotalus,* are venomous snakes unique to the New World. Our knowledge of them has been increased by innumerable observers from the time of the earliest explorers to the present. The state of that knowledge has been reviewed in detail by Howard K. Gloyd (1940, reprinted in 1978) and Lawrence M. Klauber (1956, partially revised and reprinted in 1972). Both of these monographs, Gloyd's *The Rattlesnakes Genera Sistrurus and Crotalus* and Klauber's *Rattlesnakes: Their Habits, Life Histories and Influence on Mankind,* compliment each other. Combined they represent a most thorough, scholarly description of rattlesnake anatomy, taxonomy, phylogeny, distribution, and natural history. Although the pace of accumulation of new information and understanding of the rattlesnakes has slowed, there is still much that is not yet known and understood. Even in the area of taxonomy, where much of our knowledge is well established and stable, there are still gaps, especially with regard to populations of South American and Mexican rattlesnakes.

Recent investigations into the Mexican rattlesnake fauna have enriched our knowledge of these poorly understood rattlesnakes. A recent publication *The Natural History of Mexican Rattlesnakes* (Armstrong and Murphy 1979), is an excellent source of behavioral and habitat information. It is to Klauber (1972) that we refer the reader for complete synonymies and taxonomic descriptive analysis of the 31 recognized species and the 70 recognized subspecies. Since that authoritative report, six newly described subspecies have withstood critical examination and are recognized as follows:

1. *Crotalus ruber lorenzoensis* (Radcliffe and Maslin, 1975).
2. *Crotlaus willardi obscurus* (Harris, 1974). This subspecies was conventionally described by Harris and Simmons (1976); however, the name was occupied by Harris (1974). (See Smith et al., 1975.)
3. *Sistrurus ravus brunneus* (Harris and Simmons, 1978; as redescribed by Campbell and Armstrong, 1979).
4. *Sistrurus ravus exiguus* (Campbell and Armstrong, 1979).
5. *Crotalus triseriatus armstrongi* (Campbell, 1979).
6. *Crotalus lepidus maculosus* (Tanner, Dixon, and Harris, 1972).

The six subspecies nominated by Hoge (1965) of South America's races of *Crotalus durissus* [(*terrificus*)] [see *Crotalus durissus* (neotropical rattlesnake)] are in need of redescriptive, comparative analysis in accordance with the rules of the *International Commission on Zoological Nomenclature* (see Klauber, 1972: pp. 35-36) and are therefore not included in Table 1. The same problem of taxonomic status applies to *Crotalus durissus trigonicus*, *Crotalus ruber monserratensis* and *Crotalus triseriatus quadrangularis* (Harris and Simmons, 1978), and *Crotalus lepidus castaneus* (Zertuche and Trevino, 1978). We also suggest that further analysis is required concerning Harris and Simmons' (1978) elevation of *Crotalus triseriatus aquilus* to species status (see McCranie and Wilson, 1979). We have presented briefly in this chapter some common characteristics of rattlesnakes, the recognized species, and their distribution and have reviewed the yield and lethal toxicity of their venoms.

Species and Distribution

The rattlesnakes are members of the family of venomous snakes known as the Crotalidae (or, as preferred by some taxonomists, Crotalinae—a subfamily of

Table 1 Recognized Rattlesnake Species and Subspecies, Original Describer, Year, and Common Name

Genus—*Crotalus,* Linnaeus, 1758

C. adamanteus Beauvois, 1799, Eastern diamondback rattlesnake

C. atrox Baird and Girard, 1853, Western diamondback rattlesnake

C. basiliscus basiliscus Cope, 1864, Mexican west coast rattlesnake

C. basiliscus oaxacus Gloyd, 1948, Oaxacan rattlesnake

C. catalinensis Cliff, 1954, Santa Catalina Island rattlesnake

C. cerastes cerastes Hallowell, 1854, Mojave Desert sidewinder

C. cerastes cercobombus Savage and Cliff, 1953, Sonoran Desert sidewinder

C. cerastes laterorepens Klauber, 1944, Colorado Desert sidewinder

C. durissus durissus Linnaeus, 1758, Central American rattlesnake

C. durissus culminatus Klauber, 1952, Northwestern neotropical rattlesnake

C. durissus terrificus Laurenti, 1768, South American rattlesnake

C. durissus totonacus Gloyd and Kauffield, 1940, Totonacan rattlesnake

C. durissus tzabcan Klauber, 1952, Yucatan neotropical rattlesnake

C. enyo enyo Cope, 1861, lower California rattlesnake

C. enyo cerralvensis Cliff, 1954, Cerralvo Island rattlesnake

C. enyo furvus Lowe and Norris, 1954, Rosario rattlesnake

C. exsul Garman, 1883, Cedros Island diamond rattlesnake

C. horridus horridus Linnaeus, 1758, timber rattlesnake ⋅

C. horridus atricaudatus Latreille, 1802, canebrake rattlesnake ⋅

C. intermedius intermedius Troschel, 1865, Totalcan small-headed rattlesnake

C. intermedius gloydi Taylor, 1941, Oaxacan small-headed rattlesnake

C. intermedius omiltemanus Gunther, 1895, Omilteman small-headed rattlesnake

C. lannomi Tanner, 1966, Autlan rattlesnake

C. lepidus lepidus Kennicott, 1861, Mottled rock rattlesnake

C. lepidus klauberi Gloyd, 1936, banded rock rattlesnake

C. lepidus maculosus Tanner, Dixon, and Harris, 1972, Durango rock rattlesnake

C. lepidus morulus Klauber, 1952, Tamaulipan rock rattlesnake

C. mitchellii mitchellii Cope, 1861, San Lucan speckled rattlesnake

C. mitchellii angelensis Klauber, 1963, Angel de la Guarda Island speckled rattlesnake

C. mitchellii muertensis Klauber, 1949, El Muerto Island speckled rattlesnake

Table 1 (Continued)

C. mitchellii pyrrhus Cope, 1866, Southwestern speckled rattlesnake

C. mitchellii stephensi Klauber, 1930, panamint rattlesnake

C. molossus molossus Baird and Girard, 1853, Northern blacktail rattlesnake

C. molossus estebanensis Klauber, 1949, San Esteban Island rattlesnake

C. molossus nigrescens Gloyd, 1936, Mexican blacktail rattlesnake

C. polystictus Cope, 1865, Mexican lance-headed rattlesnake

C. pricei pricei Van Denburgh, 1895, Western twin-spotted rattlesnake

C. pricei miquihuanus Gloyd, 1940, Eastern twin-spotted rattlesnake

C. pusillus Klauber, 1952, Tancitaran dusky rattlesnake

C. ruber ruber Cope, 1892, red diamond rattlesnake

C. ruber lorenzoensis Radcliffe and Maslin, 1975, San Lorenzo Island diamond rattlesnake

C. ruber lucasensis Van Denburgh, 1920, San Lucan diamond rattlesnake

C. scutulatus scutulatus Kennicott, 1861, Mojave rattlesnake

C. scutulatus salvini Gunther, 1895, Huamantlan rattlesnake

C. stejnegeri Dunn, 1919, long-tailed rattlesnake

C. tigris Kennicott, 1859, tiger rattlesnake

C. tortugensis Van Denburgh and Slevin, 1921, Tortuga Island diamond rattlesnake

C. transversus Taylor, 1944, cross-banded mountain rattlesnake

C. triseriatus triseriatus Wagler, 1830, central plateau dusky rattlesnake

C. triseriatus aquilus Klauber, 1952, Queretaran dusky rattlesnake

C. triseriatus armstrongi Campbell, 1979, Armstrong's dusky rattlesnake

C. unicolor Van Lidth de Jeude, 1887, Aruba Island rattlesnake

C. vegrandis Klauber, 1941, Uracoan rattlesnake

C. viridis viridis Rafinesque, 1818, prairie rattlesnake

C. viridis abyssus Klauber, 1930, Grand Canyon rattlesnake

C. viridis caliginis Klauber, 1949, Coronado Island rattlesnake

C. viridis cerberus Coues, 1875, Arizona black rattlesnake

C. viridis concolor Woodbury, 1929, midget faded rattlesnake

C. viridis helleri Meek, 1905, southern Pacific rattlesnake

C. viridis lutosus Klauber, 1930, Great Basin rattlesnake

C. viridis nuntius Klauber, 1935, Hopi rattlesnake

C. viridis oreganus Holbrook, 1840, northern Pacific rattlesnake

Table 1 (Continued)

C. willardi willardi Meek, 1905, Arizona ridgenose rattlesnake

C. willardi amabilis Anderson, 1962, Del Nido ridgenose rattlesnake

C. willardi meridionalis Klauber, 1949, southern ridgenose rattlesnake

C. willardi obscurus Harris and Simmons, 1974, New Mexican ridgenose rattlesnake

C. willardi silus Klauber, 1949, west Chihuahua ridgenose rattlesnake

Genus *Sistrurus,* Garman, 1883

S. catenatus catenatus Rafinesque, 1818, Eastern massasauga

S. catenatus edwardsii Baird and Girard, 1853, desert massasauga

S. catenatus tergeminus Say, 1823, Western massasauga

S. miliarius miliarius Linnaeus, 1766, Carolina pygmy rattlesnake

S. miliarius barbouri Gloyd, 1935, Eastern pygmy rattlesnake

S. miliarius streckeri Gloyd, 1935, Western pygmy rattlesnake

S. ravus ravus Cope, 1865, Mexican pygmy rattlesnake

S. ravus brunneus Harris and Simmons, 1978, Oaxacan pygmy rattlesnake

S. ravus exiguus Campbell and Armstrong, 1979, Guerreran pygmy rattlesnake

Source: Adapted from Klauber (1956).

Viperidae), the pit vipers, because of their facial heat sensory pits; however, the rattle is their most unique feature. The rattle, when present, distinguishes rattlesnakes from all other snakes. However, rattleless forms do exist, and the rattle is inherently lacking in a significant number of adult specimens of *Crotalus catalinensis* (Klauber, 1972: p. 33) and *C. ruber lorenzoensis* (Radcliffe and Maslin, 1975), both of which are insular forms of the Gulf of California, Baja, Mexico. Other conditions may exist in any rattlesnake population which effectively eliminate the rattle as an identifying characteristic in individuals. Neonate rattlesnakes, lacking in functional rattle segments, may appear rattleless and are often practically soundless, especially the smaller forms. Furthermore, adults in several of the smaller taxa often produce practically inaudible sound despite numerously segmented rattles. This would suggest that the warning or protective function popularly ascribed to the rattle could not be applied to these groups of rattlesnakes. Congenital defects or injury to the matrix that forms the rattle may produce rattleless individuals and also confuse those unfamiliar with rattlesnakes. All other gross anatomical features such as retractable fangs, heat-sensory pits, blotched, spotted, or banded pattern, head shape, and general body conformation are shared by other New World pit vipers as well and have led to misidentification. Generally speaking, there is no single characteristic of rattlesnakes that can be used without qualification for identification purposes.

Morphologic aberrations in rattlesnakes have been reported by Gloyd (1935, 1958), Klauber (1956), and Nickerson and Mays (1968) among others. The aberrancies most noticeably appear as various degrees of pattern and color variations, including rare albinistic and melanistic individuals. Pattern aberrancies range from near patternless to longitudinally striped specimens (Figure 1). A more common occurrence is either the fusion of a few dorsal blotches (appearing partially striped) or a vertebral division of the middorsal blotches, simulating a twin-spot appearance. See Bechtel (1978) for further information on factors influencing color and pattern and their anomalies.

The two recognized genera of rattlesnakes, *Crotalus* and *Sistrurus,* presently comprise 28 and 3 species, respectively (Table 1). The main external characteristic separating the two genera is the organization or number of head (crown) scales (Figure 2a-d). *Sistrurus* typically exhibits 9 large scales or plates over the crown region, including the supraocular scales above the eyes, a condition generally considered primitive. Variation in the number of *Sistrurus* crown scales is relatively rare, although divided parietals are common in *Sistrurus ravus brunneus* and occasionally in *Sistrurus ravus ravus* (Campbell and Armstrong, 1979). Conversely, members of the genus *Crotalus* exhibit a wide variety of crown scale organization but consistently have smaller and more numerous scales in this region than the members of the genus *Sistrurus.*

Subspecific status has been given to many variant geographic, and often ecologic populations, and the exact number, although controversial among taxonomists, is nonetheless increasing. As more studies are carried out in Mexico, Central America, and South America, the total trinomial taxa (subspecies) of rattlesnakes likely will approach 90 in number in the near future.

Mainland Mexico is not only the approximate center of rattlesnake geographic distribution, but also has the greatest variety of rattlesnakes. There are 27 species and 54 subspecies that inhabit the Mexican mainland, Baja peninsula, and the islands of the Pacific and Gulf of California. The United States follows Mexico in rattlesnake diversity, with 15 species and 33 subspecies. Only 3 closely allied rattlesnake species are found in South American countries, and only 1 species inhabits the Central American countries of Guatemala, Honduras, Yucatan, Nicaragua, and Costa Rica.

The recognized rattlesnake species and subspecies are illustrated in Figures 3-78. The photographs are of live rattlesnakes, except for *C. lannomi*, and are arranged alphabetically by species in typical poses demonstrating head and body conformation and the variety of patterns found among the rattlesnakes. Also included are distribution maps (Figures 79-85) illustrating the general locality each taxon inhabits. The distribution information is mainly from Klauber (1956-72) and also the reports of Gloyd (1940), Cliff (1954), Hoge (1965), Hoge and Romano (1971), Conant (1975), Harris and Simmons (1977, 1978), Campbell and Armstrong (1979), and in a few instances, from our own field studies.

Figure 1 Aberrant (striped) *Crotalus viridis lutosus,* from Springdale, Utah. (Courtesy of John Lancaster, Superintendent of Zion National Park, Utah.)

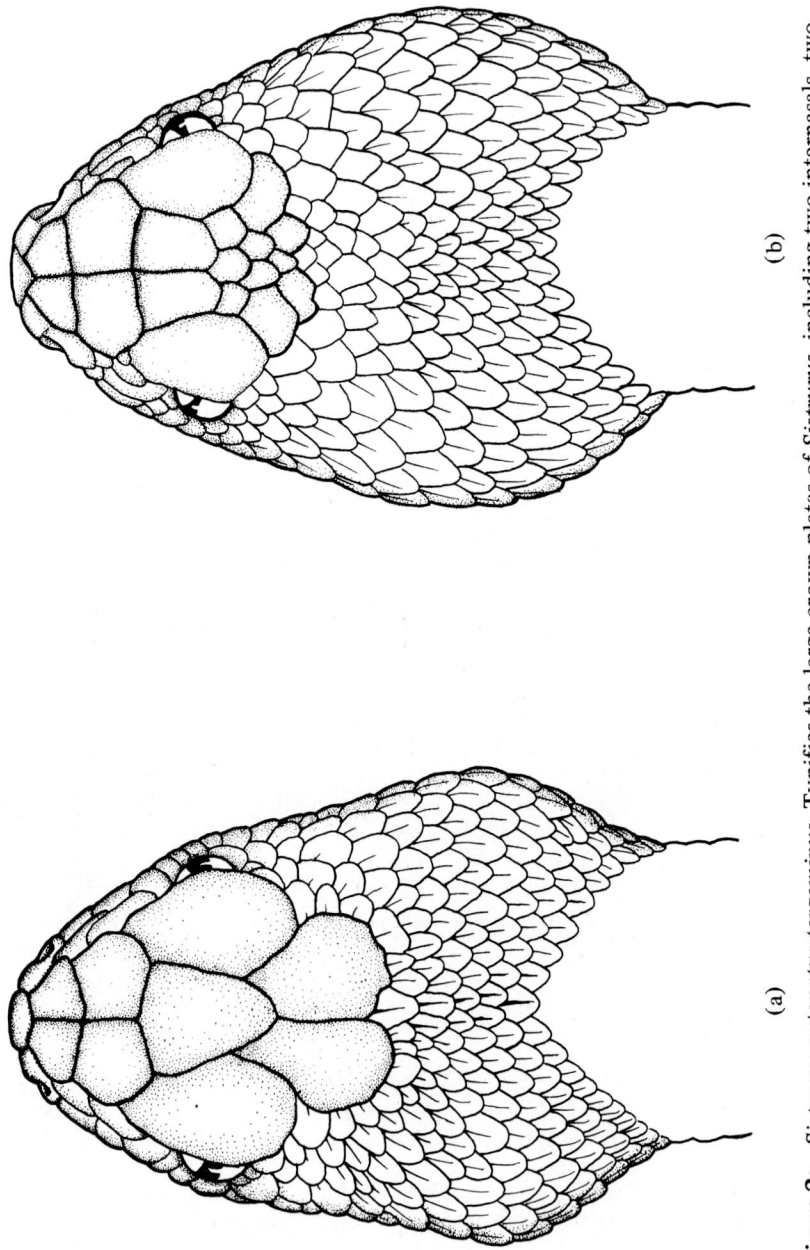

Figure 2a *Sistrurus catenatus tergeminus*. Typifies the large crown plates of *Sistrurus*, including two internasals, two prefontals, a single frontal plate, and two large parietals, in addition to the two supraocular scales.
Figure 2b *Crotalus durissus totonacus*. Typifies one of the more primitive crown scale situations in the *Crotalus* genus. The paired internasals and prefrontals are common characteristics of the *C. durissus* forms.

Venom Yield and Lethal Toxicity

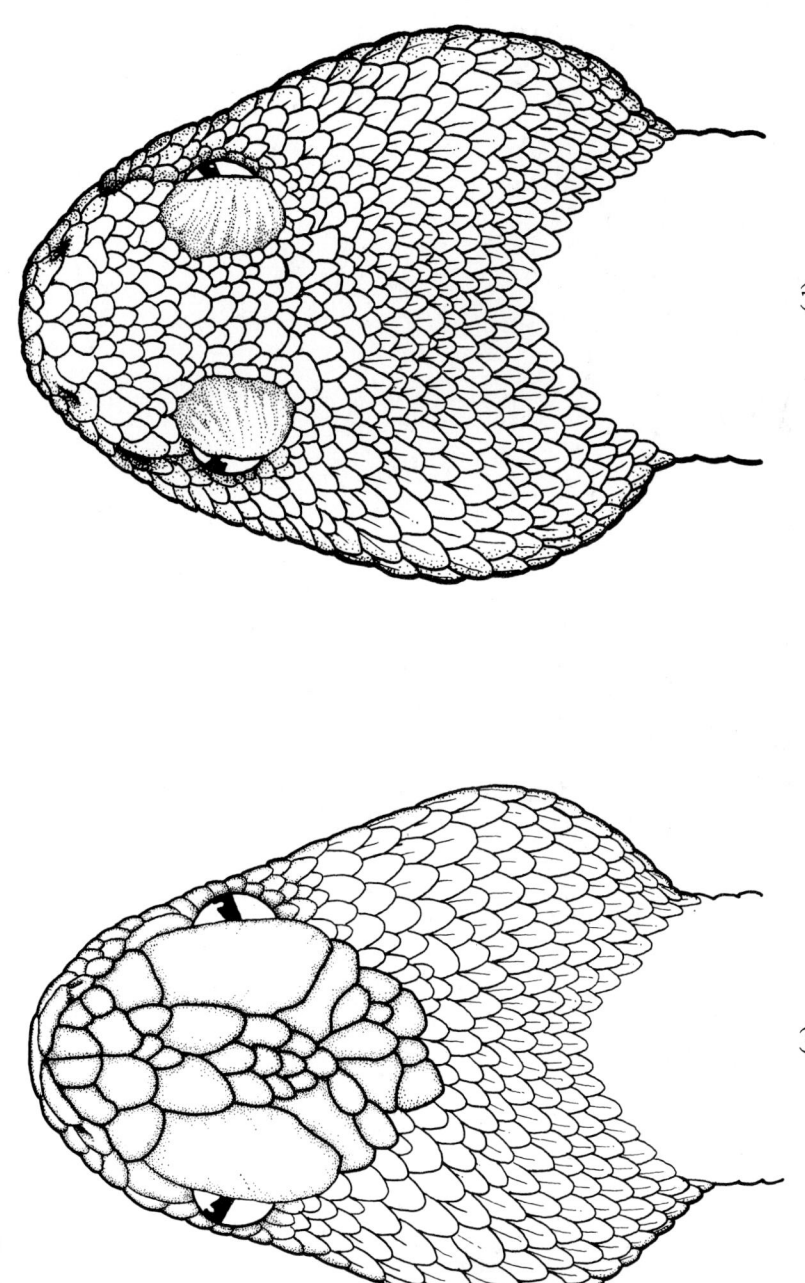

Figure 2c *Crotalus scutulatus scutulatus.* Example of a moderate subdivision of crown scales into smaller scales. In this particular *Crotalus* species, vestigal parietal scales are usually indicated by crescent-shaped scales in contact with supraoculars (posteriorly).
Figure 2d *Crotalus mitchellii pyrrhus.* Represents a *Crotalus* species in which the crown scalation has reached the extreme in exhibiting numerous small and irregularly arranged scales.

Figure 3 *Crotalus adamanteus,* Eastern diamondback rattlesnake. Specimen from Dade County, Florida. (Photo by Louis Porras.)

Figure 4 *Crotalus atrox,* Western diamondback rattlesnake. Specimen from east-central Oklahoma. (Photo by Kenneth Stockton.)

Figure 5 *Crotalus basiliscus basiliscus,* Mexican West Coast rattlesnake. Specimen from coastal Sinaloa, Mexico. (Photo by Louis Porras.)

Figure 6 *Crotalus basiliscus oaxacus,* Oaxacan rattlesnake. Specimen from highlands of central Oaxaca, Mexico. (Photo by Louis Porras.)

Figure 7 *Crotalus catalinensis*, Santa Catalina Island rattlesnake. Specimen from Santa Catalina Island, Gulf of California, Mexico. (Photo by Louis Porras.)

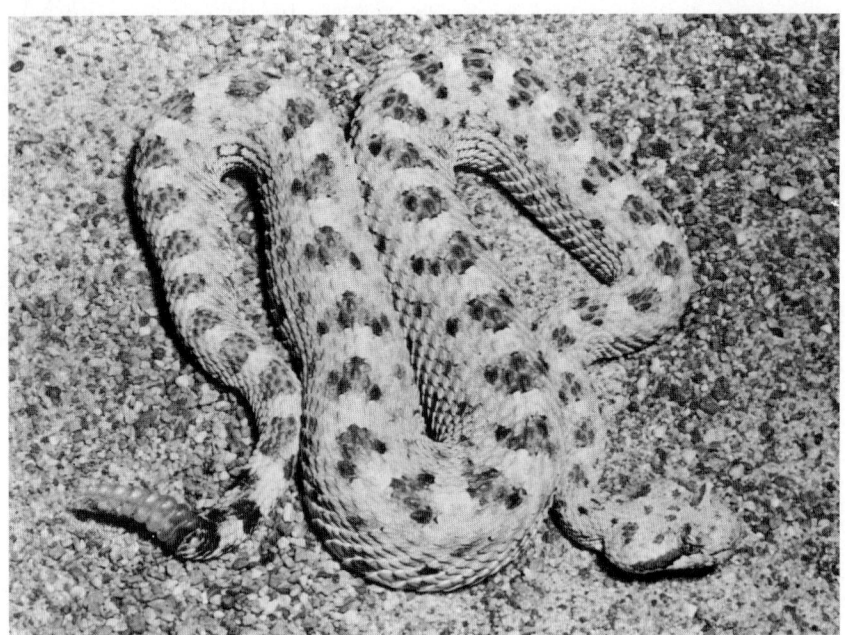

Figure 8 *Crotalus cerastes cerastes*, Mojave Desert sidewinder. Specimen from San Bernadino County, California. (Photo by John Tashjian.)

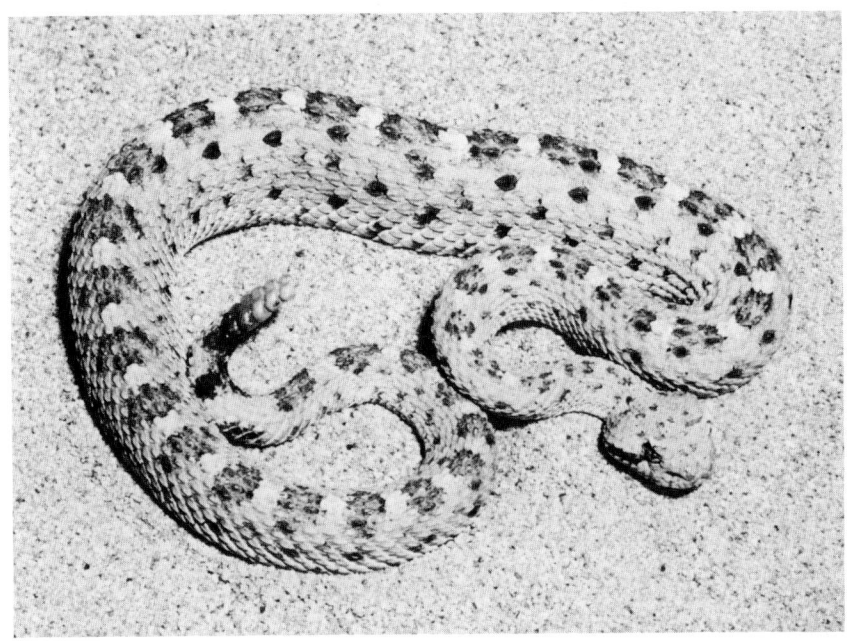

Figure 9 *Crotalus cerastes cercobombus,* Sonoran Desert sidewinder. Specimen from Pinal County, Arizona. (Photo by John Tashjian.)

Figure 10 *Crotalus cerastes laterorepens,* Colorado Desert sidewinder. Specimen from San Diego County, California. (Photo by Kenneth Stockton.)

Figure 11 *Crotalus durissus durissus*, Central America rattlesnake. Specimen from Guatemala. (Photo by Louis Porras.)

Figure 12 *Crotalus durissus culminatus*, Northwestern neotropical rattlesnake. Specimen from Michoacan, Mexico. (Photo by Louis Porras.)

Figure 13 *Crotalus durissus terrificus,* South American rattlesnake. Juvenile specimen from Venezuela. (Photo by Louis Porras.)

Figure 14 *Crotalus durissus totonacus,* Totonacan rattlesnake. Specimen from coastal Tamualipas, Mexico. (Photo by Louis Porras.)

Figure 15 *Crotalus durissus tzabcan,* Yucatan neotropical rattlesnake. Specimen from Yucatan Peninsula, Mexico. (Photo by Louis Porras.)

Figure 16 *Crotalus enyo enyo,* lower California rattlesnake. Specimen from Baja California Sur, Mexico. (Photo by Ed Cassano.)

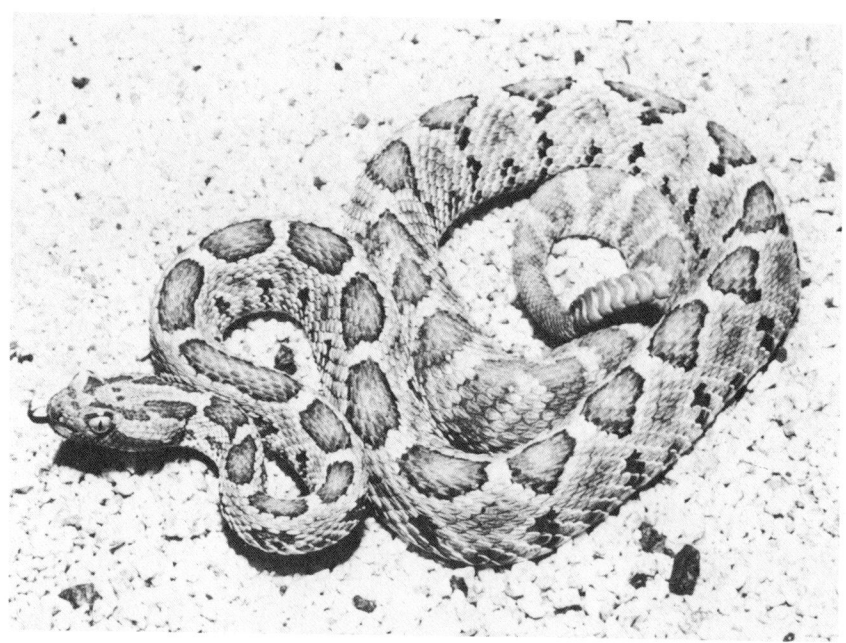

Figure 17 *Crotalus enyo cerralvensis,* Cerralvo Island rattlesnake. Specimen from Cerralvo Island, Gulf of California, Mexico. (Photo by John Tashjian.)

Figure 18 *Crotalus enyo furvus,* Rosario rattlesnake. Specimen from Baja California Norte, Mexico. (Photo by Louis Porras.)

Figure 19 *Crotalus exsul*, Cedros Island diamond rattlesnake. Specimen from Cedros Island, Pacific coast of Baja California Norte, Mexico. (Photo by Louis Porras.)

Figure 20 *Crotalus horridus horridus*, timber rattlesnake. Specimen from Scioto County, Ohio. (Photo by Louis Porras.)

Figure 21 *Crotalus horridus atricaudatus,* canebrake rattlesnake. Specimen from South Carolina. (Photo by Louis Porras.)

Figure 22 *Crotalus intermedius intermedius,* Totalcan small-headed rattlesnake. Specimen from highlands of west central Veracruz, Mexico. (Photo by Robert Simmons.)

Figure 23 *Crotalus intermedius gloydi*, Oaxacan small-headed rattlesnake. Specimen from highlands of central Oaxaca, Mexico. (Photo by Louis Porras.)

Figure 24 *Crotalus intermedius omiltemanus*, Omilteman small-headed rattlesnake. Specimen from highlands of central Oaxaca, Mexico. (Photo by John Tashjian.)

Figure 25 *Crotalus lannomi,* Autlan rattlesnake. Preserved specimen from Jalisco, Mexico. Brigham Young University Museum No. 23800. (Photo by Kenneth Stockton.)

Figure 26 *Crotalus lepidus lepidus,* mottled rock rattlesnake. Specimen from Valverde County, Texas. (Photo by Louis Porras.)

Figure 27 *Crotalus lepidus klauberi,* banded rock rattlesnake. Specimen from Magdelena Mountains, New Mexico. (Photo by Louis Porras.)

Figure 28 *Crotalus lepidus maculosus,* Durango rock rattlesnake. Specimen from extreme southwest Durango, Mexico. (Photo by Louis Porras.)

Figure 29 *Crotalus lepidus morulus,* Tamaulipan rock rattlesnake. Specimen from Gomez Farias Region, Tamualipas, Mexico. (Photo by Louis Porras.)

Figure 30 *Crotalus mitchellii mitchellii,* San Lucan speckled rattlesnake. Specimen from near San Ignacio, Baja California Sur, Mexico. (Photo by Kenneth Stockton.)

Figure 31 *Crotalus mitchellii angelensis,* Angel de La Guarda island speckled rattlesnake. Specimen from Angel de La Guarda Island, Gulf of California, Mexico. (Photo by John Tashjian.)

Figure 32 *Crotalus mitchellii muertensis,* El Muerto Island speckled rattlesnake. Specimen from El Muerto Island, Gulf of California, Mexico. (Photo by John Tashjian.)

Figure 33 *Crotalus mitchellii pyrrhus*, Southwestern speckled rattlesnake. Juvenile specimen from Washington County, Utah, U.S.A. (Photo by Louis Porras.)

Figure 34 *Crotalus mitchellii stephensi*, panamint rattlesnake. Specimen from California. (Photo by Louis Porras.)

Figure 35 *Crotalus molossus molossus,* Northern blacktail rattlesnake. Specimen from Organ Mountains, New Mexico. (Photo by Louis Porras.)

Figure 36 *Crotalus molossus estebanensis,* San Esteban Island rattlesnake. Specimen from San Esteban Island, Gulf of California, Mexico. (Photo by Louis Porras.)

Figure 37 *Crotalus molossus nigrescens,* Mexican blacktail rattlesnake. Specimen from Sierra Madre Occidental in western Zacatecas, Mexico. (Photo by Ed Cassano.)

Figure 38 *Crotalus polystictus,* Mexican lance-headed rattlesnake. Specimen from Aquascalientes, Mexico. (Photo by Louis Porras.)

Figure 39 *Crotalus pricei pricei,* Western twin-spotted rattlesnake. Specimen from highland plateau of southwest Durango, Mexico. (Photo by Ed Cassano.)

Figure 40 *Crotalus pricei miquihuanus,* Eastern twin-spotted rattlesnake. Specimen from Cerro Potosi, Nuevo Leon, Mexico. (Photo by James R. McCranie.)

Figure 41 *Crotalus pusillus,* Tancitaran dusky rattlesnake. Specimen from Sierra de Coalcoman, Michoacan, Mexico. (Photo by Louis Porras.)

Figure 42 *Crotalus ruber ruber,* red diamond rattlesnake. Specimen from Los Angeles County, California. (Photo by Kenneth Stockton.)

Figure 43 *Crotalus ruber lorenzoensis*, San Lorenzo Island diamond rattlesnake. Specimen from San Lorenzo Island, Gulf of California, Mexico. (Photo by John Tashjian.)

Figure 44 *Crotalus ruber lucasensis*, San Lucan diamond rattlesnake. Specimen from Baja California Sur, Mexico. (Photo by Louis Porras.)

Figure 45 *Crotalus scutulatus scutulatus,* Mojave rattlesnake. Specimen from western Texas. (Photo by Louis Porras.)

Figure 46 *Crotalus scutulatus salvini,* Huamantlan rattlesnake. Specimen from west central Veracruz, Mexico. (Photo by James R. McCranie.)

Figure 47 *Crotalus stejnegeri,* long-tailed rattlesnake. Specimen from southeastern Sinaloa, Mexico. (Photo by Louis Porras.)

Figure 48 *Crotalus tigris,* tiger rattlesnake. Specimen from Pima County, Arizona. (Photo by Louis Porras.)

Figure 49 *Crotalus tortugensis,* Tortuga Island diamond rattlesnake. Specimen from Tortuga Island, Gulf of California, Mexico. (Photo by Louis Porras.)

Figure 50 *Crotalus transversus,* cross-banded mountain rattlesnake. Specimen from Laguna Zempoala, Morelos, Mexico. (Photo by John Tashjian.)

Figure 51 *Crotalus triseriatus triseriatus*, central plateau dusky rattlesnake. Specimen from Laguna Zempoala, Morelos, Mexico. (Photo by Louis Porras.)

Figure 52 *Crotalus triseriatus aquilus*, Queretaran dusky rattlesnake. Specimen from San Luis Potosi, Mexico. (Photo by Louis Porras.)

Figure 53 *Crotalus triseriatus armstrongi,* Armstrong's dusky rattlesnake. Specimen from Jalisco, Mexico. (Photo by Louis Porras.)

Figure 54 *Crotalus unicolor,* Aruba Island rattlesnake. Specimen from Aruba Island, Venezuela. (Photo by Louis Porras.)

Figure 55 *Crotalus vegrandis,* Uracoan rattlesnake. Specimen from extreme northeastern Venezuela. (Photo by Louis Porras.)

Figure 56 *Crotalus viridis viridis,* prairie rattlesnake. Specimen from Park County, Wyoming. (Photo by Louis Porras.)

Figure 57 *Crotalus viridis abyssus,* Grand Canyon rattlesnake. Specimen from Marble Canyon, Coconino County, Arizona. (Photo by Louis Porras.)

Figure 58 *Crotalus viridis caliginis,* Coronado Island rattlesnake. Specimen from south Coronado Island, Pacific coast of Baja California Norte, Mexico. (Photo by John Tashjian.)

Figure 59 *Crotalus viridis cerberus,* Arizona black rattlesnake. Specimen from Pima County, Arizona. (Photo by Louis Porras.)

Figure 60 *Crotalus viridis concolor,* midget faded rattlesnake. Specimen from San Juan County, Utah. (Photo by Louis Porras.)

Figure 61 *Crotalus viridis helleri,* southern Pacific rattlesnake. Specimen from Baja California Sur, Mexico. (Photo by Louis Porras.)

Figure 62 *Crotalus viridis lutosus,* Great Basin rattlesnake. Specimen from Elko County, Nevada. (Photo by Louis Porras.)

Figure 63 *Crotalus viridis nuntius,* Hopi rattlesnake. Specimen from San Juan County, Utah. (Photo by Louis Porras.)

Figure 64 *Crotalus viridis oreganus,* northern Pacific rattlesnake. Specimen from southeastern Washington. (Photo by Louis Porras.)

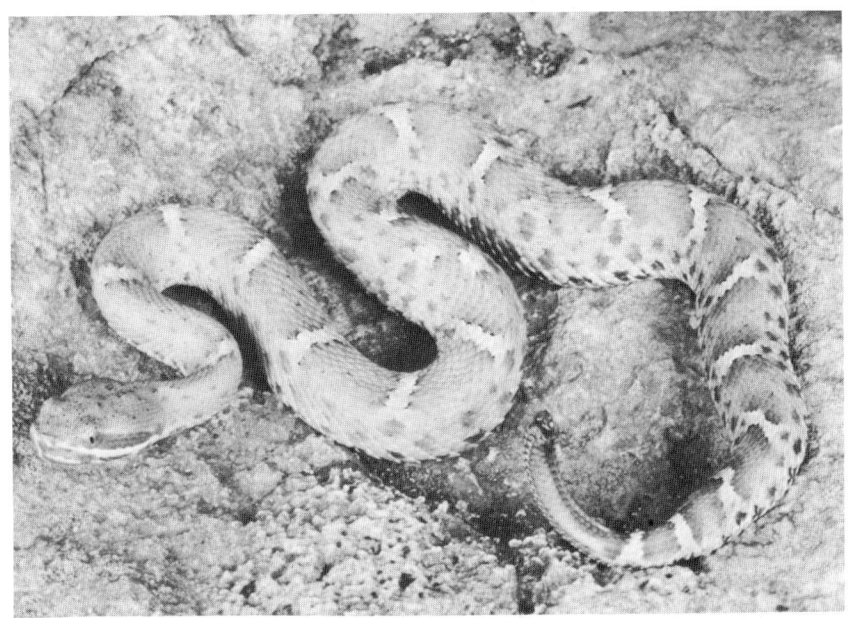

Figure 65 *Crotalus willardi willardi,* Arizona ridgenose rattlesnake. Specimen from Santa Rita Mountains, Pima County, Arizona. (Photo by Louis Porras.)

Figure 66 *Crotalus willardi amabilis,* Del Nido ridgenose rattlesnake. Specimen from Sierra del Nido, central Chihuahua, Mexico. (Photo by Louis Porras.)

Figure 67 *Crotalus willardi meridionalis*, Southern ridgenose rattlesnake. Specimen from highland plateau of southwest Durango, Mexico. (Photo by Ed Cassano.)

Figure 68 *Crotalus willardi obscurus*, New Mexican ridgenose rattlesnake. Specimen from Sierra San Luis, northwest Chihuahua, Mexico. (Photo by Louis Porras.)

Figure 69 *Crotalus willardi silus*, west Chihuahua ridgenose rattlesnake. Specimen from Arroyo Tinaja, northwest Chihuahua, Mexico. (Photo by Louis Porras.)

Figure 70 *Sistrurus catenatus catenatus*, Eastern massasauga. Specimen from Hillsdale County, Michigan. (Photo by Louis Porras.)

Figure 71 *Sistrurus catenatus edwardsii*, desert massasauga. Specimen from southern New Mexico. (Photo by John Tashjian.)

Figure 72 *Sistrurus catenatus tergeminus*, western massasauga. Specimen from Oklahoma. (Photo by Louis Porras.)

Figure 73 *Sistrurus miliarius miliarius,* Carolina pygmy rattlesnake. Specimen from Hyde County, North Carolina. (Photo by Louis Porras.)

Figure 74 *Sistrurus miliarius barbouri,* Eastern pygmy rattlesnake. Specimen from Dade County, Florida. (Photo by Louis Porras.)

Figure 75 *Sistrurus miliarius streckeri,* Western pygmy rattlesnake. Specimen from Polk County, Arkansas. (Photo by Louis Porras.)

Figure 76 *Sistrurus ravus ravus,* Mexican pygmy rattlesnake. Specimen from Huitzilac, Morelos, Mexico. (Photo by John Tashjian.)

Figure 77 *Sistrurus ravus brunneus,* Oaxacan pygmy rattlesnake. Specimen from Oaxaca, Mexico. (Photo by John Tashjian.)

Figure 78 *Sistrurus ravus exiguus,* Guerreran pygmy rattlesnake. Specimen from central Guerrero, Mexico. (Photo by John Tashjian.)

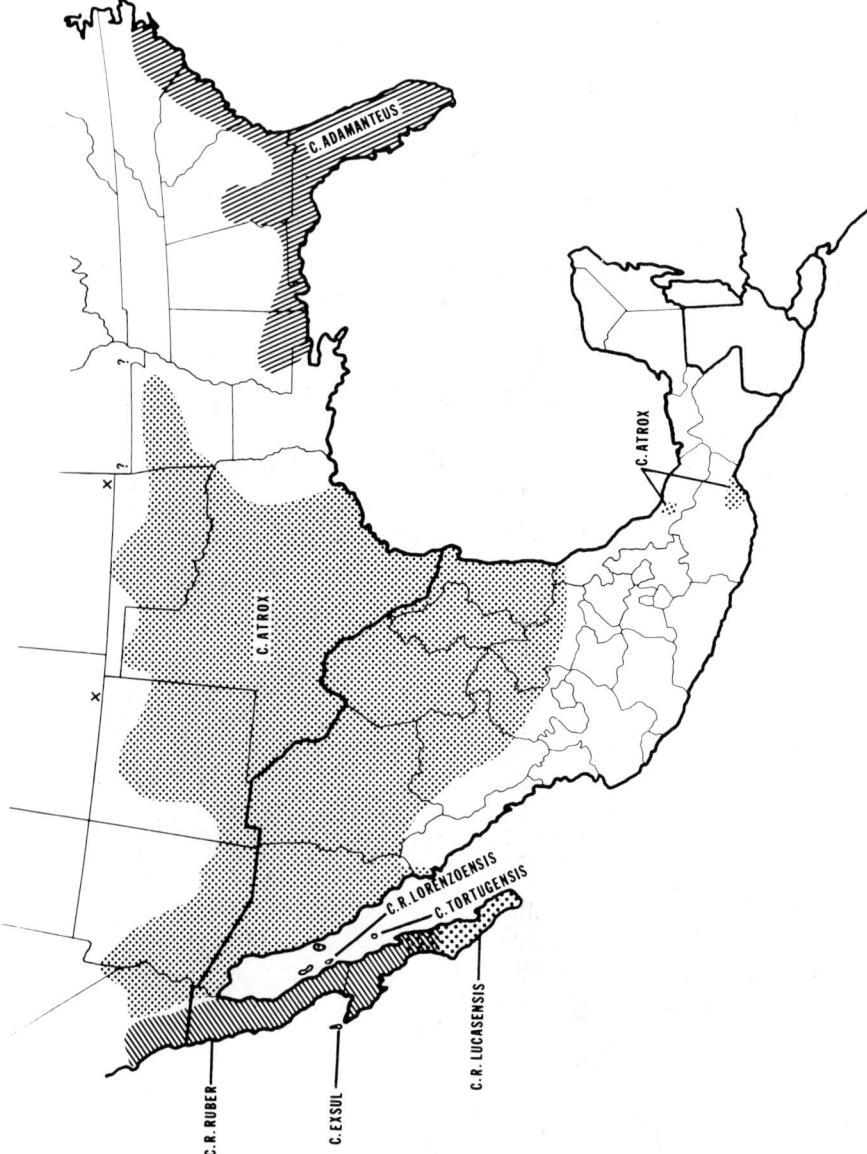

Figure 79 Distribution of the *Atrox* group: *Crotalus adamanteus, C. atrox, C. exsul, C. ruber, C. ruber lorenzoensis, C. ruber lucasensis,* and *C. tortugensis*. (Data from Gloyd, 1940; Klauber, 1972; Radcliffe and Maslin, 1975.)

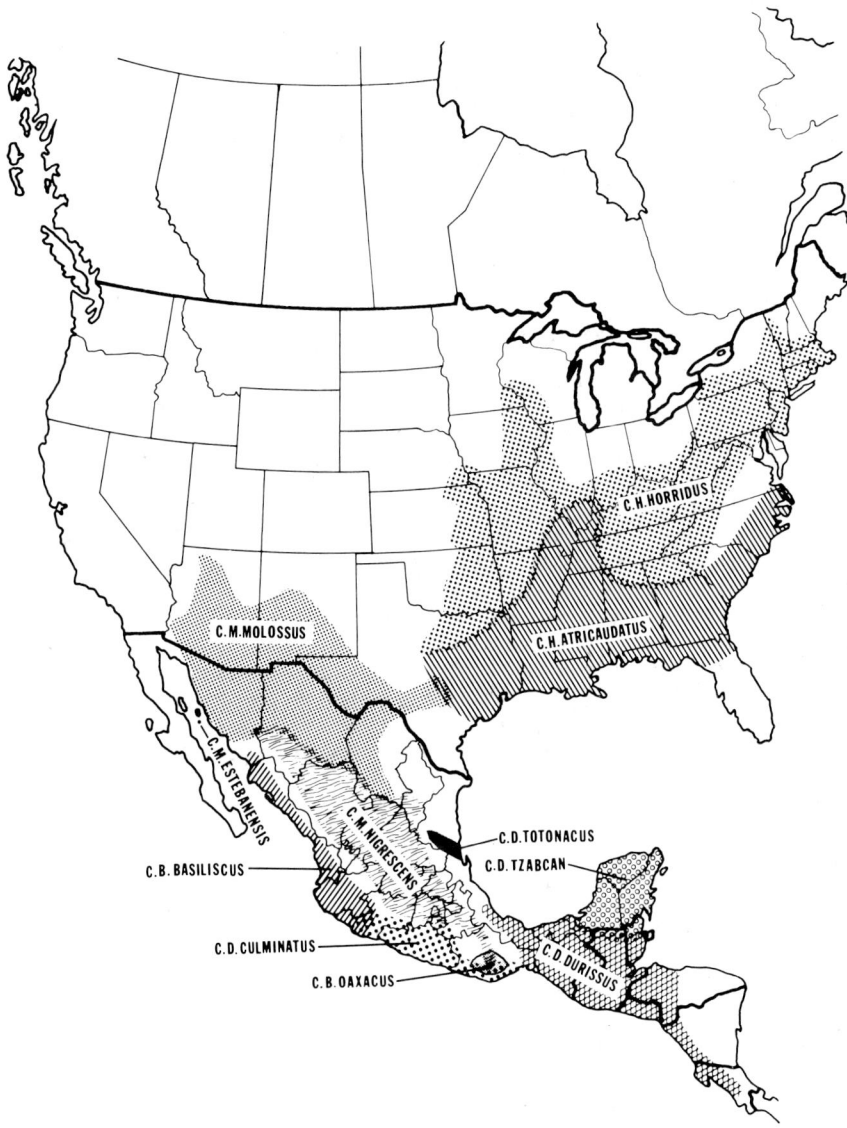

Figure 80 Distribution of North American representatives of the *Durissus* group. *Crotalus basiliscus basiliscus, C. basiliscus oaxacus, C. durissus durissus, C. durissus culminatus, C. durissus totonacus, C. durissus tzabcan, C. horridus horridus, C. horridus atricaudatus, C. molossus molossus, C. molossus estebanensis,* and *C. molossus nigrescens.* (Data from Gloyd, 1940; Klauber, 1972; Conant, 1975.)

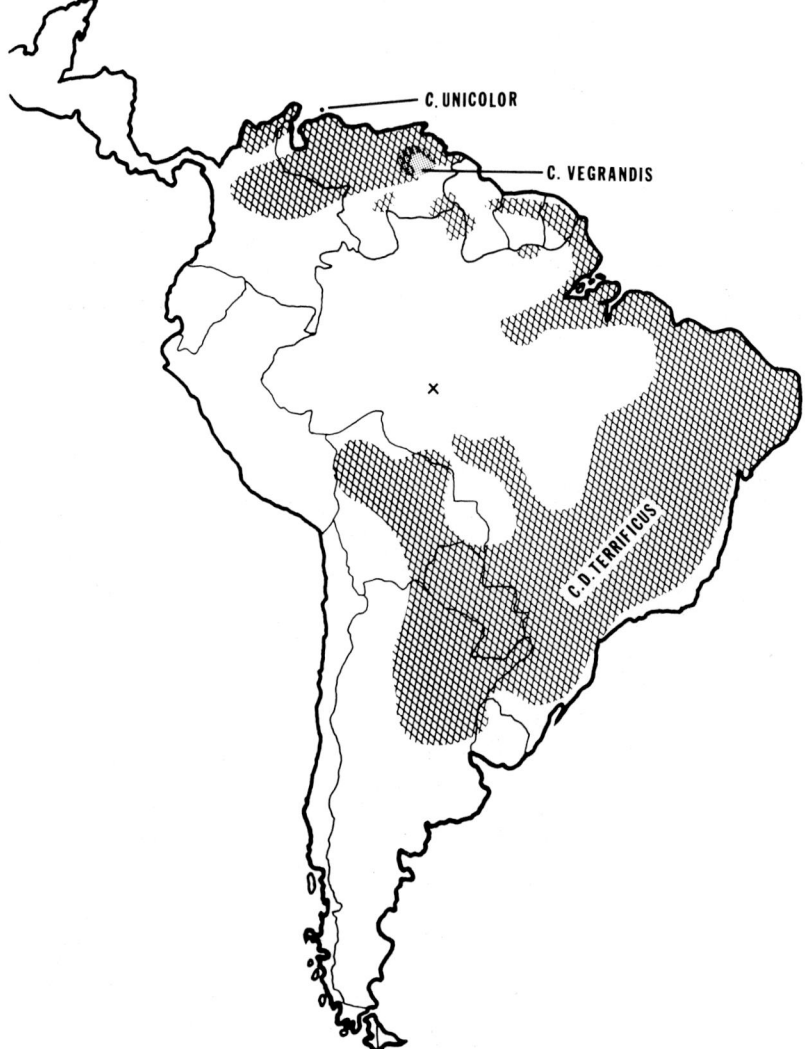

Figure 81 Distribution of the South American representatives of the *Durissus* group; *Crotalus durissus terrificus, C. unicolor,* and *C. vegrandis.* (Data from Hoge, 1965; Hoge and Romano, 1971; Klauber, 1972.)

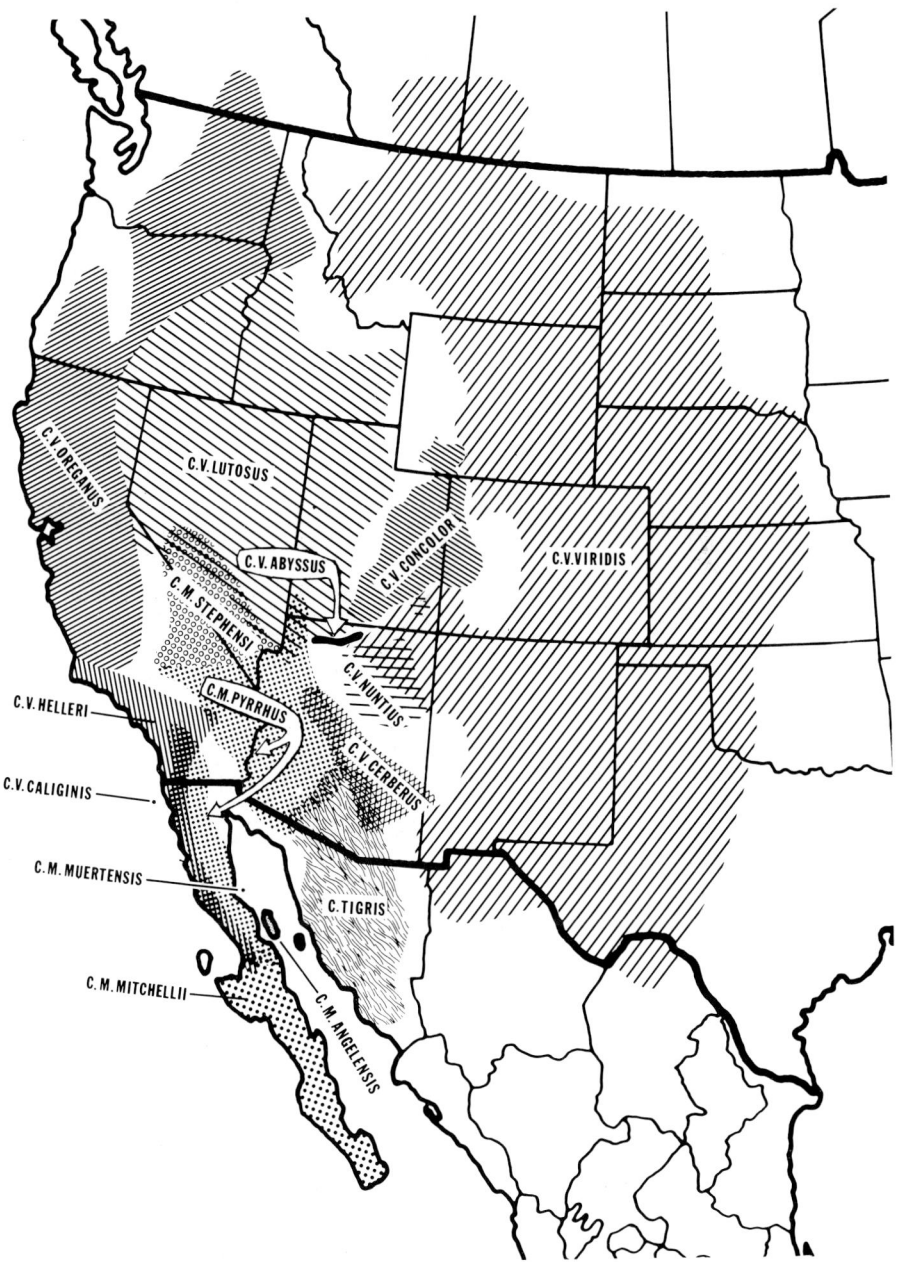

Figure 82 Distribution of the *Viridis* group; *Crotalus mitchellii mitchellii, C. mitchellii angelensis, C. mitchellii muertensis, C. mitchellii pyrrhus, C. mitchellii stephensi, C. tigris, C. viridis viridis, C. viridis abyssus, C. viridis caliginis, C. viridis cerberus, C. viridis concolor, C. viridis helleri, C. viridis lutosus, C. viridis nuntius,* and *C. viridis oreganus.* (Data from Gloyd, 1940; Cliff, 1954; Klauber, 1972; and the authors' field studies.)

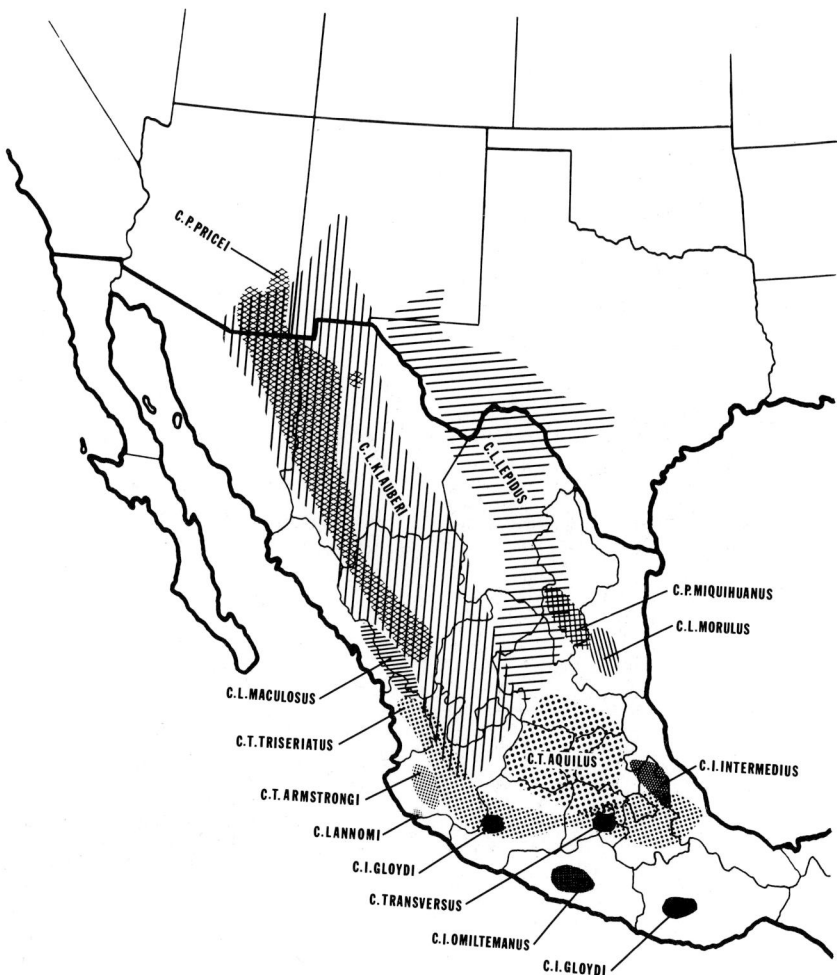

Figure 83 Distribution of *Crotalus intermedius intermedius, C. intermedius gloydi, C. intermedius omiltemanus, C. lannomi, C. lepidus lepidus. C. lepidus klauberi, C. lepidus maculosus, C. lepidus morulus, C. pricei pricei, C. pricei miquihuanus, C. transversus, C. triseriatus triseriatus, C. triseriatus aquilus,* and *C. triseriatus armstrongi.* (Data from Gloyd, 1940; Tanner, 1966; Klauber, 1972; Tanner et al., 1972; Campbell and Armstrong, 1979; Campbell, 1979.)

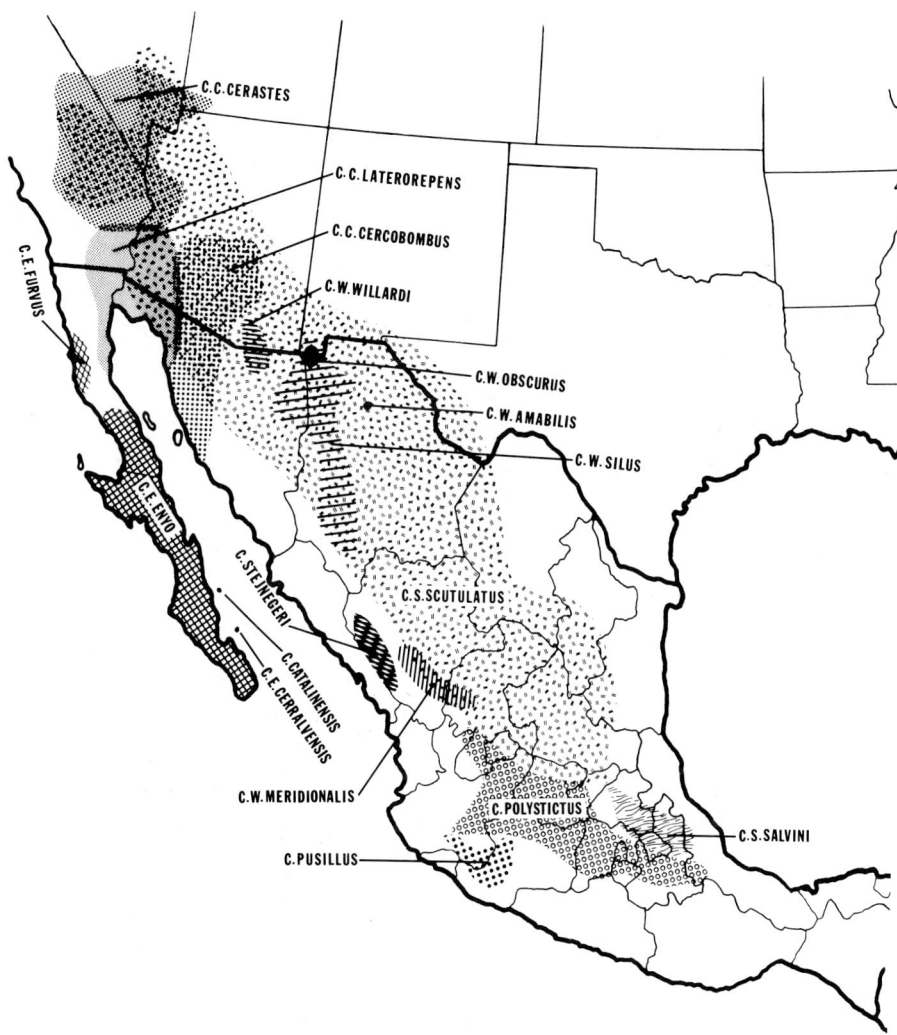

Figure 84 Distribution of *Crotalus catalinensis, C. cerastes cerastes, C. cerastes cercobombus, C. cerastes laterorepens, C. enyo enyo, C. enyo cerralvensis, C. enyo furvus, C. polystictus, C. pusillus, C. scutulatus scutulatus, C. scutulatus salvini, C. stejnegeri, C. willardi willardi, C. willardi amabilis, C. willardi meridionalis, C. willardi obscurus,* and *C. willardi silus*. (Data from Gloyd, 1940; Cliff, 1954; Klauber, 1972; Harris and Simmons, 1976.)

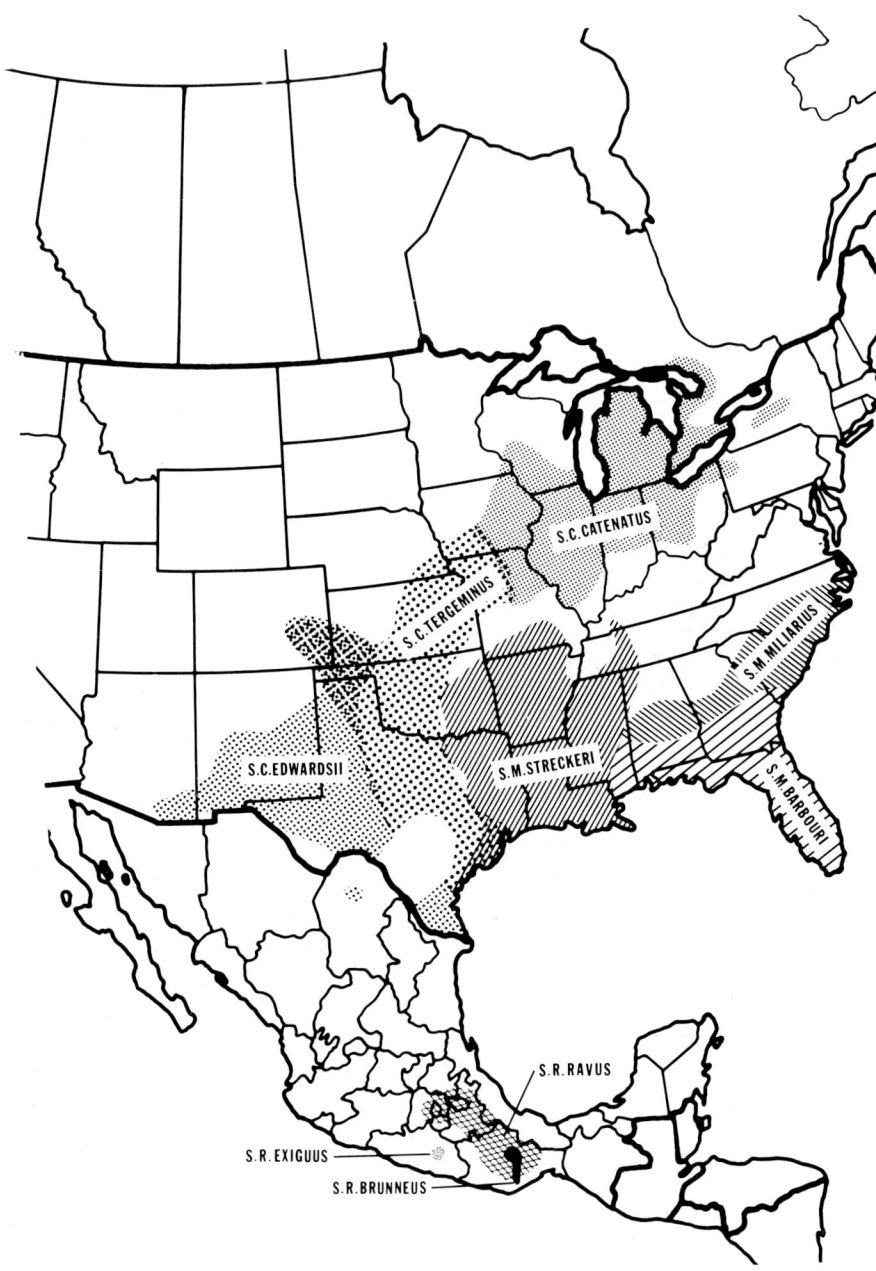

Figure 85 Distribution of the *Sistrurus* genus: *Sistrurus catenatus catenatus, S. catenatus edwardsii, S. catenatus tergeminus, S. miliarius miliarius, S. miliarius barbouri, S. miliarius streckeri, S. ravus ravus, S. ravus brunneus,* and *S. ravus exiguus.* (Data from Gloyd, 1940; Klauber, 1972; Conant, 1975; Harris and Simmons, 1978; Campbell, 1979; Campbell and Armstrong, 1979.)

VENOM YIELD AND LETHAL TOXICITY

Introduction

Venom research, including research on rattlesnake venoms, is often motivated by the need to understand the action of venoms following evenomation and the need to develop adequate medical therapies for treatment of envenomation. Rattlesnake venoms are complex mixtures of toxic proteins, with multiple biochemical, immunological, pharmacologic, and pathological effects as reviewed in other chapters of this volume. The actions of rattlesnake venoms and the adequate medical therapy following envenomation depend primarily on the relative toxicity and the quantity of venom injected; therefore, knowledge of rattlesnake venom yields and toxicities is important to both research and clinical practice.

Human envenomation by rattlesnakes may induce local tissue injury, systemic effects, morbidity, and/or death in close correlation with the yield and toxicity of the specific venom. The *lethal* toxicity is only one aspect of venom toxicology, and adequate medical therapy for envenomation by any venomous snake depends on knowledge about the myriad of toxic activities of the specific venom involved. A full understanding of the toxicology of rattlesnake venoms can only be obtained from knowledge of the biochemistry, pharmacology, and pathology of these venoms and their toxic constituents. These are presently active areas of venom research, and the present state of knowledge is reviewed in other chapters of this book.

Venom yield and toxicity from many species/subspecies of rattlesnakes are still not well characterized. This is especially true of Mexican and Central American rattlesnakes, including *Crotalus catalinensis, C. durissus* ssp., *C. intermedius* ssp., *C. lepidus* ssp., *C. mitchellii* ssp., *C. molossus estebanensis, C. polystictus, C. pricei* ssp., *C. pusillus, C. scutulatus salvini, C. stejnegeri, C. tigris, C. tortugensis, C. transversus, C. triseriatus* ssp., *C. willardi* ssp., and *Sistrurus ravus* ssp. Many have not been studied because either they are geographically isolated and rarely collected or, from the human-clinical perspective, they are relatively unimportant. Also, there are relatively few specific studies on the effect of factors such as environmental changes, geographic distribution, seasonal changes, ontogenic changes, and habitat on the venom yield and toxicity. The information presented in this chapter is from numerous investigations specifically designed to study yield and lethal toxicity and from other venom research reports where data were collected on yield and lethal toxicity incidental to other venom research. Also, data are presented from the records of the Venom Research Laboratory, Veteran's Administration Medical Center, Salt Lake City, Utah.

The lethal toxicities reviewed in this chapter are derived solely from data using laboratory mice and are expressed as the LD_{50} of venom injected intravenously (i.v.), intraperitoneally (i.p.), intramuscularly (i.m.) and subcutaneously (s.c.), except as otherwise noted. The LD_{50} is the dose of dry venom (or dry venom protein in a few instances) required to kill 50% of the test animals, expressed in milligram of venom per kilogram of test animal weight. Differences in the data from separate reports dealing with identical subspecies may be attributable to the experimental bioassay, to the number, age, sex, and strain of the mice, to the effects of temperature and other stress factors on the mice, as well as to the methods used in computation of the LD_{50} values. We have limited the subject of this review to the factors affecting venom quantity and quality specifically.

Venom Apparatus

The ability of various animal forms to envenomate reaches a high level of specialization in the snakes of the Viperidae and Crotalidae families. The venom apparatus of all rattlesnakes (Crotalidae) studied has the same basic structure (Figure 86) and includes provisions for generating, storing, and injecting venom. The anatomical features of the venom apparatus are bilateral.

Venom Glands

The main venom glands of rattlesnakes are situated in the temporal regions of the head and are somewhat almond shaped. They appear similar in shape, gross anatomy, and histology to the venom glands of other viperids (Kochva and Gans, 1966; Kochva, 1978). The muscles compressing the venom glands are separately controlled on each side of the head and function independently from the biting action. Consequently, snakebite does not necessarily result in envenomation. The quantity of available venom released may vary from none to nearly 100%, depending on several factors, including the stimuli motivating the snake's behavior.

Gland size and venom yield are directly correlated with genetically and ontogenically determined head size and conformation and with the overall length of various rattlesnakes. The main venom gland is made up of tubules lined with several epithelial cell types, but the principal cell type is the columnar, secretory cell representing at least 80% of the total cell population (Kochva, 1978). These cells synthesize the toxic proteins and polypeptides of venom and secrete venom constituents contained in secretory vesicles into the lumina of the tubules (Warshawsky et al., 1973; Kochva 1978). The tubules store venom and connect with a central storage lumen leading to a single venom duct (Figure 86). This duct transports the venom through a separate accessory gland before entering the fang sheath pocket, where the venom is directed into and through the fang.

The rate of venom protein synthesis depends on the amount of venom stored in the gland and is also likely affected by temperature, cell conditions, nutrition, and other factors. In glands full of venom, the majority of the secretory epithelial cells assume a cuboidal shape, with relatively little cytoplasm. The available

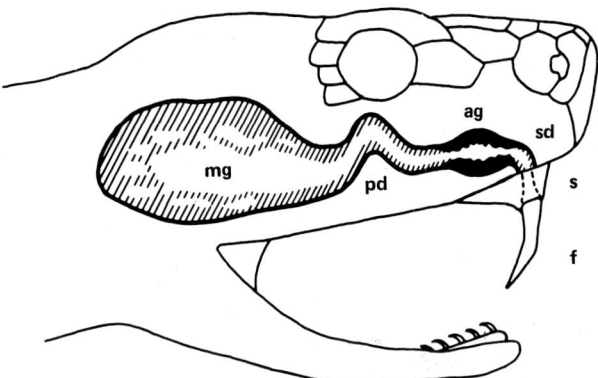

Figure 86 Schematic drawing of the venom apparatus of *Crotalidae*. Venom is synthesized and stored in the main venom gland (mg) and transported by the primary duct (pd) through the accessory gland (ag) and secondary duct (sd) which exits into the fang sheath pocket (s), diverting the venom into and through the fang (f). The loop in the primary duct accommodates the movement of the fang. All of these features occur bilaterally.

literature suggests that venom glands, once full of venom, will cease net synthesis of venom proteins and remain inactive for an indefinite time (Kochva and Gans, 1965; Shaham and Kochva, 1969; Glenn et al., 1973; Glenn and Straight, 1979). Even in surgically isolated venom glands, the majority of the secretory cells remain viable for at least 4½ years and will resume venom production if the gland is emptied of venom (Glenn and Straight, 1979). Removing venom from the gland triggers the reactivation of the secretory cells, and within hours the cuboidal cells appear columnar, with a marked incease in cytoplasm. Subsequently, within a few days, the cells reach maximal columnar height and venom secretory activity. The cell height progressively decreases to the cuboidal stage over the next few weeks. Venom secretory activity diminishes, but continues, until the gland tubules are full, which may require 30-60 days.

The cyclic process of venom synthesis in rattlesnakes has been studied most extensively in *Crotalus durissus terrificus* (DeLucca and Imaizumi, 1972; Warshawsky et al., 1972; DeLucca et al., 1974). The peak of protein synthetic activity, measured through the incorporation of ^{14}C amino acids into proteins, occurs at 4 days after milking. This peak could possibly be shifted by many factors, including age, season, temperature, and nutrition, and may vary among rattlesnake species. Available data from *Vipera palaestine* indicates that the peak of synthetic activity for the venom proteins and the peak for the cellular proteins are different (Oron and Bdolah, 1973). Also, specific venom enzymes and

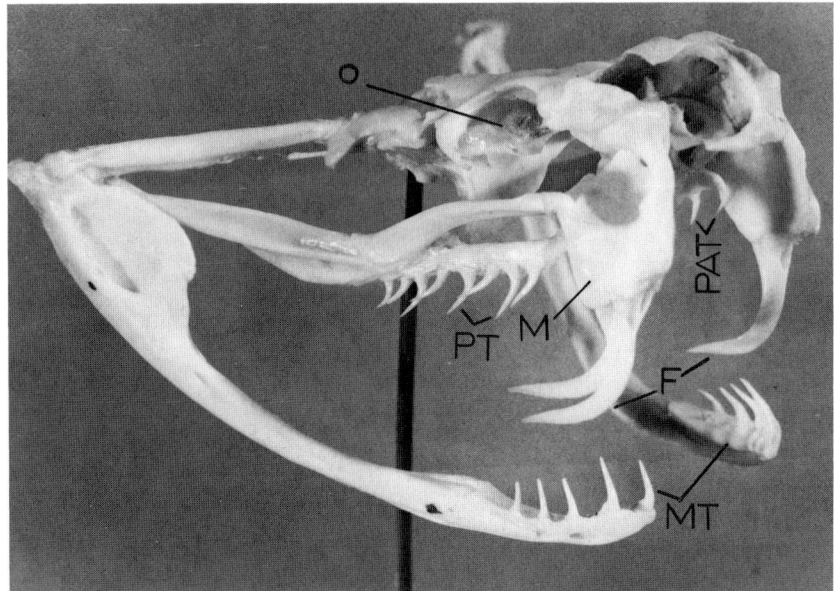

Figure 87 Skull of *Crotalus adamanteus* (156 cm specimen). Illustrates the fang length in comparison to the pterygoid, palatine, and mandibular teeth and also the delicate bone structure of the skull. The mandibular arrangement facilitates the engulfing of large prey and the opening of the mouth at approximately $180°$ for envenomating the victim. F = fang, M = maxillary, MT = mandibular teeth, O = eye orbit, PT = pterygoid teeth, and PAT = palatine teeth.

and other toxic proteins may have different, independent rates of synthesis and secretion (see Kochva, 1978; Oron et al., 1978) which could affect the venom lethal toxicity.

In addition to the main venom gland, there is also a small accessory gland (Figure 86). The function of this gland is not clear. It may secrete mucoid substances which are added to the venom for some purpose (Gennaro et al., 1963; Gans and Kochva, 1965). Recently, Ruzic and Russell (1978) reported that the accessory gland secretions of *Crotalus atrox* were only slightly lower in lethal toxicity than the crude venom and main venom gland secretions.

Fangs

The retractable or mobile fang arrangement of rattlesnakes is a distinct feature of all vipers (Viperidae) and pit vipers (Crotalidae) (Figure 87). The functional fangs are anchored in bilaterally shortened maxillaries, which rotate the fangs downward and forward about $90°$ from a resting position against the roof of the

Venom Yield and Lethal Toxicity

mouth to their biting position with the mouth open. Even newborn rattlesnakes normally have functional fangs in their maxillary sockets. The development, length, and structure of *Crotalus* and *Sistrurus* fangs have been thoroughly described by Klauber (1939, 1956). Fang shape, curvature, and relative length vary considerably among rattlesnake species. The larger species tend to have thicker, stronger, and longer fangs. Fang curvature, as measured by the external angle defined by Klauber (1939), varies from 44 to 88° interspecifically.

Fang length, defined as the straight line distance from the lower end of the upper lumen to the point of the fangs, increases as the snake ages through a periodic shedding and replacement process. Two different descriptions of fang regeneration and replacement (Mitchell, 1861; Tomes, 1877) were used in the literature until Klauber (1939) substantiated the work of Tomes (1877) as follows. On each side of the snake's head are short maxillaries, each containing two sockets. These two sockets are alternatively used to hold the active fang, while the immature replacement fangs, five to eight in progressive stages of development, are channeled in a staggered line behind the maxillary sockets. The most mature fang is always situated behind the adjacent vacant socket. The replacement fang advances to the vacant socket where it becomes rigidly seated, and the lateral old fang drops out soon after, leaving its socket vacant; the nearest immature fang will later advance to this vacant socket. There is a short period when both the old and new fangs are functional just prior to the shedding of the old fang (see Figure 87). There appears to be no synchronism in fang replacement between the two maxillae, and it is unknown if replacement is hastened by premature loss of an active fang.

The longest mature fangs are found in large specimens of *Crotalus adamanteus,* followed closely by *Crotalus atrox* and *Crotalus ruber ruber.* The maximal fang length for large *C. adamanteus* is approximately 22 mm (Klauber, 1939, 1956). Among the smaller rattlesnake species, *C. polystictus* is exceptional in having relatively long fangs (7-10 mm) in proportion to the overall length of the adult. The fangs of adult *Crotalus lepidus klauberi* and *Crotalus pricei* are notably short by comparison (3.2 to 3.7 mm).

Factors Affecting Venom Yield

Expert, indeed, will be the field collector who can catch a snake [rattlesnake], particularly one of the more nervous species, without the loss of some venom.
L. M. Klauber (1956)

This statement by Klauber points out an important problem involved with obtaining accurate venom yield data from freshly captured rattlesnakes, although in several species this would not be a problem and accurate yield data from freshly captured specimens can be obtained. Activities that occur from the bush to the beaker, including capture, bagging, transporting, housing, and handling for venom extraction, and many other factors may significantly affect yield data and must be considered and understood before evaluating yield information.

Much of the literature on rattlesnake venom yield is not adequate in the description of handling and collection methods, control of experimental and environmental factors, and recording of the vital statistics of the specimens involved.

Methods for Venom Collection

Three basic milking or venom extraction techniques have been used on rattlesnakes. All three techniques require restraining the snake by hand or with some restraining device. *Voluntary* extraction allows the snake to bite, once or any number of times, into a rubber covered vessel, expressing venom voluntarily. *Manual* extraction techniques require that finger pressure be applied to the venom glands to more or less forcibly milk out the venom. *Electrical* extraction methods employ the application of electricity to the maxilary region, thereby stimulating contraction of the muscles surrounding the venom gland and expulsion of the venom.

Each of the above methods are useful and appropriate depending on the intended purpose of the study and how the information is to be applied. In the case of obtaining maximum or average yield data, manual or electrical extraction methods are better utilized than voluntary methods, which rely totally on the snake's willingness to cooperate and other factors not easily controlled (see Johnson, 1938; Belluomini, 1968; Glenn et al., 1972; Ditada et al., 1978). Whatever the extraction method employed, it is reasonable to assume that significant yield variation will occur depending on the method. Therefore, in obtaining and presenting venom yield data, it is essential to control and carefully describe the collection method.

Venom Extraction Frequency

The venom yield values of freshly captured specimens are widely dispersed. Even captive yield values generally have a high dispersion and do not fit normal probability curves, providing problems in statistical analysis (Klauber, 1956; Glenn et al., 1972). Klauber (1956) reported that captive rattlesnakes averaged only one-half the venom yields of newly captured specimens. However, he suggested that well-fed, adequately housed, captive specimens would yield as much venom as the newly captured specimens. *Crotalus ruber lucasensis* specimens produced only 90% of their initial (freshly captured) dry venom yields following 54 days without food (Klauber, 1956). A group of *C. atrox* produced only 48% of their initial dry venom after 38 days. Bolaños (1972) also reported that captive *Crotalus durissus* (*durissus*) produce less venom than freshly captured specimens (see Table 11). Conversely, Glenn et al. (1972) found that 40 captive *C. atrox*, housed individually at 75-80°F without food, produced more venom than when freshly captured if rested 2-4 weeks.

Obviously, captive conditions would influence the yields obtained from any species. However, in Klauber's previously mentioned investigation on *C. ruber lucasensis* venom yield, the yields continually increased over a 54 day period. Additionally, Glenn et al. (1972) found that *C. atrox* yields were greatly reduced if 14 day extraction cycles were continuously used. Amaral (1928) also noted that recently captured crotalids yield less venom than when properly confined for 2 or 3 weeks. From these data, plus the present knowledge of venom gland secretory activity, it would appear that venom extraction frequency plays a major role in yield statistics.

Development, Growth, and Aging

Venom yield increases dramatically as rattlesnakes grow to their maximum genetically controlled size range. Venom yields correlate closely with the development of venom glands, which is associated with head growth specifically and related generally to overall growth as the snake ages. Venom yield data have been reported as adult or juvenile yields: neither description adequately defines the size of the specimen. For the purposes of this discussion, neonate rattlesnakes are defined as less than 3 months of age; juveniles less than 2 years. Male rattlesnakes in general reach sexual maturity between 2 and 4 years of age, and females between 3 and 5 years (Klauber, 1937; Glissmeyer, 1951). These age groups are correlated with specific size ranges (length) for each species and subspecies of rattlesnakes.

Yield related to length has been used by several investigators to demonstrate the effect of normal development on venom yield. Data relating the maximum and/or average venom yield to overall length of a specimen for any given species is considered useful in evaluating the possible seriousness of envenomation.

Venom Yields in Neonates and Juveniles There has been much discussion concerning the envenomation capabilities of the newborn venomous snake (see Stadleman, 1928). Stadleman (1928) even experimented with self-inflicted envenomation by forcing a newborn (6 hr) copperhead (*Agkistrodon contortrix* ssp.) to bite him. The results convinced him "to conduct all future research solely with mice." The question of whether snakes are venomous at birth seems no longer in doubt, however, quantitative neonatal yield data are relatively scant and are lacking for the majority of the rattlesnake species. Nonetheless, some interesting findings have been reported.

Venom gland cellular activity begins in the middle to late embryonic developmental stage of *Vipera palaestinae* (Fein, et al., 1971). These authors obtained "no biochemical evidence of the exact point of initiation of the synthesis of any specific secretory product(s)"; however, they remarked that the venom gland's central and tubular lumina are filled with venom less than 2 weeks posthatching. Unfortunately, no similar studies have been reported for any rattlesnakes which would prove useful in evaluating their venom yield capabilities at birth.

The watery or thinner nature of neonate *C. atrox* venom has received comment (Johnson et al., 1968; Allen and Maier, 1941). Klauber (1956) found that compared with adult venoms, juvenile venoms do in fact exhibit low specific gravities and low solid residue content per unit volume after dessication. This, in part, may reflect the relative state of hydration of the animals in question. At least the specimen's state of hydration is a factor to consider in collecting venom yield data. Klauber (1956) collected more neonatal and juvenile rattlesnake venom yield data than all other investigators combined. Using manual venom extraction techniques, he milked 416 specimens of juvenile *Crotalus viridis viridis* ($\approx$300 mm). From regression lines determined from yield data of over 1,000 *C. viridis viridis* of all lengths, Klauber computed that *C. viridis viridis* juveniles (of this approximate length) from Colorado would produce 6.2 mg per snake, and 4.6 mg per snake would be obtained from similar juveniles from South Dakota. However, the actual yields he obtained from 170 Colorado and 72 South Dakota *C. viridis viridis* averaged only 0.5 and 1.26 mg per snake, respectively: far below the expected values. Another South Dakota group of 174 *C. viridis viridis*, slightly longer than the above mentioned group, yielded only 2.37 mg per snake. Other investigators reported that venom yields of 11 *C. viridis viridis* between 9 and 15 days of age totaled only 12 mg dry weight (1.1 mg per snake) (Fiero et al., 1972). Venom subsequently collected at 79 days of age yielded 29 mg (2.64 mg per snake).

Klauber (1956) also collected yield data from 135 juvenile *C. atrox* with dry weight yields averaging only 2.2 mg per snake. In a smaller study Reid and Theakston (1978) and Theakston and Reid (1978) correlated venom dry weight yields with the age and weight of 3 *C. atrox* juveniles over a 20 month period. They collected their yield data by allowing the snakes to voluntarily bite into a covered petri dish at monthly intervals, from 2 to 22 months of age. Venom yields (dry weight) averaged less than 10 mg per snake until the specimens reached 4 months of age. Yields gradually increased, averaging approximately 150 mg per snake by 22 months of age. Minton (1957) had previously conducted similar experiments with a single *C. atrox* specimen. Beginning at 14-16 months of age, 19 consecutive monthly milkings resulted in a 660% increase in monthly venom production rates (40-264 mg), compared with only an approximate 100% increase in length (22 1/4 to 43 1/4 in.).

Venom Yield and Lethal Toxicity 65

An average dry weight venom yield of 3.33 mg per snake was obtained from 9 neonate siblings of *Crotalus horridus atricaudatus* at 5 days of age (Minton, 1967). At 28 days of age, the 9 siblings more than doubled their venom yield to 7.55 mg per snake, whereas their length had increased only approximately 16%.

Recently, venom was collected from 7 captive-born neonate siblings of *C. ruber ruber* (Baja), 5 days of age (born 9/7/79), by manually applied pressure to the venom glands (Glenn and Straight, unpublished). A total of 0.19 ml, representing 16.1 mg of dry venom following lyophilization, resulted in a dry weight yield of 2.3 mg per snake and a concentration of total solids of only 84 mg/ml. This concentration was remarkably low compared with venom from adult snakes.

The largest rattlesnake species at birth is *C. adamanteus,* measuring from 14 to 17 in. (35.6 to 43.2 cm) in length and weighing from 46 to 77 g (Glenn et al., 1975). Allen and Maier (1941) collected 270 mg of "clear and colorless" dried venom from 20 *C. adamanteus* specimens, 1 week old, representing an average yield of 13.5 mg per snake. These data demonstrate the higher yield capabilities of neonate *C. adamanteus* as compared with other species.

Although these various studies contain a number of differences, including venom collection techniques, period of fasting previous to venom collection, temperatures at which specimens were maintained, etc., the results consistently indicate that venom yield in neonate rattlesnakes is lower than expected based on body length (Table 2).

Venom Yields in Adults   Studies referring to adult venom yields actually encompass a wide range of rattlesnake lengths in their data, especially in the larger forms of rattlesnakes. *Crotalus atrox* and *C. adamanteus* adults may range from $\simeq$75-225 cm (from 2 1/2 to 7 1/2 ft) in length, resulting in a wide range in venom yield data for the "adult" group. Conversely, the smallest forms of rattlesnakes are sexually mature at about two-thirds of the length they ultimately reach (Klauber, 1937); thus the adult venom yield data are confined to narrower limits and, therefore, are more generally representative of adults in these forms.

The adolescent stage of rattlesnake development is a period of rapid growth (Klauber, 1937). This is accompanied by venom yields that increase exponentially in *C. atrox* (Minton, 1957; Reid and Theakston, 1978), and in *C. horridus atricaudatus* (Minton, 1967), with possible seasonal fluctuations. However, in exceptionally large individuals, venom yield curves may increase approximately linearly in the medial maturation years and may increase exponentially again at the upper end of the growth range in *C. atrox* (Klauber, 1956; Glenn et al., 1972) and in *C. ruber ruber, C. ruber lucasensis,* and *C. viridis viridis* (Klauber, 1956).

Table 2 Venom Yield of Neonate and Juvenile Rattlesnakes

Number of specimens	Species	Age or length	Average yield (dry weight, mg)	Reference
20	Crotalus adamanteus	1 week	13.5[a]	Allen and Maier (1941)
3	Crotalus atrox	2–8 months	<11[a]	Reid and Theakston (1978)
135		Juvenile	2.2	Klauber (1956)
9	Crotalus horridus atricaudatus	5 days	3.3[a]	Minton (1967)
10		<45 days	<10[a]	Johnson et al. (1968)
7	Crotalus ruber ruber (Baja, Mexico)	5 days	2.3[a]	Glenn and Straight, unpublished
170	Crotalus viridis viridis (Colorado)	~300 mm	0.5	Klauber (1956)
72	(South Dakota)	~300 mm	1.26	Klauber (1956)
174	(South Dakota)	Slightly longer than 300 mm	2.37	Klauber (1956)
11		9–15 days	1.1[a]	Fiero et al. (1972)
11		79 days	2.64[a]	Fiero et al. (1972)

[a]Captive born.

Venom Yield and Lethal Toxicity

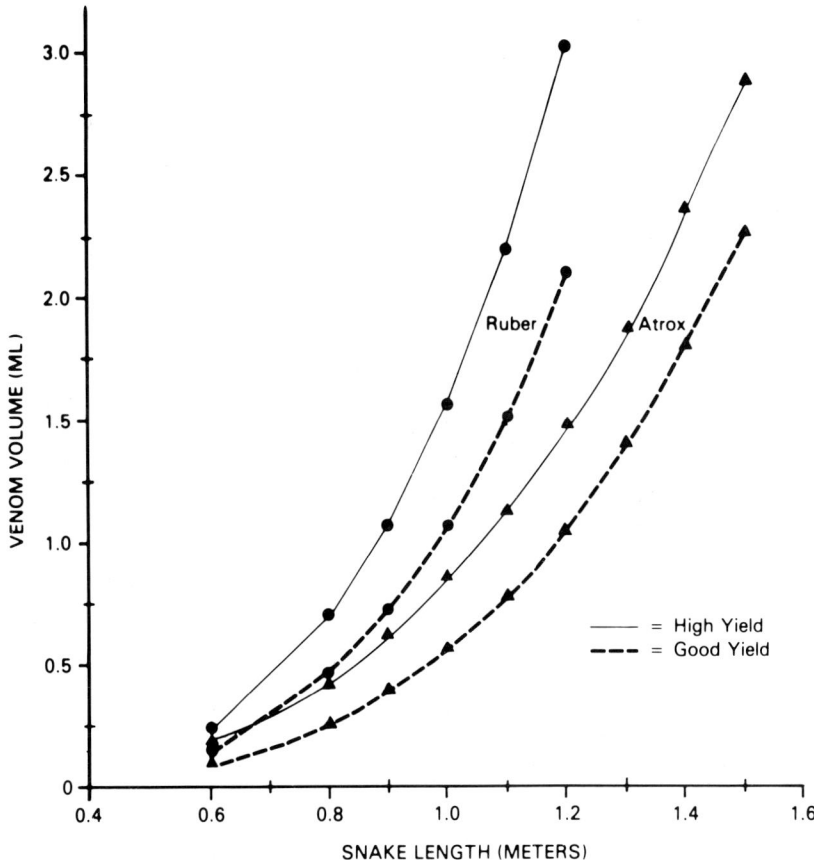

Figure 88 Correlation of venom yield (volume) with length in *Crotalus ruber ruber* and *Crotalus atrox*, demonstrating the exponential relationship of venom volumes to snake length. (Data from Klauber, 1956.)

Venom yields related to growth and development in rattlesnakes may vary markedly for a given species owing to geographic location, especially if the subspecies is disbursed over a wide range. For example, significant maximal and average size differences do occur among certain populations of *C. atrox, C. ruber ruber, C. viridis viridis,* and possibly other subspecies from geographically speareated areas.

George (1930) reported that the average yield for a given size snake of a species varies directly with the cube of the length of the snake. However, Klauber (1956; see Figure 88) demonstrated that rattlesnake venom yield increases exponentially when compared with growth and development. Yield increases proportionally faster than the length, square or cube of the length, and cube of the head length, but increases proportionally slower than weight. It is evident that ontogeny is the most significant factor affecting venom yields of rattlesnakes.

Sexual Dimorphism

Male rattlesnakes differ from females in scalation, including the range of ventral and subcaudal scale numbers, and most males also display faster, and ultimately 10-12% greater, growth (length) (Heyrend and Call, 1951; Klauber, 1937, 1956). The single exception is *Crotalus cerastes*: females significantly grow faster and attain greater length than males (Klauber, 1937: p. 27).

When comparing the venom yields of each sex, for individuals of equal length, no significant difference has been reported except in *Crotalus ruber lucasensis*, where the female venom yield is greater (Klauber, 1956). *Crotalus cerastes* females would also likely yield more venom than males of equal size. Therefore, the fact that males generally produce greater venom yields seems directly related to their faster growth and development and ultimately larger size. The effect of sexual dimorphism on venom yield is probably not important considering all the other variables involved, and we would agree with Klauber (1956) that sexual dimorphism has a relatively insignificant effect on venom yield. However, behavioral sexual dimorphisms might be important in voluntary venom delivery, although no data are presently available.

No sexual dimorphism in rattlesnake venom protein content or general toxicity has been reported. Recent investigations found no significant difference in the venom protein content or intraperitoneal venom lethal toxicity of male and female *Crotalus viridis concolor* and *Crotalus scutulatus scutulatus* (Glenn and Straight, 1977, 1978).

Environmental and Seasonal Effects

Rattlesnakes occupy a variety of habitats with diverse climates geographically distributed from southern Canada southward to northern Argentina. Many species often inhabit seemingly inhospitable life zones; however, they have adapted to specific microhabitats within these zones. Besides a decrease in venom yield owing to recent use of venom for feeding or predator defense, various environmental factors may affect venom yield, including photoperiod, seasonal moisture, activity, dormancy, ecdysis, and gravidity. However, there is little evidence as to how much these factors affect venom yield.

Of all the natural environmental factors which might influence venom yields, in any rattlesnake species, temperature is likely the most important. All physiological processes of ectothermic reptiles are affected by temperature, and one would suspect critical changes to occur in venom secretory function as well.

Venom Yield and Lethal Toxicity

The importance of temperature effects on snake venom secretory rates and yields is seen in Kochva's (1960) investigations with another viperid, the palaestine viper (*Vipera palaestinae*). He designed experiments specifically to investigate the effects of temperature on venom yield. The results indicated that the venom yields of captive snakes were directly related to the temperatures at which the specimens were maintained. Little, if any, venom was produced at 13.5°C, but venom production gradually increased as temperature was increased up to 26.5°C. This corresponded to seasonal yield variation in wild freshly captured specimens, where greater yields were produced in the summer than in the winter and spring. Also, by artificially heating captive snakes in the winter months yields remained high. This would indicate that temperature, in part, is responsible for seasonal yield variation. Similar temperature studies with rattlesnakes have yet to be reported, although one would expect similar results.

Houssay (1923) noted that winter temperatures greatly reduced venom yields of *C. durissus,* and similar temperature effects were reported for *C. durissus terrificus* specimens housed outdoors, as compared with specimens housed indoors, in Sao Paulo, Brazil (Belluomini, 1968). Although no temperatures were reported, the snakes artificially heated indoors survived longer and thereby produced more venom than specimens maintained outdoors in open pits (from Belluomini's data, p. 114). Also, the seasonal variation in the monthly venom yield of a single captive *C. atrox* specimen reported by Minton (1957) suggests an effect of temperature, especially since the temperature range reported for the 18 month study period ranged from 62 to 95°F. Yield declined in the winter months and increased appreciably in the spring; however, monthly temperature ranges were not reported. These data suggest the value of accurately controlling and reporting temperature in any captive venom yield investigations, a practice that has not been adhered to in the past. In this respect, much yield data are derived from reports where yield is only incidently recorded and included in research focused on other objectives.

Factors Affecting Venom Lethal Toxicity

Animal Models

The resistance or sensitivity of animals to rattlesnake venoms is highly variable, depending on the test animal model involved. Animal models used to test venom toxicity have included bacteria, insects, amphibians, reptiles, birds, and many different mammals. Although a few publications have extrapolated human lethal toxicity doses from these animal model data, there is little scientific basis for such extrapolations.

To review the numerous reports concerning the toxicity of rattlesnake venoms in various animal forms is beyond the scope of this chapter. Generally, within the groups of the more common laboratory animals, the order of sensitivity to rattlesnake venom lethal toxicity is pigeons > rabbits > mice > rats. Pigeons were used extensively as animal models for toxicity testing during the earlier investigative reports; however, laboratory bred albino mice are presently the most common test subjects. Although rattlesnake venom lethality differs significantly between pigeons and mice, the venoms from various rattlesnake species do show a similar relative order of lethality in both test animals (see Githens and Wolff, 1939a-c).

Ontogenic Effects on Venom Lethal Toxicity

We have previously discussed the significant quantitative increases in venom yield that occur throughout the rattlesnake's growth, development, and aging process. There is evidence that ontogeny also affects qualitative changes in venom lethal toxicity. Theakston and Reid (1978) reported remarkable changes in venom composition of 3 juvenile *C. atrox* between 2 and 22 months of age. Monthly venom extractions were used, and the intravenous lethal toxicity was highest (LD_{50} 1 mg/kg) at 2 months of age, gradually decreasing until 13 months of age (LD_{50} 5 mg/kg), where the lethal toxicity leveled off. Proteolytic activity, on both casein and gelatin, rose steadily for 12 months and stabilized thereafter. Dramatic changes also occurred in other biologic activities of the juvenile venoms during the 21 month study. The results of Minton (1957, 1975), who obtained venom from a single *C. atrox* specimen over a 19 year period, are similar to the findings of Theakston and Reid (1978). The subcutaneous lethality of this specimen's venom decreased 2.4-fold over the 19 year period. Also, the subcutaneous lethal toxicity of venom from *C. viridis viridis* juveniles, 2 weeks to 3 months of age, are also approximately twice as toxic as adults (see Table 22) (Fiero et al., 1972).

The pooled venoms from juvenile *C. horridus horridus* specimens, 3-18 months of age, exhibit lower protease activity (casein) and higher lethal toxicity than venom from adults (Bonilla et al., 1973). However, Minton (1967) found that pooled neonate (5 days old) venom from 9 siblings of *C. horridus atricaudatus* was approximately 10 times less toxic than venom collected at 182 days of age: i.p. LD_{50} 2.5 mg/kg versus 0.26 mg/kg, respectively. Both interperitoneal and subcutaneous LD_{50} steadily increased from birth to 182 and 271 days of age, respectively. Notably, all the juvenile venoms from 5 to 370 days of age were significantly higher in lethal toxicity than the parental adult female (see Table 13). The above reports are reasonably consistent and indicate that venom quality is affected by ontogeny in the species investigated. It would be interesting to know if juvenile *C. durissus terrificus* and *C. scutulatus scutulatus* venoms that have well-defined major lethal toxins, crotoxin and Mojave toxin, also are more toxic than adult venoms.

The advantage of juveniles producing venoms of greater lethal toxicity possibly counterbalances the low yields of juveniles; however, the toxicity may also coincide with environmental parameters or ontogenic changes in food prey. For example, young rattlesnakes of a given species may feed on insects or lizards early in life and rodents or birds as adults. The venom composition of juveniles therefore may be suited for killing and digesting different prey than the venom of adults. Whatever the reason for the ontogenic changes in the toxicities of rattlesnake venoms, they are important considerations in toxicology investigations and envenomation therapy.

Geographic and Individual Variation in Lethal Toxicity

Linnean systematics in rattlesnake classification does not include venom composition and other physiological characteristics as criteria for taxonomic designation. In several instances, categorizing rattlesnake lethal toxicities by species or subspecies does not reflect the total state of knowledge concerning venom lethality. For example, geographic differences in venom composition and lethal toxicities are notable in *C. durissus terrificus* (see Gonçalves and Deutsch, 1956; Hoge, 1965; Rodriguez et al., 1974; Rodriguez and Scannone, 1976) and *C. scutulatus scutulatus* (Glenn and Straight, 1978) as they are presently classified. The venoms of *C. durissus terrificus* specimens from northeastern Brazil are missing the basic polypeptide toxin, crotamine, whereas specimens from southern Brazil, Argentina, and Venezuela (var. *C. durissus cumanensis*: Hoge, 1965) possess this venom component. The extreme difference in *C. scutulatus scutulatus* venom lethal toxicity is due to some north central Arizona specimens lacking this species' major lethal toxin, Mojave toxin (see Chapter 4). Other taxa that may exhibit substantial geographic or population variation in venom lethal toxicity are *C. lepidus klauberi* (Table 14) and *C. horridus atricaudatus* (Table 13). In this regard, present taxonomic status of several rattlesnake forms may be incomplete as suggested by venom toxicology results.

Included among the many factors that might provide discrepancies among investigative reports is the individuality of living organisms, including rattlesnakes. Many, if not the majority of the lethal toxicities reported, are derived from "pooled" venoms, especially if the venoms are commercially supplied. Thus, any individual (or geographic) variation occurring within a subspecies would not be noted, nor would it interfere, with the ultimate objective of the venom's experimental use. The establishment of individual variation in lethal toxicities has received little attention. Minton (1953) tested seven adult *Crotalus horridus horridus* specimens (from Indiana) individually, and found the mouse i.p. LD_{50} values ranged from 1.48 to 7.80 mg/kg. The most toxic and least toxic venom samples came from specimens captured at the same locality on the same day. The individual variation in intraperitoneal LD_{50} values of 17 adult

C. viridis concolor specimens from the same den site was 0.13 to 0.45 mg/kg protein (Glenn and Straight, 1977). Also, 28 individually tested specimens of *C. scutulatus scutulatus* from southwestern Arizona and southern California produced intraperitoneal LD_{50} values ranging from 0.13 to 0.54 mg protein/kg (Glenn and Straight, 1978).

Theakston and Reid (1978) found that considerable ontogenic-related changes occurred in the venom yield and toxicity of three individual *C. atrox* juveniles, reporting significant individual variation among the specimens in fibrinogenolytic, hemorrhagic, and phospholipase activities, and found an approximate twofold variation in individual lethal toxicities. The individual variability of venom lethality should not affect the individual's prey-killing ability since such overkill doses are obviously injected. However, the individual variability may affect the severity of the symptoms in human or large animal envenomations.

The above findings are important in that the clinical symptoms encountered in human envenomation may be influenced not only by the subspecies and its age but also by the geographic region of the specimen involved and possibly by the individual envenomer.

Venom Yield and Lethal Toxicity by Species/Subspecies

Considering all the previously discussed factors affecting venom yield and toxicity data, the problems of evaluating the variety of published reports are evident. However, the information gained from the combined data of these reports lend greatly to our better understanding the role of venom yield and toxicity in the medical aspects of rattlesnake envenomation. Commonly, in medical management of snakebite, questions immediately arise concerning the species and its size and the toxic qualities and quantity of the specific venom involved. Therefore, knowledge regarding venom yield capabilities and lethal toxicity of each rattlesnake species or subspecies is of practical value. Maximum venom yields reflect the highest yield recorded from an individual of a species or subspecies, and the average yield indicates extrapolated data from a selected group of individuals of a given taxon. Maximum yields are isolated numbers and occur infrequently. Average yield data are for comparative analysis of the various taxa only and do not represent the average amount of venom delivered in any individual rattlesnake envenomation. Owing to the variable water content in venoms, emphasis is placed on the *dry weight venom yield* and not the liquid volume recorded.

There are several problems in ascertaining the accuracy of venom yields for any rattlesnake from literature values. For example, Amaral (1928) categorized the average venom yields of 12 North American rattlesnake varieties as young, adult, old, and exceptional specimens. No specific lengths or ages were designated and the number of specimens was not reported in 6 of the 12 species listed. The data were compiled from "the extraction of venom from several thousand specimens." Amaral's authroitative report is to be given credit for separating the data into 4 separate groups, even though they are not adequately defined. He also succeeded in demonstrating the intraspecific differences in venom yields resulting from ontogenic trends. Also, Klauber's data (1956), which are cited throughout this section, are typical and often quoted data; however, the data should be critically examined. The method for calculating the average yield value requires several assumptions. It may not be realized generally that this yield information was in fact calculated and does not represent the actual yield data obtained for many of the taxa. The calculations were inserted to make sense out of combining large, medium, and small categories into an average adult yield value. In several instances, the calculations actually represent large adult yields. Medium length snakes were calculated as 50% of the large, whereas the small specimens which contributed to the total yield were left unspecified as to how they fit into the calculations of average yield. Specimens milked a second time were calculated at 50% of their initial length values and also were included in the formula. There clearly is a need for further comparative yield research involving most of the taxa reported; however, because of Klauber's expertise and impressive specimen numbers, these data are the best available.

The following species and subspecies accounts include lethal toxicity information expressed as milligram of dry venom per kilogram of body weight required to kill 50% of the animals (mice) involved. All recorded data are mouse data unless otherwise specified. Some of the older citations utilized minimum lethal dose (MLD) statistics, which represent the lowest venom dose required to kill an animal and are designated. Parentheses enclosing the subspecies designation in the table listings indicate the investigators recorded the species only in their report. When only the species is listed, the subspecies is uncertain, even from information contained from the report cited.

Crotalus adamanteus (Eastern Diamondback Rattlesnake)

Venom Yield *Crotalus adamanteus* (Figure 3), a monotypic species, is the largest of all rattlesnakes in several aspects including body length and bulk, head dimensions, and fang length. Occasionally, individuals may exceed 213 cm (7 ft) in total length. The highest maximal dry weight yield for this species is likely to exceed 1 g. However, the maximum yield reports are 864 mg (Stadleman, 1929) and 848 mg (KLauber, 1956) (Table 3); neither are well documented. Stadleman's 864 mg yield, from a 5 1/2 ft specimen, was calculated as follows. The pooled dry weight of venom collected from 9 specimens (4.438 g solids) was divided by the pooled total volume (20.5 ml), and then the resultant gram-percent solids (21.6%) was applied equally to each individual specimen's recorded volume to obtain dry weight yield. This method incorrectly assumes that

Table 3 *Crotalus adamanteus* Venom Yield

Number of specimens or extractions	Venom yield (dry weight, mg) Average	Maximum	Reference
?	240 (young)	–	Amaral (1928)
?	300 (adult)	–	Amaral (1928)
?	600 (old)	–	Amaral (1928)
?	750 (exceptional)	–	Amaral (1928)
9	492	864	Stadleman (1929)
108	415	–	Allen and Maier (1941)
23	604	–	Allen and Maier (1941)
22	606	848	Klauber (1956)

no significant variance in percent of solids occurs between individuals; such is not the case. For example, when 66 individual *C. adamanteus* venom samples were collected and processed separately (Table 4), the lyophilized dry weight/liquid volume ranged from 130 to 273 mg/ml. The only other data on individual *C. adamanteus* variation in venom concentration indicated a 240-300 mg/ml range among 9 "not previously handled" specimens (Deichmann et al., 1958). Reports on total solids from pooled *C. adamanteus* venom ranged from 216 to 286 mg/ml (Amaral, 1928; Stadleman, 1928; Stadleman, 1929; Allen and Maier, 1941; Glenn et al., 1972).

Table 4 *Crotalus adamanteus* Venom Yield

Number of specimens (total extractions)	Total length (cm)	Venom yield (mg) Dry weight average (range)	Protein average (range)
6 (18)	120-130	265 (120-426)	233 (103-388)
9 (27)	135-145	376 (176-643)	324 (148-566)
7 (21)	150-155	446 (243-683)	372 (209-540)

Note: Venom was collected by electrical stimulation (Glenn et al., 1972) at 30-35 day intervals (3 venom extractions from each specimen) from 22 commercially supplied individuals from south Florida. Specimens were caged separately and maintained at 74-82°F (mean, 77°F), under 12 h diurnal lighting. Venom was centrifuged at 1000 g for 20 min, and the supernatant was lyophilized. All venom was individually collected, processed, and calculated. Biuret protein (Gornall et al., 1949) of dry individual venoms ranged from 73 to 94% (mean, 86%), and lyophilized dry weight per liquid volume ranged from 130 to 273 mg/cc (mean, 213 mg/cc).

Venom Yield and Lethal Toxicity

Table 5 *Crotalus adamanteus* Venom Lethal Toxicity

Mouse LD$_{50}$ (mg/kg)				
i.v.	i.p.	i.m.	s.c.	Reference
–	2.0[a]	–	–	Githens and Wolff (1939c)
1.54	–	–	–	Criley (1956)
–	1.67	–	14.55	Minton (1956)
1.7-2.0	–	–	–	Gingrich and Hohenadel (1956)
–	–	10.7	–	Deichmann et al. (1958)
1.68	1.89	–	–	Russell and Emery (1959)
–	3.75[b]	–	–	Hall and Gennaro (1961)
1.65-1.75[c]	–	–	–	Russell and Eventov (1964)
1.95-2.06[c]	–	–	–	Russell and Eventov (1964)
2.49[b]	–	–	–	Vick et al. (1966)
1.58[b]	–	–	–	Vick (1971)
2.36	–	–	–	Friederich and Tu (1971)
2.24[d]	–	–	–	Bonilla et al. (1971)
1.33[d]	2.22[d]	–	–	Kocholaty et al. (1971)
–	2.7[d]	–	–	Glenn and Straight (1978)
–	–	–	11.4[e]	Broad et al. (1979)

[a]Minimum lethal dose.
[b]LD$_{100}$.
[c]Dessicated at room temperature for 5 days (= 1.65-1.75) and lyophilized for 24 h (= 1.95-2.06) and tested at yearly intervals for 6 years.
[d]Venom protein weight basis.
[e]These authors report an s.c. LD$_{50}$ of 7.7 mg/kg for this venom when a 1.0% BSA solution is used as suspension instead of normal saline medium.

Klauber's maximum yield of 848 mg (1956) is extensively cited in the literature; however, the manner in which this figure was derived is not included in his report. Klauber's 848 mg and Stadleman's 864 mg figures are not unreasonable, but close examination of their publications demonstrate the problems encountered with evaluating published yield information.

The average dry weight yield recorded by Allen and Maier (1941) from 108 presumably adult specimens was 415 mg per snake (328-604 mg). One group of 23 large specimens averaging 4 1/2 ft produced 13.90 g (50 cc) of dry venom, an average of 604 mg per snake. These authors also presented venom volume data from 15 separate groups of *C. adamanteus* containing from 5 to 52 specimens. A total of 732.5 cc of venom was collected from 416 specimens, averaging 1.76 cc (0.83 to 3.46 ml) per snake. An additional 6 ft specimen produced 4.5 ml of venom.

Klauber's (1956) dry weight yield per fresh adult for *C. adamanteus* was listed at 666 mg per snake, which he considered too high. Stadleman (1929) reported an average yield of 492 mg from 9 specimens, 3.0 to 5 ft 5 in. in length.

Venom Lethal Toxicity The lethal toxicity of *C. adamanteus* is comparable to and even greater than many of the smaller forms of rattlesnakes. Thus, combined with its enormous yield capabilities, *C. adamanteus* is placed high on the list of dangerous species. A small molecular weight basic protein toxin has been isolated from the venom of this species. The toxin has a marked effect on neuromuscular transmission and induces death by pronounced ischemia of heart muscle and subsequent myocardial failure (Bonilla et al., 1971, 1972).

C. adamanteus venom has a consistently higher lethal toxicity than *C. atrox* venom in mice, dogs, and monkeys (Vick, 1971). Russell and Eventov (1964) found no significant decrease in lethal toxicity over a 6 year period with either dessicated or lyophilized *C. adamanteus* venom stored in an airtight vial in the dark at 5° [sic] (Table 5). The lethal toxicity data (LD_{50} values) of the numerous reports cited in Table 5 fall into a narrow range: the i.v. LD_{50} values range from 1.33 to 2.36 mg/kg and the i.p. LD_{50} values range from 1.67 to 2.71 mg/kg. The lethal toxicity is significantly lower when the venom is injected i.m. or s.c. (Table 5).

Crotalus atrox (Western Diamondback Rattlesnake)

Venom Yield This monotypic species (Figure 4) occasionally approaches 213 cm (7 ft) in total length. A maximum dry yield of 1145 mg (3.9 cc), reported by Klauber (1956) from a 1655 mm specimen, is the highest yield documented for any rattlesnake species. Klauber also speculated that exceptionally large specimens of this species and *C. adamanteus* "should occasionally produce well over 1000 mg and even up to 1500 mg of dried purified venom." Museum records show that this high yield was obtained from an individually collected and processed sample (L. M. Klauber Library, San Diego Natural History Museum, *Venom,* vol. 3, Lot No. 414).

Theakston and Reid's (1978) data on three captive-born *C. atrox* specimens indicated that venom yield rapidly increases during the first 2 years growth. Monthly extracted venom yields reached 150 mg at 1 year of age (70 cm), and 275 mg was produced by one 20-month-old individual between 92 and 100 cm in length.

Venom Yield and Lethal Toxicity

C. atrox venom volume data has been reported by Klauber (1956) (Figure 88) and Glenn et al. (1972). The maximum volume recorded by the latter authors was 3.1 cc from a 165 cm individual. The concentration of venom (weight of solids per liquid volume) in 40 adult *C. atrox* was 240-275 mg/cc, and the biuret protein, as percentage of dry weight, was proportionally high at 92-97%, gradually increasing during 14 weeks of captivity (Glenn et al., 1972). Allen and Maier (1941) reported a 221 mg/cc concentration of solids and Klauber (1956) noted that the concentration of solids of 294 mg/cc in the record yield specimen was "unusually high." Jiminez-Porras (1961) also demonstrated that *C. atrox* venom solids concentration varies considerably: the concentration of solids ranged from 234 to 357 mg/ml in 16 individually collected samples.

The average adult yield from *C. atrox* is highly variable, owing to the extreme range of adult length (from ~60 to over 220 cm). Although a monotypic species, the easterly forms of *C. atrox* in Oklahoma, Texas, and eastern Mexico demonstrate modal squamation differences and are generally larger than the westerly forms in portions of New Mexico, Arizona, and California. Comparative head size also somewhat diminishes westerly (Klauber, 1930: p. 9) and venom yield may therefore be lower; however, these differences have not been studied. The average dry yield of 24 manually milked specimens, from 3 to 4 ft in length, was reported by Allen and Maier (1941) to be 230 mg (221 mg/cc). Klauber (1956) calculated a dry yield of 277 mg from 373 specimens of all sizes. Amaral (1928) listed *C. atrox* average venom yields as young = 90, adult = 120, old =

Table 6 *Crotalus atrox* Venom Yield

Number of specimens	Total length (cm)	Venom yield (dry weight mg)	
		Average	Maximum
21	60-70	77	151
32	80-100	191	316
17	110-130	322	481
13	140-150	365	550
11	155-160	423	688
5	162-168	596	853

Note: Initial venom yield was collected by electrical stimulation (Glenn et al., 1972) from individuals collected in southern Texas in March, 1972-1974. Specimens were not transported or housed individually but were left undisturbed for 17-23 days after arrival. Room temperature ranged from 70 to 82°F; lighting was 12 h diurnal. Venom was centrifuged at 1000 g for 20 min, and the supernatant was lyophilized. All venom was individually collected, processed, and calculated.

240, and exceptionally old specimens = 600 mg. Although the age designations are poorly defined, Amaral realized the value of demonstrating the wide variation of yields in *C. atrox* specimens. A detailed analysis of *C. atrox* venom yield according to length is provided in Table 6.

This species generally exhibits an irritable disposition and vigorously defends itself by standing its ground when cornered and elevating its head high in a lateral "S" posture; large individuals may strike near the knee region from level ground. Combining these traits with its high venom yield capabilities and toxic qualities places this snake among the most dangerous in North America. Interestingly, in experiments designed to test venom delivery, *C. atrox* specimens consistently delivered more venom than *C. adamanteus* (Gennaro et al., 1961).

Venom Lethal Toxicity *Crotalus atrox* venom lethal toxicity seems to be attributed to its high proteolytic and hemorrhagic activities, and no major lethal

Table 7 *Crotalus atrox* Venom Lethal Toxicity

	Mouse LD_{50} (mg/kg)		
i.v.	i.p.	s.c.	References
–	5.5[a], 13.6[a]	–	Macht (1937)
–	6.0[a]	–	Githens and Wolff (1939c)
–	–	16.11	Schoettler (1951)
–	8.42	19.25	Minton (1956)
3.5–6.3	–	–	Gingrich and Hohenadel (1956)
4.20	3.71	–	Russell and Emery (1959)
5.56[b]	–	–	Vick et al. (1966)
3.56	–	–	Friederich and Tu (1971)
3.03[b]	–	–	Vick (1971)
–	4.8–6.25	–	Glenn et al. (1972)
2.67[c]	4.26[c]	–	Kocholaty et al. (1971)
–	–	11.65–28.0[d]	Minton (1975)
1.0–5.0[e]	–	–	Theakston and Reid (1978)
–	4.5[c]	–	Glenn and Straight (1978)
–	8.0	–	Perez et al. (1978)

[a]Minimum lethal dose.
[b]LD_{99} or LD_{100}.
[c]Venom protein weight basis.
[d]Lethality range of single specimen, tested 6 times over a 19 year period.
[e]Juvenile venoms, 2–22 months of age.

Venom Yield and Lethal Toxicity

toxin has been isolated from this venom. The lethal toxicity data (Table 7) vary considerably depending on injection route and venom collection procedures. Both the proteolytic and hemorrhagic activities are affected by temperature, pH, and other factors requiring proper collection and processing procedures (see Chapter 5). Considerable variation occurs in the general toxicity, including lethal toxicity, of *C. atrox* venom depending on the age of the snake (Minton, 1975; Theakston and Reid, 1978), resulting in a reduction in lethal toxicity with age. However, this is more than offset from a clinical perspective by the enormous increase in venom yield with age (growth). Owing to the widespread distribution of this species, geographic variations may also occur in venom composition and toxicity, although this has not been investigated.

Crotalus basiliscus (Mexican West Coast Rattlesnake)

Subspecies	Approximate maximum length (cm)
C. basiliscus basiliscus (Mexican West Coast rattlesnake)	205
C. basiliscus oaxacus (Oaxacan rattlesnake)	205*

Venom Yield The impressive size (length and bulk) of this species (Figure 5 and 6) is surpassed only by *C. adamanteus* and *C. atrox*. It is possibly the largest form of the *durissus* group [Gloyd's taxonomic grouping (1940)]. Maximum yield capabilities are unknown, but based on body and head size (Klauber, 1938), yields may exceed those of Central American *C. durissus* ssp. The only report of venom yield from this species is by Klauber (1956), and the average dry yield of 297 mg was calculated from only 3 fresh adults (*C. basiliscus basiliscus*).

Venom Lethal Toxicity The LD_{50} data of Criley (1956) and Gingrich and Hohenadel (1956) illustrated in Table 8 are questionable. The lethal toxicities are well above the mode for rattlesnake intravenous LD_{50} values. Another problem with the venom lethality data of this species is that Schoettler's (1951) subcutaneous LD_{50} is lower than Githens and Wolff's (1939c) intraperitoneal LD_{50} value. Rarely does the subcutaneous lethal toxicity value of any rattlesnake venom exceed intraperitoneal values. Further investigations of *C. basiliscus* ssp. are warranted for the purpose of resolving these discrepancies as they now stand.

*Little is known about the southern race *C. basiliscus oaxacus*, but maximum lengths may equal that of the northern race, *C. basiliscus basiliscus*.

Table 8 *Crotalus basiliscus* Venom Lethal Toxicity

	Mouse LD$_{50}$ (mg/kg)		Reference
i.v.	i.p.	s.c.	
–	4.0[a]	–	Githens and Wolff (1939c)
–	–	2.8	Schoettler (1951)
13.0	–	–	Gingrich and Hohenadel (1956)
11.11	–	–	Criley (1956)

[a]Minimum lethal dose.

Crotalus catalinensis (Santa Catalina Island Rattlesnake)

No venom yield or toxicology data have been reported for this rattleless form (Figures 7 and 84). The maximum yield recorded by our laboratory is 32.6 mg from a large 72.5 cm (29 in.) male. The average initial (dry weight) yield of 5 adults, 63.8 to 83.8 cm in length, was 20.8 mg (Glenn and Straight, unpublished). The mouse intraperitoneal LD$_{50}$ dose, based on dry weight, ranged from 2.9 to 3.2 mg/kg, comparing 3 of the 5 individually collected and processed samples mentioned above. The maximum length achieved by this insular species is approximately 90 cm (36 in.).

Crotalus cerastes (Sidewinder)

Subspecies	Approximate maximum length (cm)
C. cerastes cerastes (Mojave Desert sidewinder)	59
C. cerastes cercobombus (Sonoran Desert sidewinder)	63
C. cerastes laterorepens (Colorado Desert sidewinder)	83

Venom Yield The common name sidewinder is often misapplied to several rattlesnakes in the southwest. The species (Figures 8-10) is easily recognized by the conspicuous, hornlike, pointed scales projecting above the eyes. *Crotalus cerastes laterorepens* is generally the largest of the three subspecies, and the fang length is proportionally longer in this subspecies than in *C. cerastes cerastes* (Klauber 1944: p. 107). Females of this species deviate from the general trend of

Venom Yield and Lethal Toxicity

Table 9 *Crotalus cerastes* Venom Yield

Number of specimens	Venom yield (dry weight, mg)		Reference
	Average	Maximum	
12	18	–	Amaral (1928)
9	15	–	Klauber (1928)
124	22	67	Githens (1933)
169 (one-half were adults)	22.8	–	Klauber (1944)
Several small groups (adults)	40–50	–	Klauber (1944)
Two groups (adults)	65.8–68.4	–	Klauber (1944)
119 (82 large)	33	63	Klauber (1956)

rattlesnakes in that they grow faster and attain greater ultimate length than the males. Females also are exceptional in having comparatively larger heads than males (Klauber, 1938), and venom yield differences between males and females of equal length may be significant, although this has not been investigated.

Only Linnean binomenclature has been used in reporting *C. cerastes* venom yield statistics (Table 9), except for Minton's report (1956). Klauber (1956) estimated that exceptionally large individuals (females) might produce 80 mg of dry venom.

Table 10 *Crotalus cerastes* Venom Lethal Toxicity

Subspecies	Mouse LD_{50} (mg/kg)			Reference
	i.v.	i.p.	s.c.	
C. cerastes ssp.	–	2.73[a]	–	Macht (1937)
	–	8.50[a]	–	Githens and Wolff (1939c)
	2.60	2.08	–	Emery and Russell (1963)
	2.5	–	5.5	Minton and Minton (1969)
	1.95	–	–	Vick (1971)
C. cerastes laterorepens	–	–	5.50	Minton (1956)

[a]Minimum lethal dose.

Venom Lethal Toxicity This species has received little attention as far as venom toxicology investigations are concerned. Although it is generally considered less dangerous than most rattlesnakes, serious envenomations have occurred. The lethal toxicity and yield statistics indicate that large specimens should justifiably be respected for their envenomation capabilities (Tables 9 and 10). Subspecies differences in lethal toxicities have not been investigated.

Crotalus durissus (Neotropical Rattlesnake)

Subspecies	Approximate maximum length (cm)
C. durissus durissus (Central American rattlesnake)	180
C. durissus culminatus (Northwestern neotropical rattlesnake)	170
C. durissus terrificus (South American rattlesnake)	153
C. durissus totonacus (Totonacan rattlesnake)	170
C. durissus tzabcan (Yucatan neotropical rattlesnake)	170

Hoge (1965) has proposed that six additional subspecies be assigned to the South American *C. durissus* taxa (Figures 12-15), separated from *C. durissus terrificus* (*C. durissus cascavella, C. durissus collilineatus, C. durissus cumanensis, C. durissus dryinus, C. durissus marajoensis* and *C. durissus ruruima*). The subspeciations are principally based on pattern differences, and some taxonomists have yet to recognize these subdivisions of *C. durissus terrificus* until further descriptive analysis is presented in full accordance with the rules of the International Commission on Zoological Nomenclature (see Klauber, 1972; McCranie and Wilson, 1979). It is likely that enough variation in taxonomic features may exist in such a widespread species to justify Hoge's subspeciation, at least in part. The present situation is confusing regarding classification and presentation of venom toxicology results since three recent toxicology reports have followed Hoge's taxonomic theme (Rodriguez et al., 1974; Rodriguez and Scannone, 1976; Scannone et al., 1978).

Crotalus durissus is the most widely distributed rattlesnake species in the New World (Figures 80 and 81). Subspecific forms range from the Tropic of Cancer in Mexico southward to northern Argentina. The closely allied species of the *durissus* group (Gloyd, 1940), *C. durissus, C. basiliscus, C. horridus, C. molossus, C. vegrandis,* and *C. unicolor,* blanket the entire north-south distribution of rattlesnakes (Canada to Argentina). Except for the speciation of *C.*

vegrandis and *C. unicolor, C. durissus* solely dominates the rattlesnake taxa of South America and Central America (excluding Mexico). A disjunction in *C. durissus* distribution exists in Panama, commonly referred to as the Panamanian gap.

Venom Yield Central American *durissus* may occasionally exceed 183 cm (6 ft) in length and are generally larger than South American *durissus,* except possibly Venezuelan *C. durissus terrificus* [var. *cumanensis* (Hoge)], which produce significantly greater venom yields than *C. durissus terrificus* from Brazil and Argentina (see Rodriguez et al., 1974) and closely compare to Central American *durissus* (Table 11). Hoge (1965) reported a dwarfed race (no lengths specified) of *C. durissus terrificus* (var. *ruruima*) occurring near the Venezuelan-Brazilian border, suggesting that the maximum and average yields of Venezuelan *C. durissus* can vary greatly within relatively short geographic distances.

Venom gland secretory activity investigations indicate maximal yields are not obtained from captive *C.durissus terrificus* unless extraction intervals are of 30-40 days duration (De Lucca et al., 1974). This is notable since the voluminous extractions reported by Belluomini (Table 11) include data from captive specimens milked at 15-20 day intervals.

A recent report indicates that approximately twice as much venom (*C. durissus terrificus*) is obtained by manual and electrical venom extraction (MVE and EVE, respectively) than by voluntary venom extraction (VVE) techniques (Table 11) (Ditada et al., 1978). The specimens were milked at 15 day intervals. Also, according to Houssay (1923), *C. durissus terrificus* venom quantity varies with geographic origin (Argentinian specimens produce less than Brazilian) and seasonal factors such as temperature, since monthly venom yield was greatest in the Argentine summer, diminished gradually in the fall, remained low in the winter, and increased again in the spring.

Belluomini (1968) suggested that 75% of *C. durissus terrificus* bite victims receive 50 mg (dry weight) of venom. This assumption was based on data indicating that 75% of the snakes produced a maximum quantity of 50 mg. This logic seems unwarranted as it is highly unlikely that the snakes would consistently inject their maximum venom supply into bite victims.

Bolaños (1972) reported that captive *C. durissus* from Costa Rica produce less venom than specimens milked for the first time (Table 11); this actually would depend on rest intervals between extractions and captive conditions.

Venom Lethal Toxicity Of all the crotalid venoms, *C. durissus terrificus* venom has been studied most extensively. Its neurotoxic components have been of great interest to venom toxicologists (see Chapters 4 and 5). Its major neurotoxin, crotoxin, is also present in high concentration in the closely related species *C. vegrandis* (Scannone et al., 1978) and is very similar to the Mojave toxin of *C. scutulatus scutulatus* (see Chapter 4).

Table 11 *Crotalus durissus* Venom Yield

	Venom yield (dry weight, mg)			
Subspecies	Number of extractions	Average	Maximum	References
C. durissus durissus (Honduras)	83	113	—	George (1930)
(Locality ?)	4	240	—	Githens (1933)
(Locality ?)	5	277	—	Klauber (1956)
(Costa Rica)	19 (initial yield)	159	—	Bolaños (1972)
(Costa Rica)	? (captive yield)	77	—	Bolaños (1972)
C. durissus terrificus (var. *cumanensis*) (Venezuela)	130	115	—	Rodriguez et al. (1974)
C. durissus terrificus (Argentina)	508	23	90	Houssay (1923)
(Brazil)	2,758	24	300	Schoettler (1951)
(Brazil)	2,810	35	—	Belluomini (1968)
(Brazil)	4,431	44	—	Belluomini (1968)
(Brazil)	5,039	32	—	Belluomini (1968)
(Brazil)	17,809	28	—	Belluomini (1968)
(Brazil)	587	22	—	Belluomini (1968)
(Brazil)	421	33	—	Belluomini (1968)
(Brazil)	2,068	31	—	Belluomini (1968)
(Argentina)	79	28[a]	—	Ditada et al. (1978)
(Argentina)	79	54[b]	—	Ditada et al. (1978)
(Argentina)	115	49[c]	—	Ditada et al. (1978)
C. durissus totonacus (Mexico)	1 (large)	514	—	Klauber (1956)

[a]Voluntary venom extraction (VVE).
[b]Manual venom extraction (MVE).
[c]Electrical venom extraction (EVE).

Table 12 *Crotalus durissus* Venom Lethal Toxicity

	Mouse LD_{50} (mg/kg)			
Subspecies	i.v.	i.p.	s.c.	Reference
C. durissus durissus	1.20	0.67	–	Kocholaty et al. (1968)
	0.10[a]	–	–	Vick (1971)
	1.84	0.80	–	Bolaños (1972)
	1.24[b]	0.67[b]	–	Kocholaty et al. (1971)
C. durissus terrificus (var. *cumanensis*)	0.67	1.14	–	Rodriguez et al. (1974)
C. durissus terrificus	–	0.13[c]	–	Githens and Wolff (1939c)
	–	–	0.60	Schoettler (1951)
	0.13	–	–	Criley (1956)
	0.13-0.14	–	–	Gingrich and Hohenadel (1956)
	0.14-0.21	–	–	Brazil et al. (1966)
	0.35	–	–	Friederich and Tu (1971)
	0.27[c]	0.40[b]		Kocholaty et al. (1971)
	0.28[b]	0.25[b]	1.40[b]	Glenn and Straight (1977)
	–	0.20[b]	–	Glenn and Straight (1978)
	–	0.13[b]	–	Hendon and Tu (1979)
C. durissus totonacus	–	2.5	–	Possani et al. (1980)

[a]LD_{100}
[b]Venom protein weight basis.
[c]Minimum lethal dose.

Comparative chromatographic fractionation, using Sephadex G-100 gel filtration of venom proteins from the Venezuelan rattlesnakes *C. durissus terrificus* (var. *cumanensis*) and *C. vegrandis*, shows a close similarity in protein constituents, and both venoms exhibit similar chemical and physical characteristics (Scannone et al., 1978). However, Venezuelan *C. durissus terrificus* (var. *cumanensis*) have an i.v. LD$_{50}$ of 0.67 mg/kg and an i.p. LD$_{50}$ of 1.14 mg/kg (Rodriguez et al., 1974) which are closer to *C. durissus durissus* lethal toxicities than to Brazilian *C. durissus terrificus* or *C. vegrandis*. Interestingly, these data support Klauber's opinion (1972: p. 36) that northern South American *durissus* are morphologically more similar to Central American *durissus* (across the Panamanian gap) than to Brazilian *C. durissus terrificus*. The venom of the northernmost (*durissus*) subspecies, *C. durissus totonacus*, is relatively low in lethal toxicity compared with other subspecific forms of the species (Table 12). No toxicology information is available for *C. durissus culminatus* or *C. durissus tzabcan*.

Crotalus enyo (Lower California Rattlesnake)

Subspecies	Approximate maximum length (cm)
C. enyo enyo (Lower California rattlesnake)	90
C. enyo cerralvensis (Cerralvo Island rattlesnake)	80
C. enyo furvus (Rosario rattlesnake)	80

This species (Figures 16-18) is restricted to the lower California peninsula (Baja) and several of its offshore islands, including Arbajoa, Magdalena, and Santa Magarita in the Pacific and the islands of Carmen, San José, San Francisco, Cerralvo, and Espíritu Santo-Partida in the Gulf of California (Cliff, 1954; Soule and Sloan, 1966; Harris and Simmons, 1977). *Crotalus enyo cerralvensis* is endemic to Cerralvo Island and has a proportionally smaller head than *C. enyo furvus* and *C. enyo enyo*, which may influence venom gland capacity. Other island populations may be somewhat dwarfed in size and may therefore yield less venom, although the differences might be insignificant.

The only reported venom yield information for this species is Klauber's calculated average fresh adult yield of 29 mg from 16 *C. enyo enyo* specimens (Klauber, 1956). One would expect, based on size, that the maximum yield for this species would be approximately 50 or 60 mg of dry venom.

No venom lethal toxicity reports, using mice, are available for this species.

Venom Yield and Lethal Toxicity

Crotalus exsul (Cedros Island Diamond Rattlesnake)

This species (Figure 19) is a dwarfed island form very closely related to *C. ruber*. It rarely exceeds 90 cm (36 in.), and the maximum length reported is 95 cm (Klauber, 1956). The only venom yield statistics are those of Klauber (1956), who recorded 54 mg of dry venom from a single large specimen. No venom toxicology information is available for *C. exsul*; however, because of its taxonomic relationship with *C. ruber*, venom characteristics may be similar.

Crotalus horridus (Timber Rattlesnake)

Subspecies	Approximate maximum length (cm)
C. horridus horridus (timber rattlesnake)	190
C. horridus atricaudatus (canebrake rattlesnake)	190

Pisani et al. (1972) have reevaluated the taxonomic status of *C. horridus atricaudatus* (Figure 20) and concluded that *C. horridus atricaudatus* scalation was not significantly different from *C. horridus horridus* (Figure 21), and the former no longer warranted subspecific classification. If the subspecific status of *C. horridus atricaudatus* is eliminated, any variance in venom toxicology reported would be considered a geographic or population variation. Subspecies recognition of *C. horridus atricaudatus* does aid in venom discussion, and its taxonomic removal may add further to the confusion of comparative toxicology research. The widespread distribution of *C. horridus* is conducive to intraspecific variations in venom toxicity, and there is evidence that significant variations do exist (see Table 13).

Venom Yield The maximum length for *C. horridus* or *C. horridus atricaudatus* rarely exceeds 183 cm (6 ft); in fact, specimens exceeding 150 cm in length are rarely encountered throughout most of this species' range. Venom yield information is surprisingly scant considering its widespread occurrence. The first yield data for *C. horridus* were reported by Amaral (1928). He categorized the average yields of young (60 mg), adult (90 mg), and old (180 mg) *C. horridus* specimens. Minton (1953) recorded the individual yields of 7 *C. horridus* (*horridus*) specimens from Indiana. There were 6 specimens, 40-51 in. in length, that averaged 160.5 mg per snake (104-240 mg) in volumes averaging 0.5 ml (0.34 to 0.71 ml). One 37 in. individual produced 69 mg (0.23 ml) of venom. The venom solids were in relatively high concentration in these specimens at 296-351 mg/ml.

Klauber (1956) calculated the yield of 7 large and 6 medium *C. horridus* (binomially) as 139 mg per fresh adult (actual yield, 197.1 mg per snake) and recorded the maximum yield as 229 mg for *C. horridus atricaudatus*.

As previously discussed, Minton (1967) reported a considerable effect of ontogeny on venom yield and lethal toxicity among *C. horridus atricaudatus* siblings of Arkansas parentage. By comparing this study with Theakston and Reid's report (1978) of *C. atrox* sibling juvenile yields, some interesting and important comparisons can be made. The growth (length) of both species was very similar during the first 12 months of life, each reaching 75 cm ($\cong$30 in.) in length. However, considerable differences resulted in venom yield performance between the two species at 9 and 12 months of age to 62.5 cm ($\cong$25 in.) and 75 cm ($\cong$30 in.). The average yield of the *C. horridus atricaudatus* at 9 months was 25.6 mg per snake (5 specimens) and at 12 months was 54.6 mg per snake (5 specimens). The *C. atrox* specimens (3) averaged significantly greater yields of 38 (23-54.4) and 112 mg (58.7-150) at 9 and 12 months, respectively. The venom yield of adult *C. horridus* seems substantially below that of *C. adamanteus, C. atrox,* and *C. ruber ruber* and would likely be comparable to Central American *C. durissus* forms. The maximum dry yield recorded by our laboratory for *C. horridus atricaudatus* is 244 mg (a 137.5 cm male, captive 8 years, from an unknown locality). However, our experience with *C. horridus* spp. is limited to 7 specimens and 11 extractions.

Venom Lethal Toxicity The *C. horridus* lethal toxicity data presented throughout the literature are erratic and inconsistent. These variations could conceivably be related to venom processing and aging or other factors, but geographic variances may also occur. Minton (1956, 1957) demonstrated that both *C. horridus atricaudatus* (Arkansas) and *C. horridus horridus* (Indiana) venoms vary considerably in lethality, influenced by source, ontogeny, and in the case of *C. horridus horridus,* individually. An approximate fivefold variation in LD_{50} (1.48 to 7.80 mg/kg: mean, 5.10 mg/kg) occurred between the 7 individual adult *C. horridus horridus* tested (Table 13). Our experience with *C. horridus atricaudatus* venom lethality is limited to 1 pooled venom sample and 3 individual adult specimen samples. Intraperitoneal LD_{50} values using 16-20 g mice were 0.36 (pool), 0.40, 0.29, and 0.53 mg/kg. The results Minton obtained from the *C. horridus atricaudatus* juveniles and with commercially supplied venom plus our own observations indicate that this form is capable of producing very toxic venom (Table 13). The high lethal toxicity (low LD_{50}) of at least certain populations of *C. horridus atricaudatus,* combined with its venom yield and size capabilities, places specimens of this taxon among the most dangerous venomous snakes in North America.

Venom Yield and Lethal Toxicity

Table 13 *Crotalus horridus* Venom Lethal Toxicity

	Mouse LD$_{50}$ (mg/kg)			
Subspecies	i.v.	i.p.	s.c.	Reference
C. horridus	3.09	–	–	Criley (1956)
	2.63	2.94	–	Russell and Emery (1959)
C. horridus horridus	–	5.0[a]	–	Macht (1937)
	–	4.0[a]	–	Githens and Wolff (1939c)
	–	–	24.9	Schoettler (1951)
	–	1.48[b]	–	Minton (1953)
	–	3.52[b]	–	Minton (1953)
	–	4.15[b]	–	Minton (1953)
	–	5.80[b]	–	Minton (1953)
	–	6.90[b]	–	Minton (1953)
	–	7.80[b]	–	Minton (1953)
	–	7.25	9.15	Minton (1956)
	–	–	3.0	Tu et al. (1970)
	2.57[c]	–	–	Friedrich and Tu (1971)
	1.64	2.84	–	Kocholaty et al. (1971)
	–	–	5.1[c]	Moran and Geren (1979)
C. horridus atricaudatus	–	0.72[c]	8.05[c]	Minton (1967)
	–	2.50[d]	8.93[d]	Minton (1967)
	–	1.35[d]	4.51[d]	Minton (1967)
	–	0.69[d]	3.31[d]	Minton (1967)
	–	0.26[d]	2.40[d]	Minton (1967)
	–	0.27[d]	2.25[d]	Minton (1967)
	–	0.40[d]	2.82[d]	Minton (1967)
	–	6.63[e]	15.63[e]	Minton (1967)
	–	4.0	–	Johnson et al. (1968)

[a]Minimum lethal dose.
[b]Individual snakes.
[c]Commercial venom.
[d]Juvenile venom pools listed consecutively at 5, 28, 89, 182, 271 and 370 days of age.
[e]Parental adult female of juvenile venoms above those listed in d.

Crotalus intermedius (Small-Headed Rattlesnake)

Subspecies	Approximate maximum length (cm)
C. intermedius intermedius (Totalcan small-headed rattlesnake)	60
C. intermedius gloydi (Oaxacan small-headed rattlesnake)	60
C. intermedius omiltemanus (Omilteman small-headed rattlesnake)	60

All 3 subspecies (Figures 22-24) are small forms, adult length generally ranging between 32.5 and 50 cm (13-20 in.). The only reference to venom yield or lethal toxicity of this species is by Minton (1977). Two individual *C. intermedius* (Binom) averaged only 2.2 mg per snake, and the i.v. LD_{50} of the venom after 10 years of storage was 1.58 mg/kg. The two specimens were from Rancho Benito Juarez, Oaxaca, Mexico, indicating they were *C. intermedius gloydi*. In view of its small size and venom yield, Minton did not consider *C. intermedius* dangerous, although the venom exhibits strong hemorrhagic activity.

Crotalus lannomi (Autlan Rattlesnake)

This is a species known only from one 63.8 cm specimen (road kill; see Figure 25) collected near Jalisco, Mexico (Tanner, 1966).

Crotalus lepidus (Rock Rattlesnake)

Subspecies	Approximate maximum length (cm)
C. lepidus lepidus (mottled rock rattlesnake)	80
C. lepidus klauberi (banded rock rattlesnake)	85
C. lepidus maculosus (Durango rock rattlesnake)	75
C. lepdius morulus (Tamaulipan rock rattlesnake)	75

Venom yield This small mountain species (Figures 26-29) is highly diversified in color and pattern, and many isolated populations seem to exist, suggest-

Venom Yield and Lethal Toxicity

ing possible subspeciation. It is likely that taxonomists will increase the number of subspecies of *C. lepidus* over the coming years.

The first recorded venom yield data were Amaral's (1928) 30 mg average from three specimens. Other venom yield statistics for *C. lepidus* are Klauber's (1956) calculated adult average dry yield of 10 mg per extraction (actual, 5.7 mg per extraction) from 31 *C. lepidus klauberi* specimens (10 large, 16 medium, and 5 small) and Minton's (1977) dry yields of 24, 25, and 33 mg obtained from a single *C. lepidus klauberi* specimen from Zacatecas, Mexico. Also, Githens (1933) obtained an average dry yield of 5.8 mg per extraction from 28 *C. lepidus* ssp. which is comparable to Klauber's actual data.

Venom Lethal Toxicity The only LD_{50} statistics available for *C. lepidus* forms are listed in Table 14. We have recently obtained venom samples from 4 individual adult *C. lepidus klauberi*: 2 from Zacatecas, Mexico, and 2 from the Florida Mountains, New Mexico. The i.p. LD_{50} values (16-20 g mice) were 0.38 and 0.67 mg/kg for the New Mexico specimens and 3.10 and 5.0 mg/kg for the Zacatecas, Mexico, specimens. These preliminary results would indicate a considerable variance in lethal toxicity which may be associated with geographic location of the specimens involved. We are aware of 2 human envenomations by *C. lepidus klauberi* which produced serious effects and, despite its small size, some *C. lepidus* forms should be considered quite dangerous. Conversely, the southern *C. lepidus klauberi* specimens from the Zacatecas region possibly produce venom of less toxic nature. Additional toxicology investigations of all *C. lepidus* forms are warranted.

Table 14 *Crotalus lepidus* Venom Lethal Toxicity

Subspecies	Mouse LD_{50} (mg/kg)			Reference
	i.v.	i.p.	s.c.	
C. lepidus ssp.	—	0.91[a]	—	Githens and Wolff (1939c)
C. lepidus lepidus (Texas)	—	—	11.55	Minton (1958)
C. lepidus klauberi (Zacatecas, Mexico)	9.0	—	23.95	Minton (1977)

[a]Reported as minimum lethal dose; also Klauber (1956) and Minton (1977) refer to this data as *C. l. klauberi*.

Crotalus mitchellii (Bleached Rattlesnake)

Subspecies	Approximate maximum length (cm)
C. mitchellii mitchellii (San Lucan speckled rattlesnake)	95
C. mitchellii angelensis (Angel de la Guarda Island rattlesnake)	150
C. mitchellii muertensis (El Muerto Island speckled rattlesnake	65
C. mitchellii pyrrhus (Southwestern speckled rattlesnake)	120
C. mitchellii stephensi (panamint rattlesnake)	100

Venom Yield The maximal lengths attained by the 5 subspecies (Figures 30-34) are indicators of the variable venom yield produced by *C. mitchellii*. No yield data are available for the largest form, *C. mitchelli angelensis,* which may occasionally approach 140-150 cm in length, nor is there information regarding the smallest form, *C. mitchellii muertensis,* which rarely exceeds 62 cm in length. Both of these size extremes are from endemic island subspecies. Of the mainland subspecies, *C. mitchellii pyrrhus* generally grows proportionally larger than *C. mitchellii stephensi* and *C. mitchellii mitchellii.* The latter subspecies also inhabits the Baja, Mexico, islands of Carmen, Cerralvo, Espiritu Santo, Monserrate, San Jose, and Santa Margarita (Cliff, 1954; Klauber, 1956; Soule and Sloan, 1966; Harris and Simmons, 1977). These island populations are often stunted or dwarfed (to various degrees) and therefore would presumably produce less venom than their larger mainland counterparts. Certain *C. mitchellii mitchellii* populations also exhibit smaller head proportions than, for example, *C. mitchellii pyrrhus.* Thus, the general trend in comparative venom yield should be, *C. mitchellii angelensis* > *C. mitchellii pyrrhus* > *C. mitchellii stephensi* > *C. mitchellii mitchellii* > *C. mitchellii muertensis.* Table 15 presents yield data for *C. mitchellii* and represents the only available investigations to date.

Venom Lethal Toxicity According to Githens and Wolff (1939c) and, subsequently, Russell et al. (1960), *C. Mitchellii mitchellii* produces extremely lethal venom (Table 16). Both reports deal with venoms obtained from the Klauber collection. The lethal toxicities reported are significant in that *C. mitchellii mitchellii* venom is much more toxic than *C. mitchellii pyrrhus* and *C. mitchellii stephensi* venom. However, Macht's (1937) statistics for *C. mitchellii mitchellii* venom are in disagreement with the previously mentioned reports.

Table 15 *Crotalus mitchellii* Venom Yield

Subspecies	Number of Specimens or extractions	Venom Yield (dry weight mg)		Reference
		Average	Maximum	
C. mitchellii	176	135	245	Githens (1933)
C. mitchellii mitchellii	64	33,32[a]	75,90	Klauber (1936, 1956)
C. mitchellii (pyrrhus)	? (young)	60	–	Amaral (1928)
	? (adult)	100	–	Amaral (1928)
	? (old)	160	–	Amaral (1928)
	? (exceptional)	265	–	Amaral (1928)
	95	118[b]	–	Klauber (1928)
C. mitchellii pyrrhus	98	215,227[a]	350,308[a]	Klauber (1936, 1956)
C. mitchellii stephensi	13	73[a]	129	Klauber (1936, 1956)
	5	72	80	Githens (1933)

[a]Calculated for (large) fresh adult.
[b]Actual data including snakes of all sizes (San Diego County, California).

It is possible that a population or geographic variation in venom lethal toxicity may occur within *C. mitchellii mitchellii*, especially in the northern limits of this subspecies' distribution, where the area of overlap or intergradation between *C. mitchellii mitchellii* and *C. mitchellii pyrrhus* is poorly understood. Our investigations of *C. mitchellii* specimens from the Bahia de Los Angeles mainland region and Smith Island revealed a non-subspecific taxonomic status of five individuals. Based on squamation features, these specimens are intergradal in type, but three individuals could easily be considered *C. mitchellii mitchellii*. However, the most distinguishing feature of all the specimens is their venom lethality values, which more nearly correspond to *C. mitchellii pyrrhus* (see Table 16).

Table 16 *Crotalus mitchellii* Venom Lethal Toxicity

Subspecies	Mouse i.p. LD_{50} (mg/kg)	Reference
C. mitchellii mitchellii	2.05[a]	Macht (1937)
	0.18[a]	Githens and Wolff (1939c)
	0.40 (0.25[a])	Russell et al. (1960)
	0.26[b]	Glenn and Straight, unpublished
C. mitchellii pyrrhus	3.41[a]	Macht (1937)
	2.5[a]	Githens and Wolff (1939c)
	2.4–4.0[c]	Glenn and Straight, unpublished
	2.13–2.48[d]	Glenn and Straight unpublished
C. mitchellii stephensi	5.46	Macht (1937)

[a]Minimum lethal dose.
[b]A single 75 cm adult from near Mulege, Baja, Mexico.
[c]The LD_{50} range of five individually tested adult specimens from Baja De Los Angeles (three) and Smith Island (two), Baja, Mexico.
[d]The LD_{50} range of three 70–111 cm specimens, tested individually, from Washington, County, Utah.

We presently conclude that *C. mitchellii pyrrhus* and *C. mitchellii stephensi* venom lethal toxicities fit within the general mode of rattlesnakes (i.p. LD_{50}, 2-6 mg/kg). However, *C. mitchellii mitchellii* produces one of the most lethal venoms in the New World. No venom toxicology reports are available concerning the insular subspecies *C. mitchellii angelensis* and *C. mitchellii muertensis*.

Crotalus molossus (Blacktail Rattlesnake)

Subspecies	Approximate maximum length (cm)
C. molossus molossus (Northern blacktail rattlesnake)	126
C. molossus estebanensis (San Esteban Island rattlesnake)	100
C. molossus nigrescens (Mexican blacktail rattlesnake)	110

Although no venom yield data are available for the southernmost subspecies *C. molossus nigrescens* (Figure 37), it may produce greater yields owing to its larger head dimensions than *C. molossus molossus* (Figure 35) and *C. molossus estebanensis* (Figure 36). The latter subspecies is an endemic island form, comparatively smaller in length and head size than the mainland forms, and therefore should produce comparatively less venom. Klauber (1956) obtained only 32 mg from a single large *C. molossus estebanensis* specimen. Klauber's (1956) calculated yield for (large) fresh adults of *C. molossus molossus* was 286 mg (from 64 extractions), and the maximum yield recorded was 540 mg. Amaral (1928) obtained an average dry yield of 180 mg from 5 adult *C. molossus* ssp. The only venom lethal toxicity reports (mouse) are the intraperitoneal minimum lethal doses of Macht (1937) and Githens and Wolff (1939c) which varied from 2.7 to 7.0 mg/kg, respectively. Both investigators may have obtained their venoms from the Klauber collection. Additional toxicology studies of the *C. molossus* ssp. are needed.

Crotalus polystictus (Mexican Lance-Headed Rattlesnake)

Minton (1977) justly considered this medium-sized (65-95 cm) species (Figure 38) dangerous because of its long fangs and venom yield (average yield of 3 specimens is 101.3 mg per snake). He also reported that the lethal toxicity (i.v. LD_{50}, 3.37 mg/kg; s.c. LD_{50}, 13.3 mg/kg) fits the mode of the majority of rattlesnake species and is not considered exceptional.

Crotalus pricei (Twin-Spotted Rattlesnake)

Subspecies

C. pricei pricei (Western twin-spotted rattlesnake)
C. pricei miquihuanus (Eastern twin-spotted rattlesnake)

Both subspecies (Figures 39 and 40) are small, with specimens only occasionally approaching 64 cm in length. Klauber (1956) calculated a fresh adult (large) average dry yield of 8 mg per snake from 30 specimens of *C. pricei pricei*.

No mouse LD_{50} data have been reported.

Crotalus pusillus (Tancitaran Dusky Rattlesnake)

No venom yield or toxicology information is presently available for this species (Figure 41). It is closely related to *Crotalus triseriatus* and is considered quite primitive. The longest of 18 specimens measured by Klauber (1952: p. 36) was 67.4 cm in length. Armstrong and Murphy (1979) consider this species locally common in intermittent mountain habitats in southern Jalisco and west central Michoacan, Mexico.

Crotalus ruber (Red Rattlesnake)

Subspecies	Approximate maximum length (cm)
C. ruber ruber (red diamond rattlesnake)	153
C. ruber lorenzoensis (San Lorenzo Island diamond rattlesnake)	100
C. ruber lucasensis (San Lucan diamond rattlesnake)	131

Venom Yield The venom yield capabilities of *C. ruber ruber* (Figure 42) and *C. ruber lucasensis* (Figure 44) are equaled or surpassed only by *C. atrox* and *C. adamanteus* when compared by length up to 120 cm (see Figure 88 and Tables 3, 4, 6, and 17). Both *C. atrox* and *C. adamanteus* attain greater lengths than *C. ruber* and therefore produce greater maximal yields, but the yields of *C.*

Table 17 *Crotalus ruber* Venom Yield

Subspecies	Number of specimens	Venom yield (dry weight, mg)		Reference
		Average	Maximum	
C. ruber (ruber)	? (young)	120	—	Amaral (1928)
	? (adult)	240	—	Amaral (1928)
	? (old)	450	—	Amaral (1928)
	? (exceptional)	550	—	Amaral (1928)
	142[a]	191	—	Klauber (1928)
	298	246	425	Githens (1933)
	554[b]	364	668	Klauber (1956)
C. ruber lucasensis	16	226	370	Githens (1933)
	381[b]	234	707	Klauber (1956)

[a]Specimens of all sizes.
[b]Calculated fresh (large) adult yield.

Table 18 *Crotalus ruber* Venom Lethal Toxicity

	Mouse LD$_{50}$ (mg/kg))			
Subspecies	i.v.	i.p.	s.c.	Reference
C. ruber (ruber)	–	11.0[a]	–	Githens and Wolff (1939c)
	–	–	21.25	Minton (1956)
	3.97	–	–	Criley (1956)
	3.48	–	–	Gingrich and Hohenadel (1956)
	3.51–3.91	4.22–5.11	–	Emery and Russell (1963)
C. ruber lucasensis	–	4.0[a]	–	Githens and Wolff (1939c)

[a]Minimum lethal dose.

ruber ruber and *C. ruber lucasensis* are quite impressive. Both subspecies must be considered potentially dangerous rattlesnakes.

The San Lorenzo Island subspecies, *C. ruber lorenzoensis* (Figure 43), is a typically dwarfed insular population with an inheritable tendency towards the loss of the rattle (Radcliffe and Maslin, 1975). No toxicologic information is available for this recently described subspecies. At least eight other islands of the Baja Peninsula, Mexico, are inhabited by *C. ruber ruber* (four) of *C. ruber lucasensis* (four). Venom yields may also be effectively reduced where stunted growth occurs on any of these islands.

Venom Lethal Toxicity *Crotalus ruber* ssp are taxonomically closely related to *C. atrox* in morphology, and venom lethal toxicity values are also comparable between the two species (Tables 7 and 18).

Crotalus scutulatus (Mojave Rattlesnake)

Subspecies	Approximate maximum length (cm)
C. scutulatus scutulatus (Mojave rattlesnake)	123
C. scutulatus salvini (Huamantlan rattlesnake)	100

Table 19 *Crotalus scutulatus scutulatus* Venom Yield

Specimens	Venom yield (dry weight, mg)		Reference
	Average (range)	Maximum	
182	40	110	Githens (1933)
228	77[a]	141	Klauber (1956)
16 males (60.5-110 cm (mean, 85.5)	61.6 (15-139)	139[b]	Glenn and Straight (1978)
13 females 52-87.5 cm (mean 67.5)	22.9 (8-45)	45	Glenn and Straight (1978)

[a]Calculated for fresh (large) adult

[b]110 cm male specimen from near Phoenix, Arizona; captive 2 years.

Venom Yield The only data available for *C. scutulatus scutulatus* (Figure 45) is that of Githens (1933), Klauber (1956), and Glenn and Straight (1978) (Table 19). Venom yield information is lacking for the southern (Mexico) subspecies, *C. scutulatus salvini* (Figure 46). Our experience with *C. scutulatus salvini* is limited to single extractions of a 90 cm male and 85 cm female, which produced 61 and 52 mg, respectively, following 24 day fasts.

Klauber (1930: p. 52) noted that the eastern forms of *C. scutulatus scutulatus* (eastern Arizona and north central Mexico) apparently have larger heads than those farther west (California, southwestern Arizona, and northwestern Mexico), indicating possible differences in yield capabilities.

Venom Lethal Toxicity It is generally recognized that *C. scutulatus* is one of the most dangerous rattlesnakes in North America. This is supported by its combined toxic qualities and moderate yield capabilities. However, significant variation in lethal toxicity has been reported in Arizona specimens (Glenn and Straight, 1978) (Table 20). Some, but not all, individuals from east central Arizona produce venom distinctly lower in lethality and are neutralized better by commercial antivenin. As yet, the geographic distribution of the less toxic variety is unknown and requires further investigation (also see Chapter 4). Also, systemic neurotoxic symptoms, which are often delayed, are observed with *C. scutulatus scutulatus* envenomations in the southern California and southern Arizona regions, whereas envenomations occurring in Phoenix to Flagstaff, Arizona, region may or may not produce these symptoms.

Venom Yield and Lethal Toxicity

Table 20 *Crotalus scutulatus* Venom Lethal Toxicity

Subspecies	Mouse LD_{50} (mg/kg)				Reference
	i.v.	i.p.	i.m.	s.c.	
C. scutulatus (scutulatus)	–	10.0[a]	–	–	Macht (1937)
	–	0.6[a] 0.7[a]	–	–	Githens and Wolff (1939c)
	0.14	0.14– 0.38	–	–	Emery and Russell (1963)
	0.21	0.23	–	0.31	Minton and Minton (1969)
	0.18	0.18	–	–	Kocholaty et al. (1971)
	0.18	–	–	–	Bieber et al. (1975)
	–	0.26	–	–	Hendon (1975)
	–	0.45[b]	–	–	Hendon (1975)
C. scutulatus scutulatus	0.21	0.23	–	–	Russell (1967)
	0.14	–	–	–	Pattabhiraman and Russell (1975)
	0.12– 0.18[c]	0.13– 0.54[c]	0.6– 0.9[c]	–	Glenn and Straight (1977, 1978)
	–	0.09– 0.12[d]	–	–	Glenn and Straight (1978)
	–	2.3– 3.8[e]	–	–	Glenn and Straight (1978)
	0.14	–	–	–	Pattabhiraman et al. (1978)
C. scutulatus salvini	–	0.18[f]	–	–	Glenn and Straight (1978)

[a]Minimum lethal dose.
[b]Venom stored frozen at -20°C for 1 year.
[c]Range of individual specimen venoms from California and Arizona (protein weight basis).
[d]Two individual specimen venoms from southwestern Utah (protein weight basis).
[e]Range of 8 individual specimen venoms from north central Arizona (protein weight basis).
[f]Protein weight basis.

Both *C. scutulatus scutulatus* and *C. scutulatus salvini* rank among the most toxic snake venoms in the New World. The principal lethal venom toxin in *C. scutulatus*, Mojave toxin, has biochemical and immunological similarities to crotoxin, the principal lethal venom toxin in *C. durissus terrificus, C. durissus terrificus* (var. *cumanensis*), and *C. vegrandis* (see Chapters 4 and 5). Both toxins exhibit neurotoxic activities. The less toxic variety of *C. scutulatus scutulatus* venom (venom B: Glenn and Straight, 1978) do not contain "Mojave toxin" (Alan Bieber, personal communication; see Chapter 4).

Crotalus stejnegeri (Long-Tailed Rattlesnake)

No toxicologic information has been reported concerning this small, little known Mexican rattlesnake (Figure 47). The longest recorded specimen is 63.8 cm (25.1 in.; Armstrong and Murphy, 1979). This species exhibits unusually long fangs for its size (Klauber, 1939).

Crotalus tigris (Tiger Rattlesnake)

Venom Yield The maximum length recorded for *C. tigris* (Figure 48) is 88.5 cm (34.8 in; Klauber, 1956). This species has a remarkably small head, accompanied by short, delicate fangs (Klauber, 1931). It is easily confused with *C. mitchellii,* which shares similarities in color and especially pattern. The small head dimensions of *C. tigris* are striking and appear disproportionate to body size. This feature is further corroborated by the venom yield data obtained from Klauber (1931, 1956) and Githens (1933). Although, Amaral (1928) reported an average dry yield of 60 mg from two (adult) *C. tigris.*

Githens (1933) obtained a maximum yield of 17 mg and an average yield of 10 mg per extraction from 14 fresh adults. Klauber (1931) averaged 11.6 mg per extraction from 11 large fresh adults. There were 3 additional extractions that lowered the average insignificantly to 11.0 (Klauber, 1956).

Venom Lethal Toxicity The only mouse lethal toxicity information for *C. tigris* is that of Githens and Wolff (1939c). These authors indicate that *C. tigris* produces a very toxic venom: mouse i.p. LD_{50} of 0.6 mg/kg. The venom was also extremely toxic to pigeons: 0.004 mg per 350 g (Githens and Wolff, 1939b). We suggest that *C. tigris* should be considered very dangerous, despite its low venom yield; further toxicology information on this species' venom would be useful.

Crotalus tortugensis (Tortuga Island Diamond Rattlesnake)

This species (Figure 49) is endemic to Tortuga Island in the Gulf of California, Mexico, and is closely allied to *C. atrox*. The maximum length reported is 105.8 cm (41.7 in; Klauber 1956). Klauber (1956) averaged 56 mg of dry venom from 6 large and 11 medium-sized fresh adults.

Venom Yield and Lethal Toxicity

We obtained yields of 67 mg and 103 mg from a 87.5 cm (34.5 in.) female and a 98 cm (38.6 in.) male, respectively, following 26 days of fasting, using manual extraction techniques.

No venom lethal toxicity information is presently available for this species.

Crotalus transversus (Cross-Banded Mountain Rattlesnake)

Only 12 specimens are scientifically known of this small, notably inoffensive, mountain form (Armstrong and Murphy, 1979) (Figure 50). The largest reported specimen is 45.9 cm (18.1 in.; Klauber, 1956); however, the maximum length may approach 61 cm (24 in.).

No venom yield or toxicology information is available.

Crotalus triseriatus (Dusky Rattlesnake)

Subspecies	Approximate maximum length (cm)
C. triseriatus triseriatus (central plateau dusky rattlesnake)	70
C. triseriatus aquilus (Queretaran dusky rattlesnake)	70
C. triseriatus armstrongi (Armstrong's dusky rattlesnake)	60

Crotalus triseriatus ssp. (Figures 51-53) are small, mountain rattlesnakes, quite common within their range in central Mexico (Klauber, 1952; Armstrong and Murphy, 1979). Other than a single large specimen of *C. triseriatus aquilus* that produced 36 mg of dry venom (Klauber, 1956), little is known about *C. tirseriatus* ssp. venom yield or toxicology. A 41.3 cm (16.3 in.) male *C. triseriatus triseriatus* produced manually extracted quantities of 5.9, 10.1, 8.7, and 11.6 mg of dry venom (at 2, 4, 6, and 8 months captivity, respectively; Glenn and Straight, unpublished). The mouse i.p. LD$_{50}$ of these individual venoms (dry weight) ranged from 3.7 to 4.4 mg/kg. From this meager data, *C. triseriatus* would be placed among the less dangerous rattlesnake varieties with low venom yields and low venom lethal toxicity.

Crotalus unicolor (Aruba Island Rattlesnake)

This species (Figure 54) is a very close relative of *C. durissus* forms and would likely be a subspecies of *C. durissus* if not for its insular isolation. We have been unable to locate any toxicological reports concerning *C. unicolor*. The largest reported specimen is 95 cm (37.4 in.; Klauber, 1956) and therefore should be considered dangerous, especially if venom toxicity is found to be similar to the *C. durissus* forms.

Crotalus vegrandis (Uracoan Rattlesnake)

Venom Yield This species (Figure 55) is apparently restricted to the savannas and alluvial plains of Anzoategui and Monagas, Venezuela, and is very similar taxonomically to *C. durissus* ssp. (Villalaz, 1966). The maximum length reported for this small form is 68.4 cm (26.9 in.; Villalaz, 1966). However, two *C. vegrandis* specimens at the Field Museum of Natural History (FMNH), Chicago, Illinois, measure 114 cm (male No. 210154) and 105.3 cm (female No. 210150) and presently represent record lengths for this species (personal communication, Harold Voris, FMNH). Villalaz (1966) obtained only 0.03 cc of venom per snake from 30 specimens (within 1 week of capture). The average dry yield was 6.03 mg per extraction, representing 226.3 mg/cc. Subsequently, Scannone et al. (1978) reported similar liquid venom volumes (0.02 ml per snake from 35 specimens, extracted at 25 day intervals). These data indicate that *C. vegrandis* specimens produce very low venom yields.

Venom Lethal Toxicity The venom composition results of Scannone et al. (1978) further verify the relationship of *C. vegrandis* to *C. durissus*. Also, the *C. vegrandis* venom i.v. LD_{50} of 0.2 mg/kg equals the lethal toxicity of *C. durissus terrificus* and places *C. vegrandis* among the leaders in rattlesnake lethal toxicity. This species should be considered dangerous owing to its venom composition, despite its low venom yield.

Crotalus viridis (Western Rattlesnake)

Subspecies	Approximate maximum length (cm)
C. viridis viridis (prairie rattlesnake)	175
C. viridis abyssus (Grand Canyon rattlesnake)	100
C. viridis caliginis (Coronado Island rattlesnake)	70
C. viridis cerberus (Arizona black rattlesnake)	110
C. viridis concolor (midget faded rattlesnake)	90
C. viridis helleri (southern Pacific rattlesnake)	145
C. viridis lutosus (Great Basin rattlesnake)	125
C. viridis nuntius (Hopi rattlesnake)	90
C. viridis oreganus (northern Pacific rattlesnake)	165

Venom Yield This widely distributed species has diversified into nine recognized subspecies and is the most common rattlesnake species in the western United States (Figures 56-64). Three subspecies are considered "stunted":

Venom Yield and Lethal Toxicity

Table 21 *Crotalus viridis* Venom Yield

Subspecies	Number of specimens or extractions	Venom yield (dry weight, mg) Average	Venom yield (dry weight, mg) Maximum	References
C. viridis viridis	19	50-90	—	Amaral (1928)
	523	41	135	Githens (1933)
	1676[a]	44	162	Klauber (1956)
C. viridis abyssus	1	60	—	Githens (1933)
	10[a]	97	137	Klauber (1956)
C. viridis caliginis		—	—	None
C. viridis cerberus	10	89	150	Githens (1933)
	15[a]	112	150	Klauber (1956)
C. viridis concolor	4[a]	22	—	Klauber (1956)
	13 males	18	34	Glenn and Straight (1977)
	4 females	9	13	Glenn and Straight (1977)
C. viridis helleri/ oreganus	880[a]	112	390	Klauber (1956)
C. viridis lutosus	98[a]	110	234	Klauber (1956)
	56	65	150	Githens (1933)
C. viridis nuntius	?	38	—	Klauber (1935)
	82[a]	51	72	Klauber (1956)
C. viridis oreganus	254	90	190	Githens (1933)
C. viridis oreganus/ helleri	880[a]	112	289	Klauber (1956)

[a]Calculated for (large) fresh adults.
Note: The *C. viridis helleri/oreganus* data were collected previous to the subspeciation of *C. viridis helleri* from *C. viridis oreganus*.

these are the Colorado Basin form, *C. viridis concolor,* the little Colorado Basin form, *C. viridis nuntius,* and the insular form, *C. viridis caliginis.* Venom yields of these small forms are considerably lower than the larger subspecies (Table 21), especially *C. viridis helleri, C. viridis oreganus,* and certain *C. viridis viridis* populations. Northern populations of *C. viridis viridis* attain greater length than the southern populations, and significant yield variation is expected within this subspecies. Owing to a general trend of comparatively larger head dimensions in *C. viridis helleri* and *C. viridis oreganus,* greater yields are obtained from these subspecies when compared by length than from other *C. viridis* spp. A generalized concept of venom yield among *C. viridis* subspecies is as follows: *C. viridis helleri* > *C. viridis oreganus* > *C. viridis lutosus* > *C. viridis cerberus* > *C. viridis abyssus* > *C. viridis viridis* > *C. viridis nuntius* > *C. viridis concolor* (and *C. viridis caliginis*).

Venom Lethal Toxicity An extreme subspecific variability in venom lethal toxicity occurs within the *C. viridis* forms. *Crotalus viridis concolor,* the Colorado Basin form, produces venom up to thirty times more lethal than the other *C. viridis* subspecies (Glenn and Straight, 1977; Table 22). Venom lethal toxicity of the other eight *C. viridis* subspecies are confined to reasonably close values, although the venom of *C. viridis viridis* juveniles is significantly more toxic (LD_{50} values) than the parental adults (Fiero, et al., 1972; Table 22).

Table 22 *Crotalus viridis* Venom Lethal Toxicity

Subspecies	Mouse LD_{50} (dry weight, mg/kg)				Reference
	i.v.	i.p.	i.m.	s.c.	
C. viridis viridis	–	2.05[a]	–	–	Macht (1937)
	–	1.25[a]	–	–	Githens and Wolff (1939c)
	1.20	–	–	–	Gingrich and Hohenadel (1956)
	1.61	2.25	–	–	Russell (1967)
	1.01	–	–	–	Friederich and Tu (1971)
	1.11	2.0	–	–	Kocholaty et al. (1971)
(juveniles)	–	–	–	5.5–7.10	Fiero et al. (1972)
(adults)	–	–	–	14.3–14.8	Fiero et al. (1972)
	–	2.0[b]	–	–	Glenn and Straight (1977)
C. viridis abyssus	–	4.6[a]	–	–	Macht (1937)
C. viridis caliginis	–	1.9–2.8[b]	–	–	Glenn and Straight (1977)

Venom Yield and Lethal Toxicity

Table 22 Continued

Subspecies	Mouse LD$_{50}$ (dry weight, mg/kg)				Reference
	i.v.	i.p.	i.m.	s.c.	
C. viridis cerberus	–	2.5[b]	6.0	–	Glenn and Straight (1977)
C. viridis concolor	0.28–0.48[b]	0.13–0.45[b]	0.8–1.3[b]	–	Glenn and Straight (1977)
	–	0.20	–	–	Glenn and Straight (1977)
C. viridis helleri	–	1.56	–	3.56	Minton (1956)
	1.84	–	–	–	Gingrich and Hohenadel (1956)
	1.0–1.44	1.4–1.82	–	–	Emery and Russell (1963)
	0.84	2.44	–	–	Kocholaty et al. (1971)
	1.85–2.13	–	–	–	Pattabhiraman and Russell (1973)
	2.0[b]	2.8[b]	5.1[b]	–	Glenn and Straight (1977)
	1.43	–	–	–	Pattabhiraman et al. (1978)
	1.2, 1.3	–	–	–	Schaeffer et al. (1978)
C. viridis lutosus	–	6.4[a]	–	–	Macht (1937)
	–	3.5[a]	–	–	Githens and Wolff (1939c)
	–	2.2	–	–	Russell (1967)
	–	1.9–2.2[b]	4.0–4.6[b]	–	Glenn and Straight (1977)
C. viridis nuntius	–	2.2[b]	–	–	Glenn and Straight (1977)
C. viridis oreganus	2.84	–	–	–	Gingrich and Hohenadel (1956)
	–	3.2[b]	3.6[b]	–	Glenn and Straight (1977)

[a]Minimum lethal dose.
[b]Protein weight basis. Range of lethal toxicity of individual specimen venom samples.

Crotalus willardi (Ridgenose Rattlesnake)

Subspecies	Approximate maximum length (cm)
C. willardi willardi (Arizona ridgenose rattlesnake)	60
C. willardi amabilis (Del Nido ridgenose rattlesnake)	60
C. willardi meridionalis (Southern ridgenose rattlesnake)	60
C. willardi obscurus (New Mexican ridgenose rattlesnake)	65
C. willardi silus (west Chihuahua ridgenose rattlesnake)	65

Klauber's (1956) report of 37 mg of dry venom obtained from one large adult *C. willardi willardi* is identical to Githens' data (1933) and represents the only venom yield data available for the species, specimens of which are shown in Figures 65-69. Likewise, venom toxicology information is scant for *C. willardi.* Githens and Wolff's (1939c) i.p. LD_{50} of 12.0 mg/kg (mouse) would indicate that this form may produce venom of very low lethal toxicity.

Sistrurus catenatus (Massasauga)

Subspecies	Approximate maximum length (cm)
S. catenatus catenatus (Eastern massasauga)	98
S. catenatus edwardsii (desert massasauga)	55
S. catenatus tergeminus (Western massasauga)	82

Venom Yield Sistrurus catenatus catenatus Figure 70; see also Figure 85) and *S. catenatus tergeminus* (Figure 72) are generally more robust than other *Sistrurus,* although *Sistrurus miliarius barbouri* occasionally approach these forms in both length and weight. *Sistrurus catenatus edwardsii* (Figure 71) is significantly smaller than *S. catenatus catenatus* and *S. catenatus tergeminus* and is regarded as less dangerous than the two larger subspecies.

Despite their small size, *Sistrurus* species are capable of producing serious envenomation in humans, especially *S. catenatus catenatus, S. catenatus tergeminus,* and *S. miliarius barbouri.* Venom yield data (Tables 23 and 24) indicate that these three varieties produce greater venom quantities than several species of *Crotalus,* including *C. intermedius, C. pricei, C. tigris, C. triseriatus* and *C. viridis concolor.* No venom yield information is available for *S. catenatus edwardsii.*

Venom Yield and Lethal Toxicity

Table 23 *Sistrurus catenatus* Venom Yield

Subspecies	Number of extractions or specimens	Venom Yield (dry weight, mg)		Reference
		Average	Maximum	
S. catenatus catenatus	69	14	33	Githens (1933)
	7[a]	31	–	Klauber (1956)
S. catenatus tergeminus	2[a]	37	–	Klauber (1956)

[a]Calculated for (large) fresh adults.

Table 24 *Sistrurus catenatus catenatus* Venom Lethal Toxicity

Mouse LD_{50} (mg/kg)			Reference
i.v.	i.p.	s.c.	
–	0.90	–	Githens and Wolff (1939c)
–	–	6.8	Schoettler (1951)
–	0.22	5.25	Minton (1956)

Lethal Toxicity The lethal toxicity of *S. catenatus catenatus* (Table 24) is greater than the mode of most rattlesnakes, and combined with venom yield data, this rattlesnake should be considered dangerous despite its comparatively small size. No venom toxicology information is available for *S. catenatus tergeminus* or *S. catenatus edwardsii.*

Sistrurus miliarius (Pygmy Rattlesnake)

Subspecies	Approximate maximum length (cm)
S. milarius miliarius (Carolina pygmy rattlesnake)	55
S. miliarius barbouri (Eastern pygmy rattlesnake)	80
S. miliarius streckeri (Western pygmy rattlesnake)	65

Venom Yield Of the three subspecies of *S. miliarius* (Figures 73-75) only *S. miliarius barbouri* attains lengths comparable to *S. catenatus* (*catenatus* and *tergeminus*). Despite of its name "pygmy rattlesnake," venom yields are some-

Table 25 *Sistrurus miliarius* Venom Yield

Subspecies	Number of specimens or extractions	Average venom yield (dry weight, mg)	Reference
S. miliarius	4	20	Amaral (1928)
S. miliarius barbouri	41	34.1	Allen and Maier (1941)
	13	34.6	Allen and Maier (1941)
	9[a]	18.0	Klauber (1956)

[a]Calculated for large fresh adult.

what surprising (Table 25), equaling the reported yields of *S. catenatus catenatus* and *S. catenatus tergeminus* and, as previously mentioned, surpassing the yields reported for the smaller *Crotalus* forms (*C. intermedius, C. pricei, C. tigris, C. triseriatus* and *C. viridis concolor*). No venom yield data is available for *S. miliarius miliarius* or *S. miliarius streckeri*.

Venom Lethal Toxicity The lethal toxicity of *S. miliarius barbouri* is apparently low (Table 26) in comparison with other rattlesnakes. However, *Sistrurus miliarius barbouri* can produce relatively severe local and occasionally systemic symptoms in human envenomation similar to the findings of Vick (1971), who reported *S. miliarius barbouri* venom causes severe edema, necrosis, and general tissue breakdown in dogs. No toxicology information is available for the smaller forms, *S. miliarius miliarius* and *S. miliarius streckeri*.

Table 26 *Sistrurus miliarius barbouri* Venom Lethal Toxicity

Mouse LD_{50} (mg/kg)			
i.v.	i.p.	s.c.	Reference
—	6.0[a]	—	Githens and Wolff (1939c)
—	—	24.25	Minton (1956)
12.6	—		Friederich and Tu (1971)
4.47[b]	—	—	Vick (1971)
2.80	6.84	—	Kocholaty et al. (1971)

[a]Minimum lethal dose.
[b]LD_{50}.

Venom Yield and Lethal Toxicity

Sistrurus ravus (Mexican Pygmy Rattlesnake)

Subspecies	Approximate maximum length (cm)
S. ravus ravus (Mexican pygmy rattlesnake)	70
S. ravus brunneus (Oaxacan pygmy rattlesnake)	70
S. ravus exiguus (Guerreran pygmy rattlesnake)	70

Venom Yield and Lethal Toxicity *Sistrurus ravus* subspecies (Figures 76-78) are geographically remote from the other *Sistrurus* forms and little is known concerning their venom characteristics. Minton (1977) obtained only 7 mg of dry venom and an i.v. LD_{50} of 3.17 mg/kg from a single specimen of *S. ravus* (ssp). He therefore expressed the view that this race is doubtful as a dangerous species based on its small size and venom yield. No other comparable data are available.

Comparative Lethal Capacity of Rattlesnakes

Rattlesnakes vary greatly in both the yield and lethal toxicity of their venoms as described in this chapter. All rattlesnakes are a potential hazard to man; however, some species/subspecies are capable of inflicting more serious envenomations than others.

The greatest venom yields are produced by the *Atrox* group (*Crotalus adamanteus, C. atrox,* and *C. ruber*), followed by certain species of the *Durissus* and *Viridis* groups [*C. durissus* from Mexico and Central America, *C. molossus, C. mitchellii (pyrrhus* and *stephensi),* and *C. viridis (helleri* and *oreganus)*]. Venom lethal toxicity data have been reported on only 35 of the 76 rattlesnake taxa (46%). The reported mouse LD_{50} values range from 0.13 to 12.6 mg/kg intravenously, from 0.09 to 7.0 mg/kg intraperitoneally, and from 0.3 to 28.0 mg/kg subcutaneously. Approximately 50% of the subspecies tested are capable of producing LD_{50} (i.v. or i.p.) values of less than 2.0 mg/kg. Eight subspecies produce LD_{50} (i.v. or i.p.) values of 0.3 mg/kg or less (*C. durissus terrificus, C. horridus atricaudatus, C. mitchellii mitchellii, C. scutulatus scutulatus, C. scutulatus salvini, C. vegrandis, C. viridis concolor,* and *Sistrurus catenatus catenatus*).

Substantial geographic differences in venom lethality occur in *C. lepidus klauberi, C. scutulatus scutulatus,* and possibly, *C. horridus atricaudatus.* There is also evidence that adolescent specimens of *C. atrox, C. horridus atricaudatus,* and *C. viridis viridis* produce venoms qualitatively different than adults, including greater lethal toxicity. Such geographic and ontogenic differences may also occur in other rattlesnake venoms.

Table 27 Total Mouse i.p. LD_{50} Doses Present in the Average and Maximum Venom Yields of Rattlesnakes

Species	Venom lethal toxicity[a] i.p. LD_{50} (mg/kg)		Venom yield (mg) Average	Venom yield (mg) Maximum	Total LD_{50} doses[b] Average	Total LD_{50} doses[b] Maximum
Crotalus horridus atricaudatus[c]	0.4	(0.3–0.5)	140	300	17,500	37,500
Crotalus scutulatus scutulatus[c]	0.24	(0.13–0.54)	70	150	14,583	31,250
Crotalus durissus durissus	0.71	(0.67–0.80)	200	500	14,084	35,212
Crotalus adamanteus	2.1	(1.7–2.7)	400	1000	9,524	23,810
Crotalus durissus terrificus	0.22	(0.13–0.40)	36	100	8,182	22,728
Crotalus mitchellii mitchellii	0.33	(0.26–0.40)	33	90	5,000	13,636
Crotalus durissus totonacus	2.5	(?)	(250)[d]	(600)[d]	5,000	12,000
Crotalus mitchellii pyrrhus	2.3	(2.13–2.5)	200	350	4,348	7,609
Crotalus atrox	5.0	(3.7–8.4)	400	1150	4,000	11,500
Crotalus ruber ruber	4.6	(4.2–5.1)	350	670	3,804	7,283
Crotalus ruber lucasensis	4.0	(?)	230	710	2,875	8,875
Crotalus viridis helleri	2.0	(1.4–2.8)	112	390	2,800	9,750
Crotalus viridis lutosus	2.1	(1.9–2.2)	110	240	2,619	5,714
Crotalus viridis concolor	0.25	(0.13–0.45)	13	34	2,600	6,800
Sistrurus catenatus catenatus	0.5	(0.2–0.9)	20	(50)[d]	2,000	5,000
Crotalus horridus horridus	4.8	(1.5–7.8)	140	300	1,458	3,125
Crotalus viridis oreganus	3.2	(?)	90	190	1,406	2,969
Crotalus horridus atricaudatus[e]	5.3	(4.0–6.6)	140	300	1,321	2,830
Crotalus scutulatus scutulatus[e]	3.0	(2.3–3.8)	70	150	1,167	2,500
Crotalus viridis viridis	2.1	(2.0–2.3)	44	165	1,048	3,929
Crotalus cerastes ssp.	2.4	(?)	30	80	625	1,667
Sistrurus miliarius barbouri	6.8	(?)	30	(50)[d]	221	368

[a] Values obtained from each species account in preceding section of this chapter. Numbers in parentheses are ranges.
[b] Number of mean i.p. LD_{50} values in the average and maximum yield.
[c] Most toxic variety (see preceding section of this chapter).
[d] Estimated.
[e] Less toxic variety (see preceding section of this chapter).

Table 27 compares various rattlesnake taxa in order of their decreasing lethal capacity, based on the total mouse i.p. LD_{50} doses in the average venom yield of each species/subspecies listed. Although mouse data cannot justifiably be extrapolated to humans or other animals, the list serves to demonstrate the extreme range in potential lethal capacity among the rattlesnakes. Considering that even the least toxic rattlesnakes, with the lowest average venom yields, can produce severe local and occasionally systemic symptoms in human envenomation, the capabilities of the more dangerous variety are quite impressive.

ACKNOWLEDGMENTS

We especially would like to thank Louis Porras, whose photographic contributions and helpful suggestions were of considerable value. We also appreciate the artistic work of Kenneth Stockton in preparing the distribution maps and *Crotalus* and *Sistrurus* head drawings and Kerry Matz for the venom apparatus drawing. To James R. McCranie we are indebted for critically reviewing this manuscript and providing helpful comments. We are also grateful for the valuable assistance of the following persons and institutions: Glenda McNulty, typing, and Paul Smith, Ron Orbacz, Joe Radford, Grant Minson, and Wallace Coleman, illustrations; John Ottley, Brigham Young University; Harold Voris, Chicago Field Museum of Natural History; Joseph Beraducci, Ed Cassano and Jim Bridges, Miami, Florida; John Tashjian, San Marcos, California; William Dennler, Toledo Zoo; Gordon Schuett, University of Toledo; Kenneth Stocks, Dover, Arkansas; Crawford Jackson and the library staff of the San Diego Museum of Natural History; Jim Bacon and the library staff of the San Diego Zoological Park; Lamar Farnsworth, Phillip Leonard, Robert Larson, and Mike Coffeen, Hogle Zoological Gardens; the Herpetology staffs of The Dallas Zoo, Houston Zoological Gardens, Steinhart Aquarium, and the Arizona-Sonora Desert Museum; and the Editorial Committee of the University of California Press, Berkeley, California. Research related to this chapter was supported by the Veterans Administration Medical Research Program and Hogle Zoological Gardens, Salt Lake City, Utah.

REFERENCES

Allen, R., and Maier, E. (1941). The extraction and processing of snake venom. *Copeia* 4:248-252.

Amaral, A. D. (1928). Studies on Snake venoms: I. Amounts of venom secreted by nearctic pit vipers. *Bull. Antivenin Inst. Am.* 1(4):103-104.

Armstrong, B. L., and Murphy, J. B. (1979). *The Natural History of Mexican Rattlesnakes*, E. O. Wiley and J. T. Collins (Eds.). *Univ. Kans. Mus. Natur. Hist. Spec. Publ.* (5): 88 pp.

Bechtel, H. B. (1978). Color and pattern in snakes (Reptilia, Serpentes). *J. Herpetol. 12*(4):521-532.

Belluomini, H. E. (1968). Extraction and quantities of venom obtained from some Brazilian snakes. In *Venomous Animals and Their Venoms*, vol. 1, W. Bucherl, E. E. Buckley, and V. Deulofeu (Eds.). Academic Press, New York, pp. 97-117.

Bieber, A. L., Tu, T., and Tu, A. T. (1975). Studies of an acidic cardiotoxin isolated from the venom of Mojave rattlesnake (*Crotalus scutulatus*). *Biochim. Biophys. Acta 400*:178-188.

Bolaños, R. (1972). Toxicity of Costa Rican snake venoms for the white mouse. *Am. J. Trop. Med. Hyg. 21*(3):360-363.

Bonilla, C. A., Fiero, M. K., and Frank, L. P. (1971). Isolation of a basic protein neurotoxin from *Crotalus adamanteus venom*. In *Toxins of Animal and Plant Origin*, vol. 1, A. deVries and E. Kochva (Eds.) Gordon and Breach Science Publishers, New York, pp. 343-357.

Bonilla, C. A., Abel, J. H., and Nelson, A. W. (1972). Protein toxin induced myocardial infarction *Technition Congress Proceedings* (program; abstract), New York, p. 65.

Bonilla, C. A., Faith, M. R., and Minton, S. A. (1973). L-amino acid oxidase phospodiesterase, total protein and other properties of Juvenile Timber rattlesnake (*C. h. horridus*) venom at different stages of growth. *Toxicon, 11:*301-303.

Brazil, O. V., Franceschi, J. P., and Waisbich, E. (1966). Pharmacology of crystalline crotoxin: I. Toxicity. *Mem. Inst. Butantan 33*(3):973-980.

Broad, A. J., Sutherland, S. K., and Coulter, A. R. (1979). The lethality in mice of dangerous Australian and other snake venoms. *Toxicon 17*:661-664.

Campbell, J. A. (1979). A new rattlesnake (Reptilia, Serpentes, Viperidae) from Jalisco, Mexico. *Trans. Kans. Acad. Sci. 81*(4):365-369.

Campbell, J. A., and Armstrong, B. L. (1979). Geographical variation in the Mexican pygmy rattlesnake, *Sistrurus ravus*, with the description of a new subspecies. *Herpetologica 35*(4):304-317.

Cliff, F. S. (1954). Snakes of the islands in the Gulf of California, Mexico. *Trans. San Diego Soc. Natur. Hist. XII*(5):67-98.

Conant, R. (1975). *A Field Guide to Reptiles and Amphibians of Eastern and Central North America*, Houghton Mifflin Co., Boston, Mass, pp. 230-238.

Criley, B. R. (1956). Development of a multivalent antivenin. In *Venoms*, E. E. Buckley and N. Porges (Eds.), American Association for the Advancement of Science, Washington, D.C., Pub. 44, pp. 373-380.

Deichmann, W. B., Radomski, J. L., Farrell, J. J., MacDonald, W. E., and Keplinger, M. L. (1958). Acute toxicity and treatment of intoxications due to *Crotalus adamanteus* (rattlesnake venom). *Am. J. Med. Sci. Aug.*:204-207.

DeLucca, F. L., and Imaizumi, M. T. (1972). Synthesis of ribonucleic acid in the venom gland of *Crotalus durissus terrificus* (Ophida, Reptilia) after manual extraction of the venom. *Biochem. J. 130*:335-342.

DeLucca, F. L., Haddad, A., Kochva, E., Rothschild, A. M., and Valeri, V. (1974). Protein synthesis and morphology changes in the secretory epithelium of the venom gland of *Crotalus durissus terrificus* at different times after manual extraction of venom. *Toxicon 12*:361-368.

Ditada, I. E., Martori, R. A., Doucet, M. E., and Abalos, J. W. (1978). Venom yield with different milking procedures. In *Toxins: Animal, Plant and Microbial*, P. Rosenberg (Ed.). Pergamon Press, Elmsford, N.Y., pp. 3-7.

Emery, J. A., and Russell, F. E. (1963). Lethal and hemorrhagic properties of some North American snake venoms. In *Venomous and Poisonous Animals and Noxious Plants of the Pacific Region*, H. L. Keegan and W. V. MacFarlane (Eds.). Pergamon Press, Elmsford, N. Y., pp. 409-413.

Fein, A., Bdolah, A., and Kochva, E. (1971). Developmental pattern of enzyme secretion in the embryonic venom glands of *Vipera palaestinae (Ophida, Reptilia). Develop. Biol. 24*:520-532.

Fiero, M. K., Siefert, M. W., Weaver, T. J., and Bonilla, C. A. (1972). Comparative study of juvenile and adult prairie rattlesnake (*Crotalus viridis viridis*) venoms. *Toxicon 10*:81-82.

Friederich, C., and Tu, A. T. (1971). Role of metals in snake venoms for hemorrhagic esterase and proteolytic activities. *Biochem. Pharmacol. 20*:1549-1556.

Gans, C., and Kochva, E. (1965). The accessory gland in the venom apparatus of viperid snakes. *Toxicon 3*:61-63.

Gennaro, J. F., Leopold, R. S., and Merriam, W. M. (1961). Observations on the actual quantity of venom introduced by several species of crotalid snakes in their bite. *Anat. Rec. 139:*303.

Gennaro, J. F., Callahan, III, W. P., and Lorincz, A. F. (1963). The anatomy and biochemistry of a mucus-secreting cell type present in the poison apparatus of the pit viper *Ancistrodon piscivorus piscivorus. Ann. N.Y. Acad. Sci. 106*:463-471.

George, I. D. (1930). Notes on the extraction of venom at the serpentarium of the Antivenin Institute at Tela, Honduras. *Bull. Antivenin Inst. Am. IV*(3): 57-59.

Gingrich, W. C., and Hohenadel, J. C. (1956). Standardization of polyvalent antivenin. In *Venoms*, E. E. Buckley and N. Porges (Eds.). American Association for the Advancement of Science, Washington, D.C., Pub. 44, pp. 381-385.

Githens, T. S. (1933). Data of Mulford Biological Laboratories. In *Venom II* (Klauber Library), Herpetology Dept. San Diego Mus. Natur. Hist., San Diego, Calif.

Githens, T. S., and Wolff, N. O'C, (1939a). The polyvalency of crotalidic antivenins. I. The influence of the composition of polyvalent antigens. *J. Immunol. 37*(1):33-39.

Githens, T. S., and Wolff, N. O'C. (1939b). The polyvalency of crotalidic antivenins. II. Comparison of polyvalent crotalidic antivenin with monovalent *Crotalus durissus durissus* antivenin. *J. Immunol.* 37(1):41-45.

Githens, T. S., and Wolff, N. O'C. (1939c). The polyvalency of Crotalidic antivenins. III. Mice as test animals for study of antivenins. *J. Immunol.* 37(1): 47-51.

Glenn, J. L., and Straight, R. C. (1977). The midget faded rattlesnake (*Crotalus viridis concolor*) venom: Lethal toxicity and individual variability. *Toxicon* 15:129-133.

Glenn, J. L., and Straight, R. C. (1978). Mojave rattlesnake *Crotalus scutulatus scutulatus* venom: Variation in toxicity with geographical origin. *Toxicon* 16: 81-84.

Glenn, J. L., and Straight, R. C. (1979). Venomoid snakes: A discussion. *Second Annual Reptile Symposium of Captive Propagation Husbandry*, Case Western Reserve University, published by Catoctin Mtn. Zoo, Thurmont, Md., pp.69-78.

Glenn, J. L., Straight, R. C., and Snyder, C. C. (1972). Yield of venom obtained from *Crotalus atrox* by electrical stimulation. *Toxicon* 10:575-579.

Glenn, J. L., Straight, R., and Snyder, C. C. (1973). Surgical technique for isolation of the main venom gland of viperid, crotalid and elapid snakes. *Toxicon* 11:231-233.

Glenn, J. L., Straight, R. C., and Snyder, C. C. (1975). Crotalidae reproduction data. *Utah Herpetol. League J.* 2(1):15-20.

Glissmeyer, H. R. (1951). Egg production in the Great Basin rattlesnake. *Herpetologica* 7:24-27.

Gloyd, H. K. (1935). Some aberrant color patterns in snakes. *Pap. Mich. Acad. Sci. Arts Lett.* 20:661-668.

Gloyd, H. K. (1940). *The Rattlesnakes, Genera Sistrurus and Crotalus.* Chicago Acad. Sci. Spec. Publ. 4, 266 pp. (Reprinted in 1978, Society for the Study of Amphibians and Reptiles.)

Gloyd, H. K. (1958). Aberrations in the color patterns of some crotalid snakes. *Bull. Chicago Acad. Sci.* 10(12):185-195.

Gonçalves, J. M., and Deutsch, H. F. (1956). Ultracentrifugal and zone electrophoresis studies of some crotalidae venoms. *Arch. Biochem. Biophys.* 60: 402-411.

Gornall, A. G., Bardawill, G. J., and David, M. M. (1949). Determination of serum proteins by means of the biuret reaction. *J. Biol. Chem.* 177:751-754.

Hall, H. P., and Gennaro, J. F. (1961). The relative toxicities of rattlesnake (*Crotalus adamanteus*) and cottonmouth (*Ancistrodon piscivorus*) venom for mice and frogs. *Anat. Rec.* 139:305-306.

Harris, Jr., H. S. (1974). The New Mexican ridge-nosed rattlesnake. *Nat. Parks Conserv. Mag.* 48(3):22-24.

Harris, Jr., H. S., and Simmons, R. S. (1976). The paleogeography and evolution of *Crotalus willardi*, with a formal description of a new subspecies from New Mexico, United States. *Bull. Maryland Herpetol. Soc.* 12(1):1-22.

Harris, Jr., H. S., and Simmons, R. S. (1977). A preliminary account of insular rattlesnake populations, with special reference to those occurring in the Gulf

of California and off the Pacific coast. *Bull. Maryland Herpetol. Soc. 13*(2): 92-110.
Harris, Jr., H. S., and Simmons, R. S. (1978). A preliminary account of the rattlesnakes with the descriptions of four new subspecies. *Bull. Maryland Herpetol. Soc. 14*(3):105-211.
Hendon, R. A. (1975). Preliminary studies on the neurotoxin in the venom of *Crotalus scutulatus* (Mojave rattlesnake). *Toxicon 13*:477-482.
Hendon, R. A., and Tu, A. T. (1979). The role of crotoxin subunits in tropical rattlesnake neurotoxic action. *Biochim. Biophys. Acta 578*:243-252.
Heyrend, F. L., and Call, A. (1951). Growth and age in western striped racer and Great Basin rattlesnake. *Herpetologica 7*:28-40.
Hoge, R. A. (1965). Preliminary account on neotripical Crotalinae (Serpentes Viperidae). *Mem. Inst. Butantan 32*:109-184.
Hoge, R. A., and Romano, S. A. (1971). Neotropical pit vipers, sea snakes, and coral snakes. In *Venomous Animals and Their Venoms*, vol. 2, W. Bucheral and E. E. Buckley (Eds.). Academic Press, New York, pp. 211-293.
Houssay, B. A. (1923). Quantities de venin fournies par les serpente venimeux de L'Argentine. *Comp. Rend. Soc. Biol. 89*:449-451.
Jimenez-Porras, J. M. (1961). Biochemical studies on venom of the rattlesnake *Crotalus atrox atrox. J. Exp. Zool. 148*:251-258.
Johnson, B. D., Hoppe, J., Rogers, R., and Stanhke, H. L. (1968). Characteristics of venom from the rattlesnake *Crotalus horridus atricaudatus. J. Herpetol. 2*(3-4):107-112.
Johnson, C. M. (1938). A new method for stripping venomous snakes. *Am. J. Trop. Med. 18*(4):385-386.
Klauber, L. M. (1928). The collection of rattlesnake venom. *Bull. Antivenin Inst. Am. II*(1):11-18.
Klauber, L. M. (1930). Differential characteristics of southwestern rattlesnakes allied to *Crotalus atrox. Bull. Zool. Soc. San Diego 6*:1-73.
Klauber, L. M. (1931). *Crotalus tigris* and *Crotalus enyo,* two little known rattlesnakes of the southwest. *Trans. San Diego Soc. Natur. Hist. VI*(24):353-370.
Klauber, L. M. (1935). A new subspecies of *Crotalus confluentus*, the prairie rattlesnake. *Trans. San Diego Soc. Natur. Hist. VIII*(13):75-90.
Klauber, L. M. (1936). *Crotalus mitchellii,* the speckled rattlesnake. *Trans. San Diego Soc. Natur. Hist. VIII*(19):149-184.
Klauber, L. M. (1937). A statistical study of the rattlesnakes: The growth of the rattlesnake. *Occ. Pap. San Diego Soc. Natur. Hist.* (3):1-56.
Klauber, L. M. (1938). A statistical study of the rattlesnakes: Head dimensions. *Occ. Pap. San Diego Soc. Natur. Hist.* (4):1-53.
Klauber, L. M. (1939). A statistical study of the rattlesnake: Fangs. *Occ. Pap. San Diego Soc. Natur. Hist.* (5):1-61.
Klauber, L. M. (1944). The sidewinder, *Crotalus cerastes,* with description of a new subspecies. *Trans. San Diego Soc. Natur. Hist. 10*(8):91-126.

Klauber, L. M. (1952). Taxonomic studies of the rattlesnakes of mainland Mexico. *Bull. Zool. Soc. San Diego* (26):1-143.

Klauber, L. M. (1956). *Rattlesnakes: Their Habits, Life Histories and Influence on Mankind*, vols. I and II, Univ. Calif. Press, Berkeley, 1476 pp. (The 1956 edition was partially revised in 1972, 1533 pp.)

Kocholaty, W. F., Goetz, J. C., Ashley, B. D., Billings, T. A., and Ledford, E. B. (1968). Immunogenic response of the venoms of fer-de-lance, *Bothrops asper asper*, and La Cascabella, *Crotalus durissus durissus*, following photooxidative detoxification. *Toxicon* 5:153-158.

Kocholaty, W. F., Ledford, E. B., Daly, J. G., and Billings, T. A. (1971). Toxicity and some enzymatic properties and activities in the venoms of Crotalidae, Elapidae and Viperidae. *Toxicon* 9:131-138.

Kochva, E. (1960). A quantitative study of venom secretion by *Vipera palaestinae*. *Am. J. Trop. Hyg.* 9:381-390.

Kochva, E. (1978). Oral glands of the Reptilia. In *Biology of the Reptilia*, vol. 8, chapter 2, C. Gans, (Ed.). Academic Press, New York, pp. 43-161.

Kochva, E., and Gans, C. (1965). The venom gland of *Vipera palaestinae* with comments on the glands of some other viperines. *Acta Anat.* 62:365-401.

Kochva, E., and Gans, C. (1966). Histology and histochemistry of venom glands of some crotaline snakes. *Copeia* 3:506-515.

Macht, D. I. (1937). Comparative toxicity of sixteen specimens of *Crotalus* venom. Proc. Soc. Exp. Biol. Med. 36(4):499-501.

McCranie, J. R., and Wilson, L. D. (1979). Commentary on taxonomic practice in regional herpetological publications. *Herpetol. Rev.* 10(1): 18-21.

Minton, S. A. (1953). Variation in venom samples from copperheads (*Agkistrodon contortrix mokeson*) and timber rattlesnakes (*Crotalus horridus horridus*). *Copeia* 4:212-215.

Minton, S. A. (1956). Some properties of North American pit viper venoms and their correlation with phylogeny. In *Venoms*, E. E. Buckley, and N. Porges (Eds.). American Association for the Advancement of Science, Washington, D.C., Pub. 44, pp. 145-151.

Minton, S. A. (1957). Variation in yield and toxicity of venom from a rattlesnake (*Crotalus atrox*). *Copeia* 4:265-268.

Minton, S. A. (1958). Observations on the amphibians and reptiles of the Big Bend region of Texas. *Southwest. Natural.* 3:28.

Minton, S. A. (1967). Observations of toxicity and antigenic makeup of venoms from juvenile snakes. In *Animal Toxins*, F. E. Russell and P. R. Saunders (Eds.). Pergamon Press, Elmsford, N.Y., pp. 211-222.

Minton, S. A. (1975). A note on the venom of an aged rattlesnake. *Toxicon* 13:73-74.

Minton, S. A. (1977). Toxicity of venoms from some little known Mexican rattlesnakes. *Toxicon* 15:580-581.

Minton, S. A., and Minton, M. R. (1969). In *Venomous Reptiles*. Scribner's New York, pp. 208-219.

Mitchell, S. W. (1861). Researches upon the venom of rattlesnakes. *Smithsonian Contrib. Knowl. 1*:16-29.

Moran, J. B., and Geren, C. R. (1979). A comparison of biological and chemical properties of three North American (Crotalidae) snake venoms. *Toxicon 17* (3):237-244.

Nickerson, M. A., and Mays, C. E. (1968). More aberrations in the color patterns of rattlesnakes (genus *Crotalus*). *Wasmann J. Biol. 26*(1):125-131.

Oron, U., and Bdolah, A. (1973). Regulation of protein synthesis in the venom gland of viperid snakes. *J. Cell. Biol. 56*:177-190.

Oron, U., Kinamon, S., and Bdolah, A. (1978). Asynchrony in the synthesis of secretory proteins in the venom gland of the snake *Vipera palaestinae*. *Biochem. J. 174*:733-739.

Pattabhiraman, T. R., and Russell, F. E. (1973). A lethal protein from the venom of the southern Pacific rattlesnake *Crotalus viridis helleri*. *Proc. West. Pharmacol. Soc. 16*:107-110.

Pattabhiraman, T. R., and Russell, F. E. (1975). Isolation and purification of the toxic fractions of Mojave rattlesnake venom. *Toxicon 13*:291-294.

Pattabhiraman, T. R., Russell, F. E., and Whigham, H. (1978). Some chemical and physiopharmacological properties of fractions from the venoms of *Crotalus viridis helleri* and *Crotalus scutulatus scutulatus*. In *Toxins: Animal, Plant and Microbial,* P. Rosenberg (Ed.), Pergamon Press, Elmsford, N.Y., pp. 211-222.

Perez, J. C., Haws, W. C., Garcia, V. E., and Jennings, III, B. M. (1978). Resistance of warm-blooded animals to snake venoms. *Toxicon 16*:375-383.

Pisani, G. R., Collins, J. T., and Edwards, S. R. (1972). A re-evaluation of the subspecies of *Crotalus horridus*. *Trans. Kans. Acad. Sci. 75*(3):255-263.

Possani, L. D., Sosa, B. P., Alagon, A. C., and Burchfield, P. M. (1980). The venom from the snakes *Agkistrodon bilineatus taylori* and *Crotalus durissus totonacus*: Lethality, biochemical and immunological properties. *Toxicon 18*: 356-360.

Radcliffe, G. W., and Maslin, T. P. (1975). A new subspecies of the red rattlesnake, *Crotalus ruber,* from San Lorenzo Sur Island, Baja California Norte, Mexico. *Copeia 3*:490-493.

Reid, H. A., and Theakston, R. D. G. (1978). Changes in coagulation effects by venoms of *Crotalus atrox* as snakes age. *Am. J. Trop. Med. Hyg. 27*(5):1053-1057.

Rodriguez, O. G., and Scannone, H. R. (1976). Fractionation of *Crotalus durissus cumanensis* venom by gel filtration. *Toxicon 14*:400-403.

Rodriguez, O. G., Scannone, H. R., and Parra, N. D. (1974). Enzymatic activities and other characteristics of *Crotalus durissus cumanensis* venom. *Toxicon 12*: 297-302.

Russell, F. E. (1967). Pharmacology of animal venoms. *Clin. Pharmacol. Ther. 8*: 849-873.

Russell, F. E., and Emery, J. A. (1959). Use of the chick in zootoxicologic studies on venoms. *Copeia 1*:73-74.

Russell, F. E., and Eventov, R. (1964). Lethality of crude and lyophilized *Crotalus* venom. *Toxicon* 2:81-82.

Russell, F. E., Emery, J. A., and Long, T. E. (1960). Some properties of rattlesnake venom following 26 years storage. *Proc. Soc. Exp. Biol. Med. 103*: 737-739.

Ruzic, N., and Russell, F. E. (1978). Studies on the main and accessory venom glands of *Crotalus atrox*. *Period. Biol. 80*(Suppl. 1):55-58.

Scannone, H. R., Rodriguez, O. G., and Lancini, A. R. (1978). Enzymatic activities and other characteristics of *Crotalus vegrandis* snake venom. In *Toxins: Animal, Plant and Microbial*. P. Rosenberg (Ed.) Pergamon Press, Elmsford, N. Y., pp. 223-229.

Schaeffer, R. C., Carlson, R. W., Whigham, H., Weil, M. H., and Russell, F. E. (1978). Acute hemodynamic effects of rattlesnake, *Crotalus viridis helleri*, venom. In *Toxins: Animal, Plant and Microbial*, P. Rosenberg (Ed.). Pergamon Press, Elmsford, N. Y., pp. 383-391.

Schoettler, W. H. A. (1951). Toxicity of the principal snake venoms of Brazil. *Am. J. Trop. Med. Hyg. 31*:489-499.

Shaham, N., and Kochva, E. (1969). Localization of venom antigens in the venom gland of *Vipera palaestinae* using a fluorescent-antibody technique. *Toxicon* 6:263-268.

Smith, H. M., Smith, R. B., and Sawin, H. L. (1975). The authorship and date of publications of *Siren intermedia* (Amphibia Caudata). *Great Basin Natural. 35*(1):100-102.

Soule, M., and Sloan, A.J. (1966). Biogeography and distribution of the reptiles and amphibians on islands in the Gulf of California, Mexico. *Trans. San Diego Soc. Natur. Hist. 14*(11):137-156.

Stadleman, R. E. (1928). The poisoning power of the newborn copperhead with case report. *Bull. Antivenin Inst. Am. II*(3): 67-69.

Stadleman, R. E. (1929). Some venom extraction records. *Bull. Antivenin Inst. Am. II*(1):29.

Tanner, W. W. (1966). A new rattlesnake from western Mexico. *Herpetologica* 22(4):298-302.

Tanner, W. W., Dixon, J. R., and Harris, Jr., H. S. (1972). A new subspecies of *Crotalus lepidus* from western Mexico. *Great Basin Natural. 32*(1):16-24.

Theakston, R. D. G., and Reid, H. A. (1978). Changes in the biological properties of venom from *Crotalus atrox* with aging. *Period. Biol. 80*(Suppl. 1): 123-133.

Tomes, C. S. (1877). On the development and succession of the poison-fangs of snakes. *Philos. Trans. 166*:377-385.

Tu, A. T., Homma, M., Hong, B-S., and Terrill, J. B. (1970). Neutralization of rattlesnake venom toxicities by various compounds. *J. Clin. Pharmacol. 10* (5):323-329.

Vick, J. A. (1971). Symptomology of experimental and clinical envenomation. In *Neuropoisons: Their Pathophysiological Actions*. vol. 1, L. L., Simpson, (Ed.). Plenum, New York, pp. 71-86.

Vick, J. A., Ciuchta, H. P., and Manthei, J. H. (1966). Pathophysiological studies on ten snake venoms. In *Animal Toxins,* F. E. Russell and P. R. Saunders (Eds.). Pergamon Press, Elmsford, N.Y., pp. 269-282.

Villalaz, A. R. L. (1966). *Crotalus vegrandis* Klauber: Redescripcion y distribucion. *Mem. Inst. Butantan 33*(3):725-734.

Warshawsky, H., Haddad, A., Goncalves, R. P., Valeri, V., and De Lucca, F. L. (1972). Protein synthesis and secretion by the venom gland of the rattlesnake *Crotalus durissus terrificus* as revealed by electron microscopic radioautography after ^{3}H-tyrosine injection. *Anat. Rec. 172*:422.

Warshawsky, H., Haddad, A., Goncalves, R. P., Valeri, V., and De Lucca, F. L. (1973). Fine structure of the venom gland epithelium of the South American rattlesnake and radioautographic studies of protein formation by the secretory cells. *Am. J. Anat. 138*:79-120.

Zertuche, J. J., and Trevino, C. H. (1978). Una nueva subspecie de *Crotalus lepidus* encontrada en nuevo leon. *II Congreso Nacional De Zoologic.* Rusumenes Univ. Autonoma De Chapingo, Monterrey, N. L., Mexico.

2
Physiological and Pharmacological Effects of Rattlesnake Venoms

BARBARA J. HAWGOOD
University of London, London, England

Introduction   121

Effects on the Cardiovascular System   122
Arterial Blood Pressure • Cardiac Function • Release of Autacoids • Hemostasis

Effects on the Respiratory System   132
Reflex Stimulation • Motor Paralysis

Effects on the Somatic Nervous System and Skeletal Muscle   136
Neuromuscular Transmission • Skeletal Muscle

Effects on Other Systems   149
Disturbance of Motor Coordination and Convulsions • Visceral Smooth Muscle and the Autonomic Nervous System

Conclusion   152

References   153

INTRODUCTION

Clinical reports of rattlesnake poisoning list a large variety of symptoms which are presented to the clinician who is treating bites (see Chapters 6 and 7). This is probably a reflection both of the diversity of species within the rattlesnake family and of the diversity in number and concentration of components within each venom. The aim of this chapter is to examine the reported effects on physiological systems of whole rattlesnake venoms as well as their purified components in order to assess how they affect humans and animals. Detailed reviews up to 1976 have been published (Tu, 1977; Lee, 1979a) and a full coverage of this literature will not be attempted.

EFFECTS ON THE CARDIOVASCULAR SYSTEM

Arterial Blood Pressure

A marked fall in arterial blood pressure is a common characteristic of the action of rattlesnake venoms after intravenous (i.v.) injection into anesthetized animals. An immediate fall was observed with doses as low as 20-50 µg/kg of the venoms of *Crotalus atrox, Crotalus horridus,* and *Crotalus viridis helleri* (Witham et al., 1953; Halmagyi et al., 1965; Schaeffer et al., 1978b). *C. durissus terrificus* venom showed a variation of this pattern with a triphasic response of an abrupt, transient fall in arterial pressure followed by a short duration pressor effect and a longer-lasting hypotension (Brazil et al., 1966a). Spontaneous recovery of arterial pressure occurred with these very low doses of venom, and the hypotension has been designated as *Crotalus* collapse by Halmagyi and co-workers (1965). A single rapid intravenous dose (of the order of 0.5 mg/kg) of many rattlesnake venoms produces a condition of profound and protracted hypotension termed *Crotalus* venom shock which, in anesthetized animals, usually leads to death within hours. Subcutaneous injection of *C. horridus* (or *C. atrox*) venom gave a pattern of response similar to that of a lower intravenous dose (Essex and Markowitz, 1930). Slow intravenous infusion of venom for 20 min or longer has been used to simulate the effect of a bite in studies of the mechanisms underlying venom shock and the efficacy of measures of treatment (Witham et al., 1953; Halmagyi et al., 1965; Carlson et al., 1975).

Hypotension owing to such infusions of either *C. horridus* or *C. atrox* venoms (0.06 to 0.7 mg/kg) was associated with an immediate fall in systemic arterial resistance and a later fall in cardiac output. Extensive vasodilation of the vasculature of skeletal muscle during the active fall in blood pressure was suggested by the associated invariable increase in the volume of the hindlimb and by the fact that hypotension still occurred in eviscerated dogs. Initially, splanchnic visceral volume fell but, after 5-10 min, extreme expansion resulted (Essex and Markowitz, 1930). Mean pulmonary arterial pressure and arterial resistance remained unchanged during acute *Crotalus* venom collapse induced by an infusion of 0.02 mg/kg *C. atrox* venom. At the higher doses which led to venom shock, variable changes in pulmonary arterial pressure were observed such as a slight rise in pressure followed by a fall, hypotension, and a hypertension which was associated with the presence of multiple emboli in the lung vasculature (Essex and Markowitz, 1930; Witham et al., 1953; Halmagyi et al., 1965).

Venom shock coincided with a marked fall in serum protein, a rise in blood hemoglobin content, and numerous petechiae which, in the case of an infusion of *C. viridis helleri* venom (1.5 mg/kg in 30 min) into rats, amounted to a reduction in the blood volume index by 35%, in plasma volume index by 46%, and in the red cell mass index by 22%. These changes were associated with an increased rate of disappearance of radioiodinated human serum albumin, indicating a more than twofold increase in vascular permeability (Carlson et al., 1975). However, pulmonary edema was not a part of the autopsy picture following *C. horridus, C. atrox,* or *C. viridis helleri* envenomation. Metabolic acidosis and a high blood-lactate concentration were also characteristics of severe shock in rats which had received 2.0 mg/kg *C. viridis helleri* venom in 30 min. However, if an infusion of 5% serum albumin in the amount calculated to restore the blood volume was started immediately after the end of the venom infusion, it was able to restore normal circulation, correct the defects, and lead to survival (Schaeffer et al., 1978a). Colloidal fluid replacement also restored the blood pressure following *C. horridus* envenomation and shock in dogs (Essex and Markowitz, 1930); however, actions on multiple physiological systems are probable since dextran administration in sheep shocked with *C. atrox* venom accentuated the accompanying hypoventilation, although no signs of circulatory overload were evident (Halmagyi et al., 1965). A precocious effect of *C. durissus terrificus* venom on the renal circulation has also been reported (Rovere et al., 1978).

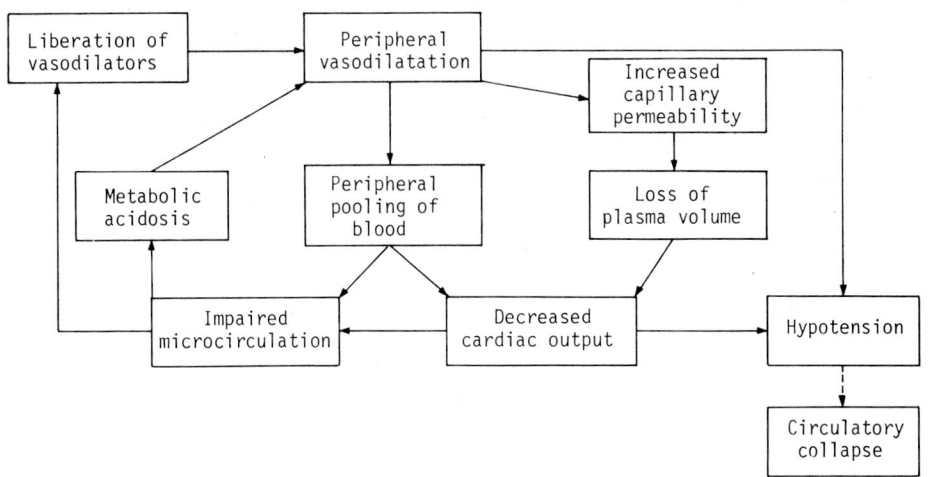

Figure 1 Some of the interactive factors involved in the genesis of shock. (Adapted from Bowman and Rand, 1980.)

Figure 1 shows both the interrelationship between the various circulatory conditions which have been reported during envenomation by rattlesnakes and how their interlocking nature can lead to a vicious circle with the onset of circulatory collapse and irreversible shock. Homeostatic mechanisms which come into play to counteract the fall in blood pressure are initiated by changes in baroreceptor activity and involve excitation of the sympathetic system, including release of norepinephrine and epinephrine from the adrenal medulla. They may override the effect of weak or transitory stimuli (see Figure 2, after indomethacin). However, if the disturbing agent is persistent, increased sympathoadrenal stimulation may exacerbate the impairment of flow in the microcirculation. Thus, in protracted shock following *C. horridus* envenomation, a restoration of systemic resistance, or even an increase to quite high levels, was associated with severely depressed cardiac output in dogs (Witham et al., 1953). Progressive hypotension will also reduce coronary perfusion, and the decrease in cardiac output will be compounded. The means whereby rattlesnake venoms induce hypotension and, in severe envenomation, circulatory collapse are multiple and may include a direct action on the vasculature. However, as Figure 1 shows, the liberation of vasodilators plays a central role, and the mechanisms whereby purified components may initiate this response is discussed under Release of Autacoids and Hemostasis.

Cardiac Function

Reports indicate that the rattlesnake venoms may have an effect, usually transient, on the excitability of the heart, but no report of a direct toxic action producing progressive, irreversible cardiac damage has been substantiated (Schaeffer et al., 1978b). In vivo, variable electrocardiogram (EKG) abnormalities appeared both early and late during the reaction to a slow infusion of *C. horridus* venom (Witham et al., 1953). A depression of the S-T segment and an increase in T wave amplitude suggested damage to the subendocardial area of the apex of the heart, but these changes usually waned and their functional importance was uncertain as they appeared during the initial stage of hypotension when tachycardia and a transient but increased cardiac output was recorded. In vitro, an equivalent amount of *C. horridus* (or *C. atrox*) venom did not impair the ability of the heart to expel blood in a heart-lung preparation (Essex and Markowitz, 1930). Rhythm irregularities with multiple premature systoles of left ventricular origin were present in about half the group of anesthetized dogs at variable times following the infusion of *C. horridus* venom (Witham et al., 1953) and shortly after a single intravenous injection of *C. adamanteus* venom (0.5 mg/kg) also in dogs (Vick, 1971). Sometimes the changes persisted. In a study of the effect of *C. scutulatus scutulatus* venom, protracted cardiac dysrhythmia was observed in 2 of 5 anesthetized rabbits given an intravenous

Physiological and Pharmacological Effects of Venoms

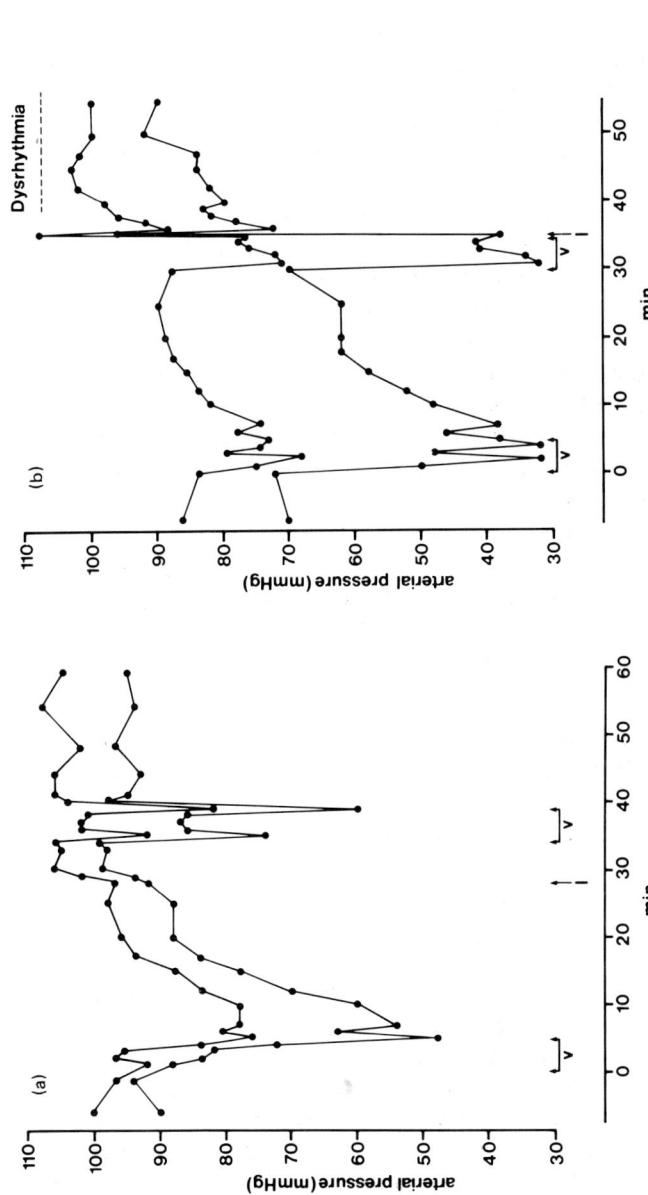

Figure 2 Effect of intravenous infusions of whole *C. scutulatus scutulatus* venom (v) (0.25 mg/kg) on the carotid arterial pressure of two anesthetized rabbits. At I a bolus intravenous injection of indomethacin (10 mg/kg) was given. (a) The indomethacin was injected prior to the second injection of venom. (b) The indomethacin was injected immediately after the second injection of venom. Injection of indomethacin solvent alone (1.9 ml of 0.13 M NaHCO$_3$) was ineffective. Control experiments showed that similarly repeated doses of the venom did not produce tachyphylaxis. In (b) cardiac dysrhythmia (---) commenced after the indomethacin injection and continued for 51 min, whereupon it spontaneously disappeared and did not return when a second injection of indomethacin was given. Prevenom mean heart rates were (a) 300 min^{-1} and (b) 330 min^{-1}, and these were not substantially altered during the course of the experiments. Anesthetic: sodium pentobarbitone, 70 mg/kg. Protracted dysrhythmia was observed in only one other experiment and commenced during the initial venom infusion prior to the injection of indomethacin. Mean heart rates fell slightly when the incidence of missed beats was of the order of 1:3.

infusion of 0.25 to 0.5 mg/kg (Figure 2b; Ghalayini, Hawgood and Whaler, unpublished observations, 1980). Mojave toxin, isolated from this venom, produced EKG abnormalities in anesthetized rabbits when given as a bolus intravenous injection of 1 mg/kg (Bieber et al., 1975), and transient dysrhythmia was observed in one anesthetized rat following a second bolus injection of 0.5 mg/kg of toxin (unpublished observations, 1980). These effects are apparently related to the presence of an intact cardiovascular system since the isolated perfused heart of the rat showed no change in rate or force of ventricular contraction at 100 μg/ml recycled Mojave toxin (Gopalakrishnakone et al., 1980).

A protein (approximately 10,000 daltons) isolated from *C. atrox* venom, when given as a bolus intravenous injection of 0.5 mg/kg into anesthetized, closed chest dogs, produced an immediate and maintained fall in systemic arterial pressure and cardiac output with no apparent change in systemic arterial resistance, suggesting a direct myocardial depressant action (Bonilla and Rammel, 1976). However, both ventricles appeared capable of maintaining adequate circulation over the recording period of 45 min. The ability of colloidal fluid replacement to restore blood circulation after shock induced by *C. viridis helleri*, *C. atrox*, and *C. horridus* venoms indicates that cardiac dysfunction is not a prime cause of hypotension. Overall, however, the evidence suggests that components of rattlesnake venoms can affect the heart, and in severe poisoning this could contribute to total collapse by facilitating the onset of myocardial weakness in the presence of a reduced coronary circulation.

Release of Autacoids

Extensive dilatation of arterioles which occurs under conditions of reduced blood flow (reactive hyperemia) is associated with the release of autacoids, and evidence is accumulating that similar mechanisms may be involved in the powerful vasodilation and increased vascular permeability which is a result of rattlesnake envenomation. The ability of colloid infusion to reverse the shock also suggests that the mediators may be transient in nature.

Bradykinin

Kininogenases which act on plasma globulins to form bradykinin have been reported in all rattlesnake venoms examined, namely *Crotalus atrox*, *C. adamanteus*, *C. durissus terrificus*, *C. horridus*, *C. basiliscus*, *C. viridis viridis*, *Sistrurus miliarius*, and *Sistrurus catenatus* (Deutsch and Diniz, 1955; Mebs, 1970; Oshima, Sato-Ohmori and Suzuki, 1969). *Crotalus* venoms also contain factors which destroy bradykinin, but overall, *C. viridis viridis* venom is particularly rich in kinin-forming activity, *C. atrox*, *C. adamanteus*, and *C. horridus* venoms are somewhat less so, and *C. durissus terrificus* venom is relatively weak (Rothschild and Rothschild, 1979).

Bradykinin has a short half-life, being inactivated by lung kininases; however, small peptides, termed bradykinin-potentiating factors, have been isolated from crotalid snake venoms, including some species of *Crotalus* (Ferreira, quoted by Rothschild and Rothschild, 1979). These preserve the action of bradykinin in the circulation (Ferreira and Vane, 1967).

In vivo, the sudden fall and subsequent recovery of systemic arterial pressure following 20 µg/kg *C. atrox* venom was found to coincide with the explosive release and rapid disappearance of plasma bradykinin. A slow venom infusion of 1.5 to 2 µg/kg min^{-1} for 30-60 min was more than sufficient to completely exhaust the plasma precursors in sheep (Margolis et al., 1965) which may form, at least in part, the basis of tachyphylaxis reported in dogs following a dose of 0.7 mg/kg *C. atrox* venom (Essex and Markowitz, 1930).

Prostaglandins

Prostaglandins of the E series reduce arterial pressure by a vasodilator action (Malik and McGiff, 1976), and the hypotensive action of *Crotalus* venoms may involve an increase in prostaglandin formation by virtue both of phospholipase A_2 and of kinin releasing activities.

Phospholipase A_2 is intimately involved in prostaglandin synthesis by providing substrate, e.g., arachidonic acid, from tissue phospholipids, substrate availability being the rate-limiting step in synthesis and release (Vogt, 1978). However, additional substrate can be provided from endogenous phospholipids by exogenous phospholipase A_2 as demonstrated by the release of prostaglandin-like substances from guinea pig lung perfused with phospholipase A_2 (*Naja naja*) venom (Damerau et al., 1975). Incubation of egg yolk with *C. durissus terrificus* and *C. adamanteus* venoms yielded an extract which induced marked hypotension when injected intravenously into anesthetized rabbits, dogs, and monkeys. As this action was completely blocked by inhibitors of prostaglandin synthetase such as indomethacin, it was attributed to the in vivo generation of prostaglandin-like substances as a result of the unsaturated fatty acids in the extract; lysolecithin was excluded (Vargaftig, 1974). Basic *C. durissus terrificus* phospholipase A_2, either alone or in complex form as crotoxin (Brazil et al., 1966a; Breithaupt, 1976), and Mojave toxin from *C. scutulatus scutulatus* venom (Bieber et al., 1975) induced acute hypotension in anesthetized animals when given by rapid intravenous injection. It has been suggested, but not proven, that this action of Mojave toxin may result from an increased generation of prostaglandin as a result of phospholipolytic activity (Gopalakrishnakone et al., 1980). It is uncertain if this mechanism could explain the tachyphylaxis which developed in the hypotensive action of crotoxin and its component phospholipase A.

Bradykinin has been shown to increase prostaglandin synthesis (Vargaftig and Dao Hai, 1972) probably by stimulation of endogenous phospholipase A_2, and medication with an inhibitor of prostaglandin synthesis prior to an intravenous injection of bradykinin curtailed the ensuing hypotension (Vargaftig, 1974). Some such mechanism would appear to be involved in the hypotension induced by *C. scutulatus scutulatus* venom as illustrated in Figure 2. In the presence of indomethacin, the hypotensive response to an infusion of venom was less, but even more striking was the rapid rate of recovery of diastolic and systolic pressures to preinjection levels. This occurred when indomethacin was given either prior to venom infusion or at the end of venom infusion in seven anesthetized rabbits (Ghalayini, Hawgood and Whaler, unpublished observations, 1980). Vargaftig (1967) has reported this effect using *Bothrops jararaca* venom.

Both prostaglandins and bradykinin increase capillary permeability to albumin, induce pooling of blood in the vasculature, and have short half-lives (Bowman and Rand, 1980). As they are inactivated by passage through the lungs, this may explain the lack of increase in lung weight following profound shock owing to *C. atrox* (Halmagyi et al., 1965) and *C. viridis helleri* envenomation (Schaeffer et al., 1978a).

Histamine and Serotonin (5-Hydroxytryptamine)

Degranulation of mast cells with histamine release was an early candidate as a mediator of rattlesnake venom shock. However, histamine (and biogenic amine) blockers such as promethazine hydrochloride failed to influence the initial rapid hypotension owing to a small standard dose of *C. atrox* venom (Halmagyi et al., 1965) and, in a monkey given a lethal injection (0.75 mg/kg) of *C. adamanteus* venom, plasma histamine levels did not increase significantly until close to death (Vick, 1971). Nevertheless, venoms from *C. adamanteus, C. atrox, C. horridus, C. durissus terrificus, C. viridis,* and *S. catenatus* have the ability to release histamine from animal tissue stores, and as these stores are only slowly depleted in envenomed animals, histamine may become important in the later stages of *Crotalus* venom shock (see the review by Rothschild and Rothschild, 1979). Originally, the phospholipase A content of venoms was considered to be the causative agent in histamine release from mast cells, although more recent studies have not supported this explanation. The original work was based on incubation of venom with an exogenous source of phospholipid such as egg yolk which allowed the formation of lysolecithin, itself cytolytic. In 1966, Rothschild showed that a fraction with phospholipase A activity isolated from *C. durissus terrificus* venom failed to release histamine from washed mast cells. Similarly, Damerau et al. (1975) demonstrated that phospholipase A_2 isolated from cobra (*Naja naja*) venom produced little mast cell degranulation in perfused guinea pig lungs and in mast cell-containing rat peritoneal suspensions, no histamine release in the perfused lung preparation, and a limited histamine release in the peritoneal suspensions; the active agent was shown to be the nonenzymic polypeptide "direct lytic factor." In certain species of *C. durissus terrificus,* the venom contains the polypeptide crotamine (mol. wt. 4880) which produces histamine release from washed cells by a process which is noncytolytic, as release was inhibited by metabolic poisons (Rothschild, 1966). *Crotalus viridis viridis* venom contains a polypeptide, myotoxin *a*, with structural and antigenic properties similar to crotamine (see Chapter 5), and crotamine-like polypeptides have been isolated from the venoms of *C. adamanteus, C. horridus horridus,* and *C. viridis helleri* venoms (Lee et al., 1973; Maeda et al., 1978). Although no direct test of the histamine-releasing activity of these substances in vivo has been reported, such an action could contribute to the induction of the state of shock.

The fraction of *C. viridis helleri* which contained the crotamine-like peptide as the major peak produced the typical features of venom shock when injected into rats at 0.5 mg/kg (Schaeffer et al., 1979b), which the authors suggested was due to a direct but transient action of a low molecular weight peptide(s) on the vasculature increasing permeability to both plasma proteins and erythrocytes. However, the lethality of this fraction was not as high as with the whole venom, and synergism with other components was suggested. It would be interesting to investigate if a transient release of histamine and/or serotonin played a role. In this context a polypeptide fraction (mol. wt. ca. 8000) from *Agkistrodon piscivorus* venom induced edema following subplantar injection into the rat paw (Bhargava et al., 1970a). The fraction was free from kininogenase activity and produced an edema of the rapid and transient type which reached a maximum after as little as half an hour and, thereafter, commenced to decline. The early phase of edema was inhibited by cyproheptadine, indicative of the release of serotonin or histamine. Intradermal injections (10 µg per spot) of *C. adamanteus* venom, *C. atrox* venom, or to a lesser extent, crotoxin also increased rat skin vessel permeability to protein by a mechanism which was blocked by pretreatment of the animal with cyproheptadine or other drug combinations of antiserotonin, antihistamine agents (Vargaftig et al., 1974). It is of interest that corticosteroid analogues have been ineffective in combating fluid loss and the onset of shock following *C. viridis helleri* envenomation (Schaeffer et al., 1979a).

The possibility that a factor in *C. durissus terrificus* venom released biogenic amines into the systemic circulation was suggested by the work of Markwardt et al. (1966b). They isolated a high molecular weight component which effected the release of both serotonin and histamine from isolated rabbit platelets by a mechanism requiring calcium ions and a viable platelet metabolism. Intravenous injection of this "amine-liberating factor" into anesthetized rabbits produced a triphasic change in blood pressure which was reproduced in part by an injection of serotonin in an amount approximating that lost from platelets. Serotonin is a potent vasoconstrictor to which pulmonary blood vessels are particularly sensitive. However, the blood vessels of skeletal and cardiac muscles are dilated, which is the basis of its short lasting hypotensive action.

In a set of interesting experiments, intravenous injection of *C. durissus terrificus* venom into anesthetized dogs produced triphasic circulatory and respiratory responses. Changes in arterial blood pressure consisted of an immediate, transient fall, a short-lasting pressor response, and prolonged hypotension. This pattern was reproduced by a high molecular weight protein, named convulxin, which was isolated from the venom. Blockade of muscarinic, nicotinic, or α-adrenergic receptors, as well as destruction of the central nervous system, failed to interfere with the progression of the circulatory changes produced by either the whole venom or convulxin (Brazil et al., 1966a; Brazil, 1972). The venom content of convulxin differs markedly between subspecies, with the venom of *C. durissus terrificus* from Santiago del Estero, Argentina, lacking convulxin (and showing only hypotension upon injection) whereas the venom of *C. durissus cascavella* contains 15% convulxin. *Crotalus durissus collineatus* venom also contains appreciable amounts of convulxin (Prado-Franceschi, personal communication, 1979). Intravenous injection of convulxin produced a rapid and profound thrombocytopenia when injected into the circulation of anesthetized rabbits. Furthermore, preincubation of convulxin with platelet-rich plasma in vitro, followed by neutralization with antiserum prior to intravenous injection, produced the powerful circulatory effects previously noted for convulxin (Prado-Franceschi et al., 1981). A detailed study of the mechanisms of action of convulxin, isolated from the venom of *C. durissus cascavella,* led to the conclusion that convulxin is a specific platelet stimulating agent which, at concentrations of 50 ng/ml and above, induces aggregation of guinea pig platelets, the release of ATP, and the synthesis of thromboxanes. Aspirin failed to interfere with platelet aggregation and release showing the activated pathway to be independent of cyclo-oxygenase. These convulxin-induced effects were Ca^{2+} dependent (Vargaftig et al., 1980). Intravenous administration of convulxin (0.3 to 3 µg/kg) to anesthetized guinea pigs induced thrombocytopenia, hypotension, and bronchoconstriction. However, the smooth muscle stimulating effects of serotonin and histamine were apparently the direct mediators of these physiological actions since pretreatment of the animal with the histamine and serotonin blockers, mepyramine and methysergide, failed to prevent both bronchoconstriction and hypotension induced by convulxin. Aspirin and indomethacin were also ineffective, suggesting that the mechanisms did not involve prostaglandin synthesis. More surprisingly, however, pretreatment of the animal with the amine blockers plus either aspirin or indomethacin suppressed the hypotension and bronchoconstriction resulting from convulxin; the final mediator in the pathway is still to be determined (Vargaftig et al., 1980).

These workers showed that, in vitro, the phospholipase A_2 subunit of crotoxin produced some thromboxane synthesis and platelet aggregation, although convulxin was 50 times more active. They believed that the pathway involved a limited source of arachidonic acid, since platelet aggregation by this phospholipase A_2 was inhibited by aspirin. Such a mechanism may provide the serotonin (and histamine) released by crotoxin in vivo as suggested by the rat skin permeability studies and may contribute to the transient hypotensive effect of this complex (see Release of Autacoids). Venom components may have more than one site of action in vivo, and a fraction from *C. durissus terrificus* venom with chemical properties similar to amine liberating factor was highly active in liberating histamine from isolated rat mast cells (Rothschild, 1966; Rothschild and Rothschild, 1979).

Of the species examined, venoms from *C. durissus terrificus* (10 µg/ml) *C. horridus horridus* (10 µg/ml), and *S. catenatus* (100 µg/ml) were all capable of releasing more than 50% of the content of stored serotonin and histamine from washed rabbit platelets, whereas those from *C. adamanteus* (100 µg/ml), *C. atrox* (100 µg/ml), and *C. viridis* (100 µg/ml) were ineffective (Markwardt et al., 1966a). It is also of interest that a dose of *C. viridis helleri* venom (50 µg/kg) which produced transient hypotension in dogs resulted in marked thrombocytopenia within 5 min of injection, probably by direct stimulation of platelet aggregation (Ruiz et al., 1980).

Hemostasis

Petechial hemorrhages are a consistent finding in severe rattlesnake venom shock, and a fall of 22% in the red cell mass index was reported following an intravenous infusion of *C. viridis helleri* venom (1.5 mg/kg, 0.5 hr) into rats (Carlson et al., 1975). However, this was not apparently a major factor in the onset of venom shock as pretreatment with antivenin prevented the appearance of petechiae but not the hypovolemia and death following a larger infusion (2 mg/kg) (Schaeffer et al., 1979a).

It has already been mentioned that Schaeffer and coworkers (1979) reported that a low molecular weight polypeptide fraction from *C. viridis helleri* venom was responsible for a direct but transient action on the vasculature to increase permeability to plasma proteins and to red blood cells (see Release of Autacoids). The situation would appear to be complex as, although gross blood was present throughout the gut lumen and serosanguinous fluid was in the abdominal cavity, petechiae were absent. The ability of a polypeptide fraction (mol. wt. ca. 8000) from *A. piscivorus* venom to induce hemorrhage and provoke edema has been demonstrated. However, the effect depended on the site as the fraction induced petechiae when applied to the surface of the dog lung, but not when injected into the rat paw, although edema did occur. It was considered likely that different factors were involved in the two actions (Bhargava et al., 1970a,b). A similarly complex situation is seen with the action of *C. adamanteus* venom which produced a marked hemorrhagic effect on the vessels of the skin but no vasculotoxic effect on the surface of the dog lung. From these studies it was concluded that hemorrhage may result from the action of different venom components depending on the species and that the effect may be highly specific for certain vascular beds (Vargaftig et al., 1974; Bonta et al., 1979). At a microscopic level, it has been shown that small amounts of hemorrhagins (mol. wt. 24,000-68,000) isolated from *C. atrox* venom disrupt the basal lamina and collagen associated with capillaries and thus lead to the passage of erythrocytes through opened intercellular junctions (Ownby et al., 1978).

Incoagulable blood was associated with transcapillary hemorrhage following *C. horridus* envenomation in dogs and, although the blood loss was relatively small, it was considered to contribute to the overall state of profound shock which developed (Witham et al., 1953). Thrombocytopenia and severe depression of plasma fibrinogen levels have been reported following experimental envenomation with *C. adamanteus* and *C. horridus horridus* venoms (Vick, 1971). However, these two symptoms are not necessarily linked as, in a study with 50 μg/kg *C. viridis helleri* venom (Ruiz et al., 1980), thrombocytopenia occurred within 5 min, whereas hypofibrinogenemia developed slowly over 4 hr. Thus different venom components may be responsible, and it is of interest that intravenous injection of a component with thrombin-like activity, isolated from *C. adamanteus* venom, produced hypofibrinogenemia but no reduction of blood platelet count in anesthetized dogs (Damus et al., 1972).

Defective hemostasis may result from the synergistic action of a number of venom factors which increase capillary and small vein fragility, particularly in the presence of increased hydrostatic pressure as a result of stasis, and which also render the blood incoagulable.

EFFECTS ON THE RESPIRATORY SYSTEM

Reflex Stimulation

Crotalus durissus terrificus venom (0.25 mg/kg) injected intravenously into anesthetized dogs produce a triphasic response consisting of an intense and transient increase in the frequency and amplitude of the respiratory movements within a few seconds of venom administration, a brief period of apnea, and finally tachypnea, which usually was very long lasting. Crotamine and crotoxin were eliminated as causative agents, but convulxin was shown to reproduce the acute changes in respiratory pattern. The effects appeared to be reflexly produced as section of both vagi prevented the apnea, whereas denervation of the carotid sinus region suppressed the initial respiratory stimulation (Brazil et al., 1966a; Brazil, 1972). It was inferred that convulxin had a direct excitatory action on the aortic and carotid chemoreceptors, and this may be mediated by serotonin released into the circulation. However, following administration of the whole venom, two of the eight anesthetized dogs with both vagi and carotid sinus nerves cut showed the usual intense respiratory stimulation, although not the subsequent apnea, and a direct excitatory action on the respiratory center was considered a possibility (Brazil et al., 1966a). No impulse activity was recorded from the phrenic nerve during apnea (Brazil, 1972), and stimulation of chemoreceptor endings in the lungs and heart was considered to initiate this inhibition. However, this is unlikely to be the sole sensory pathway as, in the Jarisch von Bezold reflex, a temporary arrest of respiration is associated with a dramatic fall in blood pressure and heart rate, whereas in envenomation, apnea was associated with a pressor response and tachycardia. Decreased baroreceptor activity may contribute to the prolonged tachypnea associated with hypotension.

Motor Paralysis

Peripheral Origin

Crotalus durissus terrificus venom causes death in most mammals by respiratory paralysis, an action attributed to the high concentration of crotoxin (60-70%) in the whole venom. However, there are marked differences in species sensitivity to crotoxin, either "native" or reconstituted from its subunits, with rats proving remarkably resistant to its lethal action (Brazil et al., 1966b; Breithaupt, 1976). Crotoxin produces flaccid paralysis owing to blockade of neuromuscular transmission, and the peripheral basis of this, in rabbits and cats, was confirmed by the continued recording of phrenic nerve discharges which increased as the amplitude of respiration became depressed and persisted for several hours after complete paralysis of the respiratory muscles when the anesthetized animals were supported by artificial ventilation (Brazil, 1966, 1972; Breithaupt, 1976). A study of the distribution of ^{131}I-labeled crotoxin produced no evidence for a central action with a very low radioactive count in brain, spinal cord, and cerebrospinal fluid (Lomba, 1969; Habermann et al., 1972). Hemorrhagic foci were absent from the meninges or brain substance of dogs killed by crotoxin (Brazil et al., 1966b).

Mojave toxin also blocks neuromuscular transmission, and in systemic intoxication of rapid onset, the cause of death in mice was respiratory paralysis. Its immunological similarity to crotoxin was confirmed by the ability of antiserum raised against crotoxin to neutralize its lethality (Gopalakrishnakone et al., 1980). The concentration of Mojave toxin in *C. scutulatus scutulatus* venom is appreciable: 1 g of lyophilized venom gave a protein yield of 190-220 mg in the first stage fraction MD-9 which contained Mojave toxin and a protein yield of 98-105 mg of MD-9v (Mojave toxin) in the second stage fractionation (A. Bieber, personal communication, 1980). With a slow rate of intoxication as from subcutaneous injection, the hypotensive actions of the whole venom and of Mojave toxin are initially apparent, although impaired respiration may hasten the onset of circulatory collapse in the terminal stages (Bieber et al., 1975; Gopalakrishnakone et al., 1980).

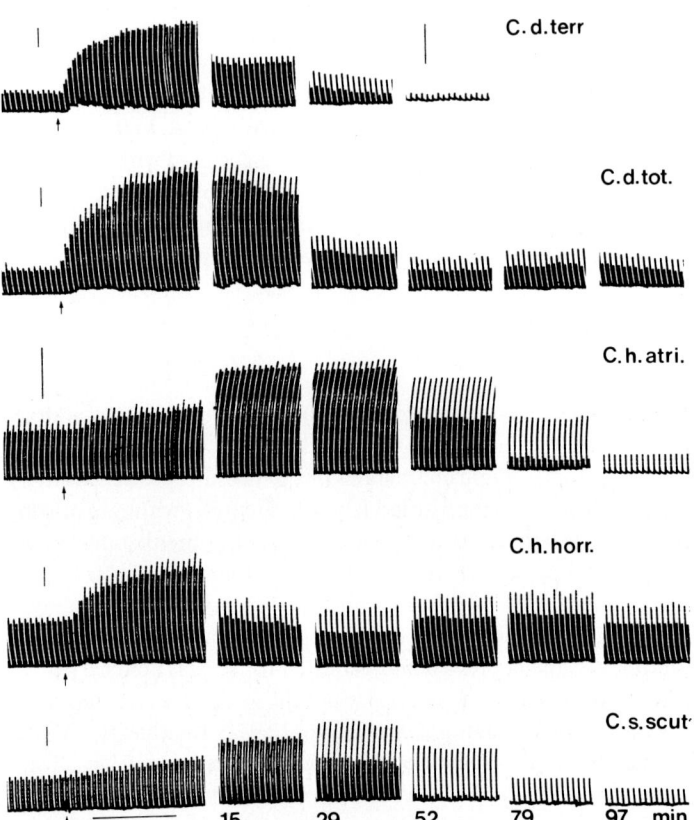

Figure 3 The effect of 5 rattlesnake venoms on the contractile response of the mouse phrenic nerve-hemidiaphragm preparation to alternatively delivered indirect (neural) and direct (muscle) stimulation. The venoms were: *C. durissus terrificus* (*C. d. terr.*) *C. durissus totonacus* (*C. d. tot.*); *C. horridus atricaudatus* (*C. h. atri.*), *C. horridus horridus* (*C. h. horr.*), and *C. scutulatus scutulatus* (*C. s. scut.*). The venom was added at the arrow to a final concentration of 10 μg/ml, and recordings are illustrated at 15, 29, 79, and 97 min. Horizontal scale, 3 min; vertical scale, 0.5 g. Stimulus parameters used in indirect stimulation were 0.05 msec, 1-2 V and in direct stimulation were 0.15 to 0.2 msec, 25 V. These parameters were supramaximal and for direct stimulation were determined in the presence of (+)-tubocurarine (4 μg/ml) during the control period. In all cases direct stimulation produced the larger of the two contractile responses. Combined stimulus frequency was 12 min^{-1}. Temperature 37°C. Krebs bicarbonate saline.

Tu (1977) suggested that crotalids possess, in varying amounts, a common lethal toxin. Evidence that *C. horridus atricaudatus* and *C. basiliscus* venoms contain a substance with antigenic similarities to crotoxin comes from the ability of these venoms to cross-react with antiserum raised against crotoxin as detected by the methods of enzyme-linked immunoassay and double immuno-diffusion (Gopalakrishnakone et al., 1981). *Crotalus horridus atricaudatus* venom (10 µg/ml) also produced blockade of neuromuscular transmission in the mouse phrenic nerve-hemidiaphragm preparation in vitro (Figure 3). In contrast, *C. horridus horridus* and *Crotalus durissus totonacus* venoms failed to show any significant cross-reaction with the anticrotoxin serum, and furthermore, no progressive blockade of neuromuscular transmission was detected at 10 µg/ml in vitro (Figure 3). This indicates a marked difference within subspecies for *C. horridus* and *C. durissus* venoms which is supported by the finding of rapid death from respiratory failure following an experimental bite by *C. horridus atricaudatus* in contrast to a survival time of more than 72 hr following the bite of *C. horridus horridus* (Vick, 1971) and a 10-fold increase in intraperitoneal (i.p.) LD_{50} in mice of *C. durissus totonacus* venom compared with *C. durissus terrificus* venom (Possani et al., 1980; Glenn and Straight, 1977). Similarly, venoms from *C. atrox* and *C. adamanteus*, as well as *C. horridus horridus*, were shown to be comparatively nontoxic at the neuromuscular junction as a complete neuromuscular blockade of the isolated guinea pig phrenic nerve-diaphragm preparation was not usually produced for several hours with venom concentrations up to 200 µg/ml (Russell and Long, 1961). There was no immunological evidence to suggest the presence of a crotoxin-like substance in the venoms of *C. atrox* and *C. viridis viridis* (Gopalakrishnakone et al., 1981), and the ability of rats to recover from severe *C. viridis helleri* envenomation (2 mg/kg) (Schaeffer et al., 1978a) testifies dramatically to the absence of crotoxin-like activity in this venom. In *Crotalus scutulatus salvini* venom the phospholipase A_2 was present in the form of an enzymic dimer, and antisera raised against this dimer cross-reacted with *C. atrox* and *C. horridus horridus* venoms but not with the venoms of *C. scutulatus scutulatus, C. durissus terrificus,* and *C. basiliscus* (Nair et al., 1979). This study, complementary to that of Gopalakrishnakone et al. (1981), suggests that rattlesnake venoms divide into at least two classes dependent on the type of complex formation of the phospholipase A_2; the physiological consequences are discussed in detail under Effects on the Somatic Nervous System.

As regards *Sistrurus* venoms, Minton (1956) presents evidence of a peripherally acting neurotoxin in *S. catenatus catenatus* venom from the profound paresis of the hindquarters of mice following intraperitoneal injection. The mouse i.p. LD_{50} of 0.22 µg/g is remarkably similar to the equivalent value of 0.25 µg/g found for *C. durissus terrificus* venom (Glenn and Straight, 1977).

Possible Central Origin

Respiratory depression of central origin was considered to occur in *C. atrox* envenomation of sheep as the alveolar hypoventilation which was observed in most animals was in contrast to the hyperventilation of other types of shock and was not relieved by restoration of the circulatory volume by dextran (Halmagyi et al., 1965). Changes in cerebral cortical activity associated with cardiac arrhythmias and dyspnea were reported in unanesthetized monkeys at about 3 hr after an intramuscular injection of a lethal dose of *C. adamanteus* venom (Vick, 1971). An injection of *C. atrox* venom gave similar results, but the exact relationship between the cortical and respiratory effects of these venoms was not clear. Electroencephalogram (EEG) changes, characterized by the sudden appearance of high voltage, slow wave activity, were indicative of cortical depression. Slow wave activity later spread to subcortical areas, but in some cases this did not occur until just prior to death. Intravenous injection of the venoms produced a similar pattern but of more rapid onset (Vick, 1971). Essex and Markowitz (1930) reported multiple diffuse and punctate hemorrhages in the brain, especially in the corpus striatum, in dogs succumbing to acute crotalin poisoning. Whether this was the case for both *C. horridus* and *C. atrox* venoms was not stated. In view of the peripheral hemorrhagic activity reported for *C. atrox* and *C. adamanteus* venoms (Ownby et al., 1978; Vargaftig et al., 1974), it would be of interest to determine if a lethal dose of these venoms induced generalized changes in permeability in the blood brain barrier, with consequential cerebral edema. The use of Evans blue to detect the passage of plasma proteins across the blood brain barrier and the distribution of radioiodinated venom would allow a distinction to be made between possible central actions and a general deterioration resulting from impairment of the cerebral circulation.

EFFECTS ON THE SOMATIC NERVOUS SYSTEM AND SKELETAL MUSCLE

Neuromuscular Transmission

Presynaptic Toxins

Crotoxin, isolated from the venom of *C. durissus terrificus,* is a complex of a basic phospholipase A_2 and an acidic nonenzymic moiety (see the review by Habermann and Breithaupt, 1978; Chapter 4). The complex produces irreversible blockade of transmission at the vertebrate neuromuscular junction by a progressive inhibition of the release of neurotransmitter in response to the nerve impulse as a result of interference with depolarization-secretion coupling (Brazil and Excell, 1971; Brazil et al., 1973; Chang and Lee, 1977; Hawgood and Smith, 1977b). As the phospholipase A_2 moiety showed a similar, though less powerful, activity (Brazil et al., 1973; Hawgood and Santana de Sa, 1979), and as the replacement of calcium by strontium ions inhibited both the phospholipase A_2 and neuromuscular blocking activities of crotoxin (Chang et al., 1977b; Hawgood and Smith, 1977b), neurotoxicity was considered to result from the hydrolytic activity of the complex on phospholipid membranes. The blockade of transmitter release could be temporarily antagonized by either increasing the external Ca^{2+} concentration (Hawgood and Santana de Sa, 1979) or adding 4-aminopyridine to the medium (Brazil et al., 1979a), which suggested that neuromuscular transmission failed before stores of transmitter were depleted. Both these procedures would increase Ca^{2+} influx into the axoplasm in response to a nerve impulse but do not differentiate as to whether the crotoxin-induced defect lay in the opening of the calcium channel or in the subsequent action of Ca^{2+} to promote exocytosis of synaptic vesicles. Crotoxin (or the homologous phospholipase A_2) also produced abnormalities in spontaneous release of transmitter as detected by the appearance of bursts of miniature endplate potentials (m.e.p.p.s) and of giant-sized m.e.p.p.s. Hawgood and Santana de Sa (1979) concluded that crotoxin induced two kinds of alterations in the structure of the presynaptic membrane, namely, the formation of transient instabilities with an increased Ca^{2+} influx and a prolonged disturbance of the calcium channel and/or the synchronized release of synaptic vesicles. Mojave toxin has a presynaptic action at end plates in the mouse diaphragm resulting in interference with depolarization-secretion coupling; propagation of the impulse into the nerve terminal may also be affected (Gopalakrishnakone et al., 1980). Figure 4 summarizes some of the stages in the formation and release of neurotransmitter which may be possible sites of toxic action.

Presynaptically active toxins have been demonstrated in the venom of some elapid species, namely, taipoxin from the venom of the Australian taipan, *Oxyuranus scutellatus scutellatus*, notexin from the venom of the Australian tiger snake, *Notechis scutatus scutatus*, and β-bungarotoxin from the venom of the Formosan krait, *Bungarus multicinctus*. Notexin, one of the two subunits of both crotoxin and β-bungarotoxin, and all three subunits of taipoxin are known to be structural homologues of either pancreatic phospholipase A_2 or its proenzyme, and the similarities in physiological action of these neurotoxins and their apparent dependence on enzymatic activity have been pointed out (Chang et al., 1977a,b). These neurotoxins induce morphologic changes in the ultrastructure of the motor nerve terminal which include an increased number of Ω-shaped profiles in the axolemma ("coated" pits), a depletion of synaptic vesicles, and progressive swelling and vacuolization of the mitochondria in the motor nerve terminals (Chen and Lee, 1970; Cull-Candy et al., 1976; Strong et al., 1977; Gopalakrishnakone and Hawgood, 1979). Some of the typical morphologic features of intoxicated motor nerve terminals are shown in Figure 5.

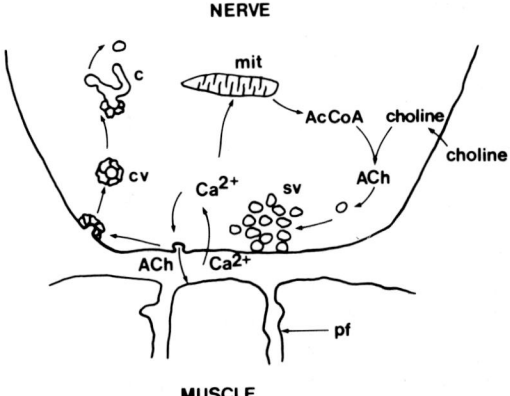

Figure 4 Schematic representation of postulated stages in the life cycle of synaptic vesicles at the motor nerve terminal. Shown are an accumulation of synaptic vesicles (SV), release of acetylcholine (ACh) by exocytosis in response to an influx of calcium ions, formation and endocytosis of a coated vesicle (CV) with incorporation into cisternae (C), synthesis of acetylcholine from acetyl-coenzyme A (AcCo A) and choline, and the refilling of empty vesicles. mit, mitochondria; pf, postsynaptic fold.

The mechanism of action at the molecular level has not been definitively established for any one of these neurotoxins, but as they are of great interest to the physiologist and biochemist who wishes to use them as a tool to probe the structure of the axon terminal membrane and the mechanism of neurotransmitter release, it is instructive to examine the known properties of the action of these elapid neurotoxins and the proposed models for their mode(s) of action. However, it is to be noted that they differ markedly in tertiary structure as the nonenzymic subunit of β-bungarotoxin is covalently bound to the phospholipase A_2 moiety and taipoxin consists of three subunits (see the review by Lee, 1979b). Incidentally, taipoxin is by far the most lethal of the presynaptic neurotoxins. A feature of the interaction of crotoxin with the membranes of erythrocytes and of microsacs derived from *Torpedo marmorata* electric organ is the dissociation of the nonenzymic subunit from the complex following the attachment of the phospholipase moiety to the membrane (Jeng et al., 1978; Bon et al., 1979). This suggests that the properties of crotoxin may lie closer to those of notexin, which consists of a phospholipase A moiety only.

In the presence of each of these neurotoxins, isolated nerve-muscle preparations show a complex pattern of changes in both the quantal content of evoked transmitter release and in the frequency of spontaneous release, which consists of three phases, namely, an initial depression, a facilitation of response, and finally, a progressive decline to complete loss of activity. At end plates in rat diaphragms poisoned with β-bungarotoxin, the changes in evoked and spontaneous release of neurotransmitter occurred in parallel (Kelly et al., 1976). However, continuous recording from end plates in frog sartorius muscles exposed to either crotoxin or notexin showed a more complex situation in that the stages of facilitation and later depression developed asynchronously in the two forms of transmitter release (Magazanik and Slavnova, 1978; Hawgood and Santana de Sa, 1979). In the mouse diaphragm intoxicated with taipoxin, m.e.p.p. frequencies never rose more than slightly above control levels except where anoxia or mechanical trauma had occurred (Kamenskaya and Thesleff, 1979). The initial depression in evoked and spontaneous transmitter release has been attributed to the effect of the binding of the neurotoxin. As shown for β-bungarotoxin, this stage is phospholipase-independent, occurring in the absence of Ca ions, whereas histidine-modified β-bungarotoxin which has lost its catalytic activity can induce only these first phase effects (Abe et al., 1977). Membranes undergoing phase transition are preferentially hydrolyzed by β-bungarotoxin, and it has been suggested that the nonenzymic moiety of the molecule acts as a recognition site, conferring a specificity of attachment of the molecule to the axolemmal membrane at a point close to the release site where it can exert its hydrolytic action (Kelly et al., 1979). Magazanik and Slavnova (1978) also suggest that, in the case of notexin, there is a direct action on the release mechanism.

However, in view of the morphologic finding of a complete reduction of the store of synaptic vesicles at severely intoxicated motor nerve terminals, this may not be the sole mechanism of action of these neurotoxins. This is consistent with the recording of low m.e.p.p. frequencies and the difficulty experienced in locating end plates in diaphragms removed from mice which were severely poisoned with crotoxin (Hawgood and Smith, 1977b). However, high m.e.p.p. frequencies were not recorded at end plates in diaphragms removed from partially intoxicated mice, and in view of the evidence suggesting a defect in release, depletion must arise from causes other than excessive release. One possibility is the internalization of the phospholipase A moiety. The membranes of internal organelles are highly susceptible to phospholipolytic attack as shown, for instance, by the disruption of synaptic vesicles by *Naja naja siamensis* phospholipase A (Heilbronn, 1972). In a morphologic study using horseradish peroxidase (HRP) conjugated β-bungarotoxin, strong binding to the plasma membrane of motor nerve terminals of the frog cutaneous pectoris muscle was observed and, interestingly, binding to some of their internal membranes including the membranes of synaptic vesicles and of mitochondria was also observed. The physiological significance of this internal binding was unclear, as it may have arisen from nonspecific fixation after rupture of the nerve terminals. However, the authors concluded that "conjugated β-bungarotoxin was selectively accumulated by nerve terminals, either by surface adsorption or pinocytotic uptake" (Strong et al., 1977). Figure 5 shows a widespread distribution of HRP reaction products at motor nerve terminals poisoned with crotoxin in vivo, and

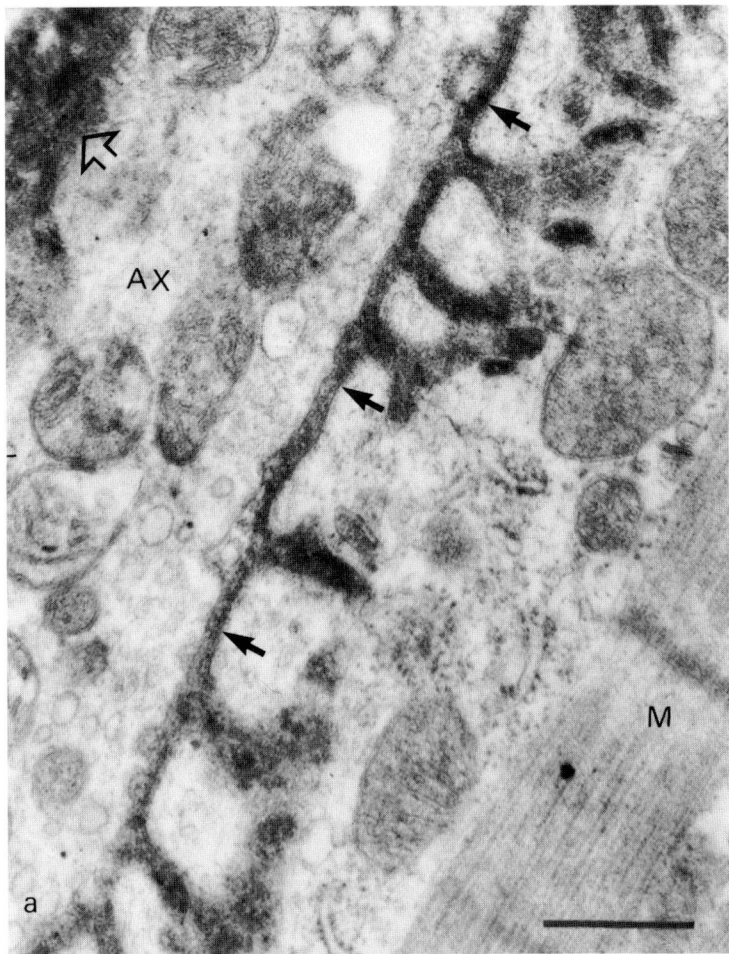

Figure 5a Neuromuscular junction in the diaphragm removed from a severely intoxicated mouse 2 hr after intravenous injection of 2 LD_{50} of reconstituted crotoxin complex. Processed by the indirect immunoperoxidase method of Daniels and Vogel (1975) using rabbit antiserum raised to the crotoxin complex and horseradish peroxidase (HRP) conjugated with anti-rabbit IgG (Miles Laboratory). The electron dense HRP reaction products fill the synaptic gap (closed arrow), the primary and secondary clefts, and are present in the region of the Schwann cell (open arrow). Stained with uranyl acetate and lead citrate. Ax, axon terminal; M, myofibrils. Calibration bar 0.9 μm. (From Gopalakrishnakone, 1979.)

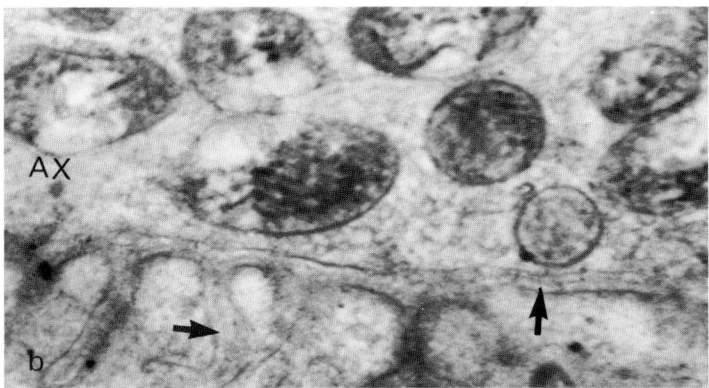

Figure 5b Neuromuscular junction in the diaphragm from a mouse 2 hr after an intravenous injection of 2 LD_{50} crotoxin complex. Processed by the indirect immunoperoxidase method as above with the omission of the antiserum. Note that no HRP reaction products are observed in the synaptic gap (closed arrows), but that the presynaptic terminal shows disrupted mitochondria and a severely reduced synaptic vesicle population as in (a). (From Gopalakrishnakone, 1979.)

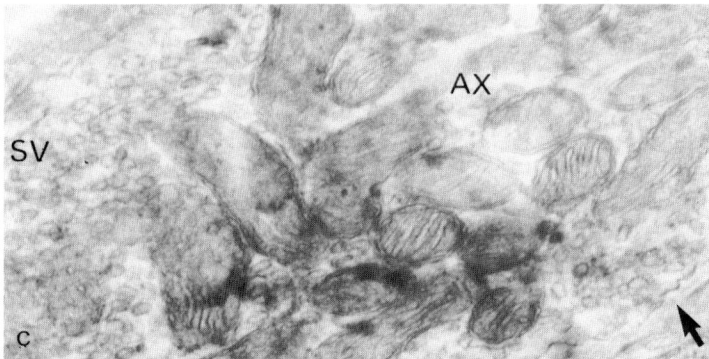

Figure 5c Neuromuscular junction in the diaphragm of a normal mouse processed as in (a). Within the axon terminal (Ax), the synaptic vesicle population (SV) and mitochondria appear normal. The synaptic gap is indicated by the closed arrow. (From Gopalakrishnakone, 1979.)

exposed to rabbit anticrotoxin serum and to anti-rabbit IgG conjugated with HRP in whole mount preparations in vitro. In addition the sarcolemmal membrane showed adsorption of crotoxin antigens (Gopalakrishnakone, 1979). No procedures were taken to determine if some of this binding could be reversed by exposure to inert, strongly binding basic proteins, so the extent of nonspecific binding is unknown. The absence of stained products from the interior of motor axon terminals in Figure 5 gives no clue as to whether or not internalization of *Crotalus* phospholipase A occurs, as in the method used, the IgG-HRP conjugate would not have been able to penetrate into the interior. No sign of axolemmal destruction was observed at end plates showing gross ultrastructural changes as a result of taipoxin poisoning, and in conjunction with electrophysiological studies, it has been suggested that the mechanism of action of this toxin involves internalization before it can affect transmitter release (Cull-Candy et al., 1976; Kamenskaya and Thesleff, 1979; Lüllman-Rauch and Thesleff, 1979). In vitro, increasing the rate of either evoked or spontaneous transmitter release reduced the considerable latency and accelerated the rate of blockade of neuromuscular transmission in the mouse diaphragm. This effect was not noted during crotoxin-induced blockade of neuromuscular transmission in the mouse diaphragm in normal Kreb's solution but was apparent if the calcium concentration was raised (Chang and Lee, 1977); the rate of blockade also showed a high temperature coefficient (Hawgood and Smith, 1977b).

An additional mode of action implicates the recycling of vesicle membrane. In mice dying from either taipoxin or notexin injections, it was noted that a reduction in the store of vesicles in motor axon terminals did not occur in the absence of impulse traffic (Cull-Candy et al., 1976). This led to the suggestion that the toxins might exaggerate vesicle disappearance during stimulation by blocking synaptic vesicle reformation through the endocytosis of coated vesicles (Figure 4). Lassignal and Heuser (1977), in a study of the effect of stimulation on end plates poisoned with β-bungarotoxin, used freeze-fracture techniques to provide evidence for such a mechanism. Stimulation led to a great increase in the proportion of nerve terminals that were covered with shallow coated pits, and histochemistry for HRP as a marker of endocytosis failed to show the usual uptake of HRP into coated vesicles.

No functional membrane grouping has yet been shown to be attacked specifically by any of these toxins, and the expression of their action would appear to be related either to an ability to internalize or to expose externally the membrane phospholipid to the catalytic site of the enzyme moiety. Thus a small species difference in membrane structure may alter the susceptibility of a tissue to attack and hence serve as the basis wherein the rat motor axon terminals are more resistant to crotoxin than are those of mice and also that mouse motor axon terminals are more resistant to notexin than are those in the frog.

Variations within vertebrate species make it difficult to transfer findings from invertebrate preparations, especially if the preparations have involved the resealing of membranes which may expose new sites. However, it is of interest that taipoxin blocked the high affinity choline transport in T sacs (which represent fragments of cholinergic nerve terminals) formed from *Torpedo marmorata* electric organ. This effect of taipoxin is presumably a minor one since histidine modification of the catalytic sites of the α and β subunits of taipoxin, which decreased the lethality to mice by 350-fold, reduced the inhibitory action on the high affinity choline transport only threefold (Fohlman et al., 1979). A different proposal comes from a study of the acetylcholine content of diaphragm strips exposed to β-bungarotoxin. An increase in tissue acetylcholine content was observed which was too large to be accounted for by the associated reduction in acetylcholine release. Since it was prevented by exchanging calcium for strontium ions in the medium, an effect on acetylcholine synthesis in response to a raised internal Ca ion concentration was tentatively suggested (Gundersen et al., 1980).

It is thus highly likely that more than one mechanism may be involved in the disruption of neurotransmitter release produced by these toxins, especially as there are reports for notexin and taipoxin that strontium ions and histidine modification of the molecule only slows, but does not completely inhibit, the onset of blockade. Limited hydrolysis of the axolemma, by affecting membrane fluidity, may also induce quite widespread effects.

Many workers have reported that crotoxin has some degree of postsynaptic action as shown by the depression of the response to exogenously applied acetylcholine (or agonist) on isolated preparations of the denervated rat and mouse diaphragm (Brazil, 1966; Hawgood and Smith, 1977a), on innervated frog sartorius muscle and guinea pig diaphragm preparations (Brazil and Excell, 1971; Brazil et al., 1979b), on the isolated electroplaque from *Electrophorus electricus*, and on microsacs of receptor-rich postsynaptic membranes from *Torpedo californica* and *Torpedo marmorata* (Hanley, 1978; Bon et al., 1979). However, no such action was demonstrated using exogenous acetylcholine on the extrajunctional acetylcholine receptors of isolated chick biventer cervicis muscle. An effect on the mouse and rat end plates was questioned from the lack of change in mean m.e.p.p. amplitude in isolated, intoxicated diaphragms (Chang and Lee, 1977). However, this latter approach would not detect small changes in postsynaptic sensitivity in view of the abnormal distribution of m.e.p.p. amplitudes with a skew to the large values which crotoxin induces (Hawgood and Smith, 1977b; Hawgood and Santana de Sa, 1979). The importance of the postsynaptic effect was usually considered to be minor in relation to the presynaptic action, although, undoubtedly, it would contribute to the complete blockade. In invertebrate systems, however, this effect may be of greater significance. That postsynaptic inhibition results from phospholipase activity was shown by the effectiveness of the homologous phospholipase A_2, and the failure of histidine-modified crotoxin, either to inhibit the response induced by cholinergic agonists on $^{22}Na^+$ efflux from preloaded acetylcholine receptor-rich microsacs from *Torpedo californica* or to inhibit depolarization of the isolated electroplaque from *Electrophorus electricus* (Hanley, 1978; Marlas, 1978; Bon et al., 1979). Binding of crotoxin to the electroplaque was phospholipase A independent, as it proceeded when calcium replaced strontium ions in the medium. However, Sr^{2+} not only arrested the inhibitory phase but induced a partial reversal of inhibition (Marlas, 1978), and this may be related to diffusion of fatty acid products from the active sites as these substances were able to mimic the action of crotoxin on microsacs (Hanley, 1978). The failure of crotoxin to interfere with the binding of either α-bungarotoxin (Hanley, 1978) or the α toxin from *Naja nigricollis* venom to acetylcholine receptors in *Torpedo* microsacs suggested that it must exert its effect through the ion channel and/or the membrane. Bon et al. (1979) put forward the concept that crotoxin "stabilizes a desensitized form of the acetylcholine receptor characterized by its high affinity for agonists." The small depression of acetylcholine sensitivity in denervated, murine hemidiaphragms induced by crotoxin (15 μg/ml) was associated with a decrease in the maximal contractile response to acetylcholine (Hawgood and Smith, 1977a), and after rapid loss of chemical excitability of the microsacs, a slow onset of destabilization was reported (Hanley, 1978). Interestingly, 4-aminopyridine antagonized the postsynaptic depression induced by crotoxin on the end plates of the isolated guinea pig diaphragm (Brazil et al., 1979a).

In vitro, basic phospholipases A_2 from *C. durissus terrificus* venom can produce both the presynaptic and postsynaptic responses of crotoxin; however, the enzyme rapidly adsorbs to surfaces, and the combination with crotapotin both reduced this nonspecific binding and increased the enzymic stability (Chang and Su, 1978). In vivo, complex formation distinctly altered the pharmacokinetic behavior of *C. durissus terrificus* phospholipase A_2 and increased the toxicity; intravital hemolysis was also reduced (Habermann et al., 1972; Habermann and Breithaupt, 1978).

Of the other presynaptic neurotoxins, notexin, but not β-bungarotoxin, produced a depression of sensitivity to acetylcholine in isolated muscle preparations (Lee et al., 1976). A neurotoxic protein from the venom of the Bulgarian viper (*Vipera ammodytes ammodytes*) which contains an acidic nontoxic component in complex with a toxic phospholipase A, at 40-50 μg/ml, showed the postsynaptic effect of a progressive reduction of m.e.p.p. amplitudes, whereas the enzyme alone, at 10 μg/ml, blocked the release of acetylcholine from frog motor nerve terminals (Blinov et al., 1979).

Heterologous Phospholipase A_2

A difference in sensitivity in the action of heterologous phospholipases A_2 on vertebrate motor nerve terminals and on excitable *Torpedo* microsacs is apparent. As with crotoxin, phospholipase A_2 from a bee (*Apis mellifica*) venom is active at very low concentrations on the microsacs (Bon, 1979), whereas at the frog neuromuscular junction, this phospholipase A_2 (5 µg/ml) did not block transmission but did prevent the inhibitory action of β-bungarotoxin. No competition for a common site was apparent (Abe and Miledi, 1978). *Apis mellifica* phospholipase A_2 as well as the acidic phospholipase A_2 from *Naja naja oxiana* venom were approximately 100-fold less potent that notexin in producing signs of disturbance of depolarization-secretion coupling and of the process of formation of new quanta at the frog neuromuscular junction (Magazanik et al., 1979). The acidic phospholipase A_2 from *Naja naja siamensis* venom was also capable of inhibiting the action of carbamylcholine on *Torpedo* microsacs by a process related to receptor desensitization (Andreasen et al., 1979).

The basis of the relative specificity of certain phospholipases A_2 for presynaptic membranes has not been fully explained; however, structural differences do exist between these enzymes, and it is interesting that attempts to inactivate the non-neurotoxic *C. adamanteus* and *C. atrox* phospholipases A_2 by modification of a critical histidine residue with *p*-bromophenacyl bromide have been unsuccessful, whereas the phospholipases A_2 showing neurotoxic activity have been so modified (Henrickson et al., 1977).

Skeletal Muscle

Toxins with Phospholipase A_2 Activity

At moderate concentrations of crotoxin (of the order of 10 µg/ml or less) which produce blockade of neurotransmitter release, no effects on the properties of isolated diaphragms from the mouse or rat were detected when measuring the contractile response to direct stimulation, the resting membrane potential, and the shape of the action potential (Chang and Lee, 1977; Hawgood and Smith, 1977b). When frog muscle was exposed to similar concentrations of crotoxin or homologous phospholipase A_2, the input resistance of the sarcolemmal membrane was unaffected, but the threshold required to initiate an action potential was raised (Hawgood and Santana de Sa, 1979). In isolated guinea pig nerve-muscle preparations, muscle excitability as shown by strength-duration curves was decreased by this dose of crotoxin only after a time lag of several hours (Brazil et al., 1979b). However, at high concentrations of crotoxin and of homologous phospholipase A_2 (30 µg/ml or more), myotoxicity was reported in mammalian isolated nerve-muscle preparations, with a reduction in contractile response accompanied by a sustained contracture (Breithaupt, 1976;

Hawgood and Smith, 1977b). No change in resting membrane potential was observed at 30 μg/ml crotoxin.

In vivo, the slowly developing myotoxic action of crotoxin and of the homologous phospholipase A_2 was readily demonstrated. In severely intoxicated mice which had received a systemic injection of crotoxin, motor nerve terminals in the diaphragm were necrotic, but the muscle fibers appeared normal (Figure 5) apart from the presence of large vacuoles in relation to the triad. However, when a sublethal dose of crotoxin or homologous phospholipase A_2 was injected into the calf muscles of one hindlimb, progressive dissolution of the internal structure of the muscle fibers was observed from 6 hr, and by 24 hr most fibers were necrotic and macrophage invasion had commenced. The ultrastructure of capillaries was normal, and the basement membrane was unaffected. Regeneration of muscle was advanced by 72 hr and was complete by 7 days (Gopalakrishnakone, 1979; Gopalakrishnakone and Hawgood, 1979). Mojave toxin produced a similar type of myonecrosis (Gopalakrishnakone et al., 1980). Certain observations point in favor of a raised intracellular Ca^{2+} concentration being involved in the mechanism of degeneration. Firstly, localized regions of overcontraction of myofibrils as well as contraction clumps in which the contractile filaments formed an increasingly homogeneous mass were apparent. Secondly, in two affected muscles considered to be degenerating slowly, dense intramitochondrial calcium deposits were observed which suggested that calcium accumulation by the mitochondria might act as a protective mechanism (Dempster et al., 1980). A rise in Ca^{2+} concentration could initiate necrosis by triggering protease activity directly or by promoting the release of lysosomal enzymes. This in turn could result from the internalization of the toxin with an action on the sarcoplasmic reticulum and/or an external action to induce a sarcolemmal defect. Mojave toxin was able to reduce the calcium sequestering activity of isolated vesicles from the sarcoplasmic reticulum (Cate and Bieber, 1976). Preliminary experiments indicated that endocytosis with uptake of HRP was present in crotoxin poisoned muscles, but it was not considered to be sufficiently prevalent to account for the extent of the muscle damage (Gopalakrishnakone, 1979).

The phospholipasic activity of crotoxin is involved in the onset of myonecrosis, as crotoxin reconstituted from histidine-modified phospholipase A_2 failed to produce necrosis when injected intramuscularly in mice (Gopalakrishnakone, 1979). Sarcolemmal defects were observed in necrotic fibers. However, in a study of notexin-induced myonecrosis, intracellular disorganization was not observed in the absence of lesions of the plasma membrane. Furthermore, immature muscles growing in vitro or regenerating in vivo were resistant to the toxin, which may reflect the lack of either an appropriate toxin binding site or an important substrate (Harris and Johnson, 1978; Harris et al., 1980). Of the other presynaptic neurotoxins, taipoxin is myonecrotic (Harris et al., 1977), but not β-bungarotoxin (Lee et al., 1976). Other phospholipases A_2 which are powerfully myotoxic include the basic enzyme from *Naja nigricollis* venom (Lee et al., 1977).

Basic Polypeptides of Low Molecular Weight

Crotamine, the basic polypeptide (mol. wt. 5000) isolated from the venom of *C. durissus terrificus* var. *crotaminicus,* produces a number of changes in the contractile properties of mammalian nerve-muscle preparations both in situ and in vitro. These changes include potentiation of the twitch amplitude, a delay in relaxation and/or after-contraction in response to a stimulus, the onset of contractures at higher doses, and the presence of spontaneous contractile activity (Cheymol et al., 1971a,b; Brazil, 1972; Chang and Tseng, 1978; Brazil et al., 1979c). Crotamine produced a rapid depolarization of the resting membrane potential of fibers of isolated rat and mouse diaphragms, an effect which was either prevented or reversed by the presence of tetrodotoxin, a specific blocker of voltage-dependent sodium channels (Chang and Tseng, 1978; Pellegrini et al., 1978). Binding of crotamine was not affected either by tetrodotoxin or by a raised Ca^{2+} concentration, but a high K^+ concentration displaced crotamine from the membrane and an increased Ca^{2+} concentration antagonized the expression of crotaminic action. Crotamine also induced a change in the sodium permeability of the sarcolemma as demonstrated by an increase in $^{24}Na^+$ influx and a decreased effective membrane resistance (Chang and Tseng, 1978). Cheymol and coworkers (1971a,b), who studied the effect of these conditions on the contracture-inducing properties of crotamine, came to the conclusion that the polypeptide interacts with membrane calcium sites to alter Na^+ or Ca^{2+} permeability. However, Chang and Tseng (1978) consider that the site of action is a regulator molecule which influences the permeability of the sodium channel and that crotamine opposes the depressant action exerted by extracellular K^+. These authors do not discuss how this effect relates to the characteristic augmentation of the twitch response, although prolongation of the active state of the muscle as a result of changes in membrane polarization plus the reported occurrence of spontaneous action potentials firing at a rate of 50-70 Hz could contribute. The resting membrane potentials recorded in these experiments fell to about -50 mV, which was close to or below the value at which action potentials could be elicited. In the absence of microelectrode penetration the depolarization may be less, since crotamine-poisoned muscles can respond to stimulation for long periods and the toxicity of crotamine is not high with an i.v. LD_{50} of 1.5 mg/kg mouse (Cheymol et al., 1971b; Brazil, 1972). Brazil et al. (1979c) point out the importance of the crotamine-induced instability of the muscle membrane with the persistence of spontaneous electrical activity and give evidence that the augmented twitch amplitude is due to a brief tetanus similar to that in myotonia. In vivo, crotamine induces spastic paralysis. Crotamine did not affect either the spontaneous or evoked release of neurotransmitter from rat motor nerve terminals, the response of guinea pig ileum to indirect stimulation, or the contraction and rhythmicity of the guinea pig atria (Chang and Tseng, 1978). Apparently, the motor nerve of the young chick is highly sensitive to crotamine, as the characteristic augmentation of the contractile response to indirect stimulation as well as spontaneous twitchings were abolished by (+)-tubocurarine. Low molecular weight basic toxins which were isolated from the venoms of *C. adamanteus, C. horridus horridus,* and *C. viridis viridis* had a similar action on this preparation (Lee et al., 1973). A polypeptide with chemical properties similar to crotamine has been isolated from *C. viridis helleri* venom (Maeda et al., 1978), and the occurrence of such molecules would appear to be relatively widespread in the venoms of different species of rattlesnakes.

Whole Rattlesnake Venoms

A crotamine-like augmentation of the contractile response to indirect and to direct stimulation of the isolated, mouse phrenic nerve-hemidiaphragm preparation was observed in the presence of venoms (10 µg/ml) from *C. durissus terrificus, Crotalus durissus totonacus, Crotalus horridus atricaudatus, C. horridus horridus,* and *C. scutulatus scutulatus* as illustrated in Figures 3 and 6. The effect was most marked with *C. durissus totonacus* venom, and although the maximal response soon declined, an underlying instability of contractile response remained as reported for the action of crotamine (Cheymol et al., 1971a).

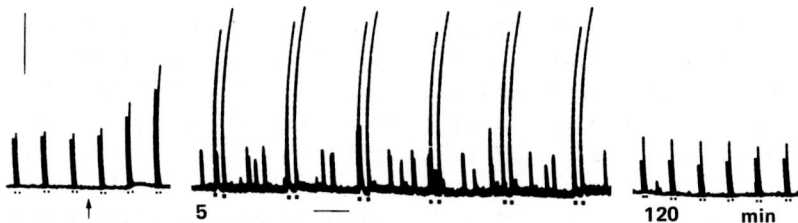

Figure 6 The contractile activity of the mouse phrenic nerve-hemidiaphragm preparation in the presence of *C. durissus totonacus* venom (10 µg/ml) which was added at the arrow. Electrical stimuli were alternately delivered to the nerve (the first stimulus of the pair) and to the muscle (· ·). Horizontal scale: 5 sec in the recording at 5 min, and 12 sec at 0 and 120 min; vertical scale: 1 g. Note that asynchronous contractile activity produced a thickening of the baseline at 5 min and that some spontaneous activity was still visible at 120 min. Stimulus parameters used in indirect stimulation were 0.05 msec, 1.2 V; in direct stimulation, 0.5 msec, 25 V. These latter parameters were determined as described in Figure 4. Combined frequency, 12 min^{-1}. Temperature $37°C$.

Spontaneous activity was observed in 2 of the 17 preparations exposed to these venoms at concentrations of 10-50 μg/ml, namely, with *C. durissus totonacus* venom (10 μg/ml, Figure 6) and with *C. horridus horridus* venom (10 μg/ml). In both cases, regularly occurring spontaneous contractions were observed which resembled fasciculations because of the firing of some motor units in short high frequency bursts. This persisted for long periods. Asynchronous spontaneous activity was also present.

In the experiments illustrated in Figure 3, venoms (10 μg/ml) from *C. durissus terrificus, C. horridus atricaudatus,* and *C. scutulatus scutulatus* produced a marked and progressive reduction of the contractile response to direct stimulation in contrast to the lack of such activity with venoms (10 μg/ml) from *C. durissus totonacus* and *C. horridus horridus.* This may be related to the presence of both crotamine-like and crotoxin-like substances in the former venoms (see Effects on the Respiratory System). However, other substances present in whole venom may affect membrane function as shown by the alterations in Na transport across the frog skin which were reported for *C. atrox* venom (Gerencser and Tu, 1977).

EFFECTS ON OTHER SYSTEMS

Disturbance of Motor Coordination and Convulsions

Systemic Injection

Both convulxin and gyroxin, proteins isolated from the venom of *C. durissus terrificus,* rapidly induce locomotor disturbances in unanesthetized animals following intravenous injection. In cats, convulxin produces loss of equilibrium, nystagmus, and convulsions, as well as gastrointestinal and respiratory disturbances. The symptoms begin a few seconds after injection and usually subside in less than 40 min. Convulxin induces seizures in mice and has an i.v. LD_{50} of 0.52 mg/kg mouse (Prado-Franceschi, 1970; Brazil, 1972). The action of convulxin to aggregate platelets with the release of biogenic amines has been discussed under Effects on the Cardiovascular System, and this property may lead either to the onset of convulsions as the result of the formation of microemboli, with subsequent transient cerebral ischemia, or to the massive release of autopharmacologic substances (Prado-Franceschi et al., 1981). Circulating amines do not readily pass the blood-brain barrier as exemplified by a brain uptake index of 2.6% for [^{14}C] serotonin relative to tritiated water (Oldendorf, 1971). Nevertheless, this value is significantly greater than the extracellular marker sucrose, and an explosive release of serotonin (and histamine) into the bloodstream may transiently raise the circulating concentration of these putative amine neurotransmitters sufficiently to allow the passage of pharmacologically active amounts into the brain.

Gyroxin (mol. wt. approx. 34,000) is a nonlethal toxin which, in mice given an intravenous injection of 5-50 μg, evoked characteristic alterations in equilibrium resulting in motor phenomena such as circling in one direction, lateral fall, rolling movements also in one direction, and short tonic seizures. Nystagmus was reported to be absent. A short lag period of about 3 min occurred before the onset of symptoms, and complete recovery was observed by 60 min. The response was not dose-dependent but rather all-or-none, and it exhibited marked tachyphylaxis even 10 and 40 days later (Barrabin et al., 1978; Seki et al., 1980). There are some resemblances to the properties of crotamine. The authors describe the syndrome as resembling that of a labyrinthic lesion and, in several cases, reported that autonomic nervous system symptoms such as pallor and tachypnea preceded the onset of the motor symptoms. It is of interest that the blood brain barrier is markedly reduced in the region of both sensory ganglia and ganglia of the autonomic nervous system as shown by the rapid passage of HRP (mol. wt. 40,000) across the capillary endothelium (Jacobs et al., 1976; Jacobs, 1977). It is an interesting possibility that gyroxin may produce a brief excitation of vestibular and other ganglia; morphologic and physiological evidence of the site(s) and mode of action are awaited.

Gross alterations in behavior such as hyperexcitability and prostration following intravenous or subcutaneous (s.c.) injection of crotamine have been considered to point to additional central effects (Habermann and Cheng-Raude, 1975). However, endothelial cells of cerebral capillaries have tight intercellular junctions and few pinocytotic vesicles, an ultrastructure which is responsible for the functional barrier to the entrance of many substances into the brain. It is difficult to envisage the passage of crotamine into the brain unless it occurs in the restricted areas where the blood-brain barrier is reduced. It is conceivable that alterations in the permeability of the blood-brain barrier may arise from enzymic degradation of membrane proteins and so promote the passage of venom components. In this context, a large intraperitoneal injection of chymotrypsin in rabbits increased the passage of penicillin into the brain (Wohlman et al., 1968). However, cerebral edema and gross neurologic changes are an invariable consequence of generalized alterations in permeability (see the review by Pardridge et al., 1975), and such irreversible changes may be demonstrated macroscopically if the animal is preinjected with trypan blue (or another vital dye) (Broman and Lindberg-Broman, 1945). No such action has yet been demonstrated for any rattlesnake venom or purified component.

Cerebral Intraventricular Injection

Intraventricular injection of crotoxin (50 μg) in cats induced clonic convulsions after a lag period of hours. These seizures, which were lethal, were in contrast to to the flaccid paralysis of systemic poisoning (Brazil et al., 1966b). A prolonged period prior to the onset of convulsions was characteristic of the central action in mice and rats, of *C. durissus terrificus* phospholipase A_2, of the relatively toxic phospholipase A_2 from *Naja nigricollis* venom, and of the relatively nontoxic phospholipase A_2 from *Hemachatus hemachatus* venom (Habermann and Cheng-Raude, 1975; Fletcher et al., 1980). Widespread hydrolysis of brain phospholipids was observed.

Crotamine (0.6 to 6 µg) induced immediate spasticity and convulsions in mice given an intralumbar injection. In rats, intraventricular injection (25µg) produced immediate hypernea and altered behavior consisting of general prostration mixed with periods of unrest. The symptoms lessened within 30 min, and recovery occurred within 6 hr (Habermann and Cheng-Raude, 1975). Although some venom components show a central action when the blood-brain barrier is circumvented, there is little evidence to suggest that they exert this effect when injected peripherally.

Visceral Smooth Muscle and the Autonomic Nervous System

Excitation of visceral smooth muscle and/or the autonomic nervous system is implicated by the symptoms of defecation, vomiting, salivation, and urination which have been reported following intravenous injections in dogs of *C. horridus* or *C. durissus terrificus* venom (Essex and Markowitz, 1930; Brazil et al., 1966b). These symptoms may arise from the action of one or more of a number of protein components. In the classic experiments of Rocha e Silva et al. (1949) the production of bradykinin was detected by contraction of isolated intestinal muscle preparations, and this response would appear to be a characteristic of the action of rattlesnake venoms owing to the presence of kininogenase (see Effects on the Cardiovascular System). Intestinal contraction, facilitation of the peristaltic reflex, and defecation are also produced by serotonin acting both directly on the D receptors of intestinal smooth muscle and on the M receptors of parasympathetic ganglion cells to release acetylcholine (Bowman and Rand, 1980). The violent intestinal contractions and salivation produced by convulxin (Brazil, 1972) may be so mediated, and it would be interesting to determine if convulxin released serotonin not only from platelets but also from the mucosal chromaffin cells of the intestine which contain most of the body's store of serotonin. Such a mechanism may also apply to *C. horridus horridus* venom, which was equally as effective as *C. durissus terrificus* venom in releasing histamine and serotonin from platelets in vitro (see Effects on the Cardiovascular System). Histamine is also known to be a ganglionic stimulant, and a dual excitatory action on smooth muscle and parasympathetic postganglionic neurones may be responsible for the onset of micturition. As regards vomiting, excitatory inputs to the vomiting center which may be activated by the venom include those from receptors in the alimentary canal tract which respond to intramuscular tension and to chemical stimuli. The depolarizing action of serotonin on the cell bodies of visceral afferents in the nodosal ganglion of the rabbit vagal trunk (Higashi, 1977) is another possible excitatory mechanism.

After intravenous injection of crotoxin (0.2 mg/kg) into dogs, defecation, vomiting, and salivation frequently occurred prior to the appearance of paralysis (Brazil et al., 1966b). These authors cite a 1938 report by Slotta and Fraenkel-Conrat that crotoxin has a stimulatory effect on the isolated guinea pig ileum. In vivo, the stimulatory action of crotoxin may be enhanced by a limited release of autacoids such as serotonin and prostaglandins. In crotoxin-injected dogs which showed myoneural blockade, crotoxin failed to affect transmission either at the cholinergic junction between pre- and postganglionic neurones of the cat superior cervical ganglion or at the adrenergic junction with the nictitating membrane (Brazil, 1966). As no further studies have been carried out on the effects of crotoxin, pointers may come from the study of other presynaptically active toxins with phospholipase A_2 activity, particularly notexin and taipoxin which are also musculotropic. Notexin was largely inactive at autonomic postganglionic synapses but induced a contraction of the longitudinal smooth muscle of the isolated guinea pig ileum which was not mediated by either cholinergic or histaminergic mechanisms. Features of the response were a spontaneous reversibility and long-lasting desensitization. This toxin also inhibits transmission in the guinea pig vas deferens and seminal vesicles by a presynaptic mechanism (Harris and Zar, 1978).

Taipoxin can attack certain peripheral cholinergic nerve terminals outside the somatic nervous system, as, following an intravenous injection into mice, ultrastructural changes were observed in the axon terminals in the adrenal medulla which were similar to those described at the neuromuscular junction (Lüllman-Rauch and Thesleff, 1979). In vitro, both taipoxin and β-bungarotoxin blocked the response of the isolated nerve-atrium preparation of the guinea pig to vagal stimulation in respect to changes both in frequency and contractility, probably by acting on the postganglionic nerve terminals to reduce the release of acetylcholine. Sympathetic stimulation was not affected (Zeigler, 1978; Muramatsu et al., 1980). There is clear evidence from several sources that these toxins can affect the autonomic nervous system and thus contribute to the overall response to envenomation.

CONCLUSION

A common feature of rattlesnake envenomation is the immediate induction of hypotension and a reduction in circulating plasma volume which may lead to shock. Evidence is accumulating that the release of autacoids by protein components is a major factor; these autacoids have a short half-life in the circulation, and the normal homeostatic mechanisms may be able to combat their action when poisoning is not severe. Many venoms also contain a more slowly acting factor which produces irreversible blockade of neuromuscular transmission, so greatly increasing the lethality of the venom. However, very few systems would appear to be immune from the venomous effects, and the synergistic action of the components is a powerful lethal device. Some of the toxic components are more widespread within the genus than originally thought, e.g., neurotoxins with phospholipase A_2 activity have been shown in the venom of the North American rattlesnake, *C. scutulatus scutulatus* as well as in the venom of the South American rattlesnake, *C. durissus terrificus,* and there is indirect evidence for its presence in at least two other North American species. The presence of low molecular weight, nonenzymic polypeptides with similarity to crotamine would appear to be quite widespread within the species; this may prove to be the case for other components such as convulxin, which was first isolated from the venom of South American species. A marked species variation in the concentration of individual purified toxins within each venom has been observed, and although the venom of relatively few species of *Crotalus* and of *Sistrurus* have so far been studied in detail, this would account for much of the individuality in the physiological and pharmacologic actions of each rattlesnake venom. Intriguing physiological problems remain such as the site and mode of action of gyroxin and the specific role of the low molecular weight polypeptides. Although these substances do not have a lethal action, their investigation may reveal interesting examples of modulation of physiological systems.

ACKNOWLEDGMENTS

I am grateful to Dr. Bernard Whaler for his many helpful suggestions during the preparation of this chapter.

REFERENCES

Abe, T., and Miledi, R. (1978). Inhibition of β-bungarotoxin action by bee venom phospholipase A_2. *Proc. R. Soc. Lond.* [*Biol.*] 200:225-230.
Abe, T., Alema, S., and Miledi, R. (1977). Isolation and characterization of presynaptically acting neurotoxins from the venom of *Bungarus* snakes. *Eur. J. Biochem.* 80:1-12.
Andreasen, T. J., Doerge, D. R., and McNamee, M. G. (1979). Effects of phospholipase A_2 on the binding and ion permeability control properties of the acetylcholine receptor. *Arch. Biochem. Biophys.* 194:468-480.

Barrabin, H., Martiarena, J. L., Vidal, J. C., and Barrio, A. (1978). Isolation and characterization of gyroxin from *Crotalus durissus terrificus* venom. In *Animal, Plant and Microbial Toxins*, P. Rosenberg (Ed.), Pergamon Press, Oxford, pp. 113-133.

Bhargava, N., Vargaftig, B. B., de Vos, C. J., Bonta, I. L., and Tijs, T. (1970a). Dissociation of oedema provoking factor of *Agkistrodon piscivorus* venom from kininogenase. In *Vasopeptides: Chemistry Pharmacology and Pathophysiology*, N. Back and F. Sicuteri (Eds.). Plenum, New York, pp. 141-148.

Bhargava, N., Zirinis, P., Bonta, I. L., and Vargaftig, B. B. (1970b). Comparison of hemorrhagic factors of the venoms of *Naja naja, Agkistrodon piscivorus* and *Apis mellifera*. *Biochem. Pharmacol. 19*:2405-2412.

Bieber, A. L., Tu, T., and Tu, A. T. (1975). Studies of an acidic cardiotoxin isolated from the venom of the Mojave rattlesnake (*Crotalus scutulatus*) *Biochim. Biophys. Acta 400*:178-188.

Blinov, N. O., Tchorbanov, B. P., Grishin, E. V., and Aleksiev, B. V. (1979). Neurotoxic action of a protein complex from the venom of the Bulgarian viper (*Vipera ammodytes ammodytes*). *Comptes rendus Acad. Bul. Sci. 32*:663-666.

Bon, C. (1979). Crotoxin: Its mechanism of action. In *Advances in Cytopharmacology*, Vol. 3: *Neurotoxins, Tools in Neurobiology*, B. Ceccarelli and F. Clementi (Eds.). Raven Press, New York.

Bon, C., Changeux, J.-P., Jeng, T. W., and Fraenkel-Conrat, H. (1979). Postsynaptic effects of crotoxin and of its isolated sub-units. *Eur. J. Biochem. 99*:471-481.

Bonilla, C. A., and Rammel, O. J. (1976). Comparative biochemistry and pharmacology of salivary gland secretions. III. Chromatographic isolation of a myocardial depressor protein (MDP) from the venom of *Crotalus atrox*. *J. Chromatogr. 124*:303-314.

Bonta, I. L., Vargaftig, B. B., and Böhm, G. M. (1979). Snake venoms as an experimental tool to induce and study models of microvessel damage. In *Handbook of Experimental Pharmacology*, Vol. 52:*Snake Venoms*, chapter 17, C. Y. Lee (Ed.). Springer-Verlag, Berlin, pp. 629-683.

Bowman, W. C., and Rand, M. J. (1980). *Textbook of Pharmacology*, 2nd ed. Blackwell Scientific Publications, Oxford, pp. 12·25, 12·34.

Brazil, O. V. (1966). Pharmacology of crystalline crotoxin. II. Neuromuscular blocking action. *Mem. Inst. Butantan. Simp. Internac. 33*:981-992.

Brazil, O. V. (1972). Neurotoxins from the South American rattlesnake. *J. Formosan Med. Assoc. 71*:394-400.

Brazil, O. V. and Excell, B. J. (1971). Action of crotoxin and crotactin from the venom of *Crotalus durissus terrificus* (South American rattlesnake) on the frog neuromuscular junction. *J. Physiol. (Lond.)*. 212:34-35P.

Brazil, O. V., Fariña, R., Yoshida, L., and de Oliveira, V. A. (1966a). Pharmacology of crystalline crotoxin. III. Cardiovascular and respiratory effects of crotoxin and *Crotalus durissus terrificus* venom. *Mem. Inst. Butantan Simp. Internac. 33*:993-1000.

Brazil, O. V., Prado-Franceschi, J., and Waisbich, E. (1966b). Pharmacology of crystalline crotoxin. I. Toxicity. *Mem. Inst. Butantan Simp. Internac. 33*:973-980.
Brazil, O. V., Excell, B. J., and Santana de Sa, S. (1973). The importance of phospholipase A in the action of the crotoxin complex at the frog neuromuscular junction. *J. Physiol. (Lond.) 234*:63-64P.
Brazil, O. V., Fontana, M. D., and Heluany, N. F. (1979a). Effect of 4-aminopyridine on the neuromuscular blockade produced by crotoxin. *Toxicon 17*: (Suppl. 1):16.
Brazil, O. V., Fontana, M. D., and Heluany, N. F. (1979b). Mode of action of crotoxin at the guinea-pig neuromuscular junction. *Toxicon 17*(Suppl. 1):17.
Brazil, O. V., Prado-Franceschi, J., and Laure, C. J. (1979c). Repetitive muscle responses induced by crotamine. *Toxicon 17*:61-67.
Breithaupt, H. (1976). Neurotoxic and myotoxic effects of *Crotalus* phospholipase A and its complex with crotapotin. *Naunyn Schmiedebergs Arch. Pharmacol. 292*:271-278.
Broman, T., and Lindberg-Broman, A. M. (1945). An experimental study of disorders in the permeability of cerebral vessels ("the blood-brain barrier") produced by chemical and physico-chemical agents. *Acta Physiol. Scand. 10*:102-125.
Carlson, R. W., Schaeffer, R. C., Whigham, H., Michaels, S., Russell, F. E., and Weil, M. H. (1975). Rattlesnake venom shock in the rat: Development of a method. *Am. J. Physiol. 229*:1668-1674.
Cate, R. L., and Bieber, A. L. (1976). Effects of Mojave toxin on rat skeletal muscle sarcoplasmic reticulum. *Biochem. Biophys. Res. Commun. 72*:295-301.
Chang, C. C., and Lee, J. D. (1977). Crotoxin, the neurotoxin of South American rattlesnake venom, is a pre-synaptic toxin acting like β-bungarotoxin. *Naunyn Schmiedebergs Arch. Pharmacol. 296*:159-168.
Chang, C. C., and Su, M. J. (1978). The mechanism of potentiation of the pre-synaptic effect of phospholipase A_2 by the crotapotin component of the crotoxin complex. *Toxicon 16*:402-405.
Chang, C. C., and Tseng, K. H. (1978). Effect of crotamine, a toxin of South American rattlesnake venom, on the sodium channel of murine skeletal muscle. *Br. J. Pharmacol. 63*:551-559.
Chang, C. C., Lee, J. D., Eaker, D., and Fohlman, J. (1977a). The pre-synaptic neuromuscular blocking action of taipoxin. A comparison with β-bungarotoxin and crotoxin. *Toxicon 15*:571-576.
Chang, C. C., Su, M. J., Lee, J. D., and Eaker, D. (1977b). Effects of Sr^{2+} and Mg^{2+} on the phospholipase A and the pre-synaptic neuromuscular blocking actions of β-bungarotoxin, crotoxin and taipoxin. *Naunyn Schmiedebergs Arch. Pharmacol. 299*:155-161.
Chen, I., and Lee, C. Y. (1970). Ultrastructural changes in the motor nerve terminals caused by β-bungarotoxin. *Virchows Arch [Zellpathol.] 6*:318-325.

Cheymol, J., Gonçalves, J. M., Bourillet, F., and Roch-Arveiller, M. (1971a). Action neuromusculaire comparée de la crotamine et du venin de *Crotalus durrissus terrificus* var *crotaminicus*. I. Sur préparations neuromusculaire in situ. *Toxicon* 9:279-286.

Cheymol, J., Gonçalves, J. M., Bourillet, F., and Roch-Arveiller, M. (1971b). Action neuromusculaire comparée de la crotamine et du venin de *Crotalus durissus terrificus* var *crotaminicus*. II. Sur préparations isolées. *Toxicon* 9: 287-289.

Cull-Candy, S. G., Fohlman, J., Gustavsson, D., Lüllman-Rauch, R., and Thesleff, S. (1976). The effects of taipoxin and notexin on the function and fine structure of the murine neuromuscular junction. *Neuroscience* 1:175-180.

Damerau, B., Lege, L., Oldigs, H.-D., and Vogt, W. (1975). Histamine release, formation of prostaglandin-like activity (SRS-C) and mast cell degranulation by the direct lytic factor (DLF) and phospholipase A of cobra venom. *Naunyn Schmiedebergs, Arch. Pharmacol.* 287:141-156.

Damus, P. S., Markland, F. S., Davidson, T. M., and Shanley, J. D. (1972). A purified procoagulant enzyme from the venom of the eastern diamondback rattlesnake (*Crotalus adamanteus*): In vivo and in vitro studies. *J. Lab. Clin. Med.* 79:906-923.

Daniels, M. P., and Vogel, Z. (1975). Acetylcholine receptor staining: Immunoperoxidase staining of α-bungarotoxin binding sites in muscle end-plates. *Nature* 254:339-341.

Dempster, D. W., Gopalakrishnakone, P., and Hawgood, B. J. (1980). Intramitochondrial calcium deposits associated with muscle necrosis induced by crotoxin. *J. Physiol.* (*Lond.*) 300:21P.

Deutsch, H. F., and Diniz, C. R. (1955). Some proteolytic activities of snake venoms. *J. Biol. Chem.* 216:17-26.

Essex, H. E., and Markowitz, J. (1930). The physiological action of rattlesnake venom (crotalin). I. Effect on blood pressure: Symptoms and post-mortem observations. *Am. J. Physiol.* 92:317-328.

Ferreira, S. H., and Vane, J. R. (1967). The detection and estimation of bradykinin in the circulating blood. *Br. J. Pharmacol.* 29:367-377.

Fletcher, J. E., Rapuano, B. E., Condrea, E., Yang, C.-C., Ryan, M., and Rosenberg, P. (1980). Comparison of a relatively toxic phospholipase A_2 from *Naja nigricollis* snake venom with that of a relatively non-toxic phospholipase A_2 from *Hemachatus haemachatus* snake venom. II. Pharmacological properties in relationship to enzymatic activity. *Biochem. Pharmacol.* 29:1565-1574.

Fohlman, J., Eaker, D., Dowdall, M. J., Lüllman-Rauch, R., Sjödin, T., and Leander, S. (1979). Chemical modification of taipoxin and the consequences for phospholipase activity, pathophysiology and inhibition of high-affinity choline uptake. *Eur. J. Biochem.* 94:531-540.

Gerencser, G. A., and Tu, A. T. (1977). Effects of *Crotalus atrox* venom on sodium transport across the frog skin. *Proc. Soc. Exp. Biol. Med.*, 156:104-108.

Glenn, J. L., and Straight, R. (1977). The midget faded rattlesnake (*Crotalus viridis concolor*) venom: Lethal toxicity and individual variability. *Toxicon* 15:129-133.
Gopalakrishnakone, P. (1979). Morphological studies on the effects of a phospholipase A_2 complex (crotoxin) from *Crotalus durissus terrificus* venom on muscle, nerve and neuromuscular junction in the mouse. Ph.D. Thesis, University of London.
Gopalakrishnakone, P., and Hawgood, B. J. (1979). Morphological changes in murine nerve, neuromuscular junction and skeletal muscle induced by the crotoxin complex. *J. Physiol. (Lond.) 291*:5-6P.
Gopalakrishnakone, P., Hawgood, B. J., Holbrooke, S. E., Marsh, N. A., Santana de Sa, S., and Tu, A. T. (1980). Sites of action of Mojave toxin isolated from the venom of the Mojave rattlesnake. *Br. J. Pharmacol. 69*:421-431.
Gopalakrishnakone, P., Hawgood, B. J., and Theakston, R. D. G. (1981). Specificity of antibodies to the reconstituted crotoxin complex, from the venom of South American rattlesnake (*Crotalus durissus terrificus*), using enzyme-linked immunosorbent assay (ELISA) and double immunodiffusion. *Toxicon 19*:131-139.
Gundersen, C. B., Newton, M. W., and Jenden, D. J. (1980). β-bungarotoxin elevates diaphragm acetylcholine levels. *Brain Res. 182*:486-490.
Habermann, E., and Breithaupt, H. (1978). Mini-review. The crotoxin complex— an example of biochemical and pharmacological protein complementation. *Toxicon 16*:19-30.
Habermann, E., and Cheng-Raude, D. (1975). Central neurotoxicity of apamin, crotamin, phospholipase A and α-amanitin. *Toxicon 13*:465-473.
Habermann, E., Walsh, P., and Breithaupt, H. (1972). Biochemistry and pharmacology of the crotoxin complex. II. Possible interrelationships between toxicity, organ distribution of phospholipase A, crotapotin and their combination. *Naunyn Schmiedebergs Arch. Pharmacol. 273*:313-330.
Halmagyi, D. F., Starzecki, G., and Horner, G. J. (1965). Mechanism and pharmacology of shock due to rattlesnake venom in sheep. *J. Appl. Physiol. 20*: 709-718.
Hanley, M. R. (1978). Crotoxin effects of *Torpedo californica* cholinergic excitable vesicles and the role of its phospholipase A activity. *Biochem. Biophys. Res. Commun. 82*:392-401.
Harris, J. B., and Johnson, M. A. (1978). Further observations on the pathological responses of rat skeletal muscle to toxins isolated from the venom of the Australian tiger snake, *Notechis scutatus scutatus*. *Clin. Exp. Pharmacol. Physiol. 5*:587-600.
Harris, J. B., and Zar, M. A. (1978). The effects of a toxin isolated from Australian tiger snake (*Notechis scutatus scutatus*) venom on autonomic neuromuscular transmission. *Br. J. Pharmacol. 62*:349-358.
Harris, J. B., Johnson, M. A., and MacDonell, C. (1977). Taipoxin, a presynaptically active neurotoxin destroys mammalian muscle. *Br. J. Pharmacol. 61*: 133P.

Harris, J. B., Johnson, M. A., and MacDonell, C. A. (1980). Muscle necrosis induced by some presynaptically active neurotoxins. In *Natural Toxins,* D. Eaker and T. Wadstrom (Eds.). *Toxicon* Suppl. No. 2, Pergamon Press, Oxford, pp. 569–578.

Hawgood, B. J., and Santana de Sa, S. (1979). Changes in spontaneous and evoked release of transmitter induced by the crotoxin complex and its component phospholipase A_2 at the frog neuromuscular junction. *Neuroscience* 4:293–303.

Hawgood, B. J., and Smith, J. W. (1977a). The presynaptic action of crotoxin at the murine neuromuscular junction. *J. Physiol. (Lond.) 266*:91–92P.

Hawgood, B. J., and Smith, J. W. (1977b). The mode of action at the mouse neuromuscular junction of the phospholipase A-crotapotin complex isolated from the venom of the South American rattlesnake. *Br. J. Pharmacol. 61*: 597–606.

Heilbronn, E. (1972). Action of phospholipase A on synaptic vesicles. A model for transmitter release? *Progr. Brain Res. 36*:29–40.

Henrickson, R. L., Krueger, E. T., and Keim, P. S. (1977). Amino acid sequence of phospholipase A_2-α from the venom of *Crotalus adamanteus. J. Biol. Chem. 252*:4913–4921.

Higashi, H. (1977). 5-hydroxytryptamine receptors on visceral primary afferent neurones in the nodose ganglion of the rabbit. *Nature 267*:448–450.

Jacobs, J. M. (1977). Penetration of systemically injected horseradish peroxidase into ganglia and nerves of the autonomic nervous system. *J. Neurocytol. 6*: 607–618.

Jacobs, J. M., MacFarlane, R. M., and Cavanagh, J. G. (1976). Vascular leakage in the dorsal root ganglia of the rat studied with horseradish peroxidase. *J. Neurol. Sci. 29*:95–107.

Jeng, T. W., Hendon, R. A., and Fraenkel-Conrat, H. (1978). Search for relationships among the hemolytic, phospholipolytic and neurotoxic activities of snake venoms. *Proc. Natl. Acad. Sci. U.S.A. 75*:600–604.

Kamenskaya, M., and Thesleff, S. (1979). Effects of taipoxin on spontaneous and evoked transmitter release from motor nerve terminals. *Toxicon 17* (Suppl. 1):80–81.

Kelly, R. B., Oberg, S. G., Strong, P. N., and Wagner, G. M. (1976). β-bungarotoxin, a phospholipase that stimulates transmitter release. In *The Synapse, Symposia on Quantitative Biology,* vol. 40. Cold Spring Harbor Laboratory, N.Y., pp. 117–125.

Kelly, R. B., von Wedel, R. J., and Strong, P. N. (1979). Phospholipase-dependent and phospholipase-independent inhibition of transmitter release by β-bungarotoxin. In *Neurotoxins, Tools in Neurobiology. Advances in Cytopharmacology,* vol. 3, B. Ceccarelli and F. Clementi (Eds.). Raven Press, New York, pp. 77–85.

Lassignal, N. L., and Heuser, J. E. (1977). Evidence that β-bungarotoxin arrests synaptic vesicle recycling by blocking coated vesicle formation. *Neurosci. Abs. 1192*:373.

Lee, C. Y. (Ed.) (1979a). *Snake Venoms Handbook of Experimental Pharmacology*, vol. 52. Springer-Verlag, Berlin, 1130 pp.

Lee, C. Y. (1979b). Recent advances in chemistry and pharmacology of snake toxins. In *Advances in Cytopharmacology*, vol. 3, B. Ceccarelli and F. Clementi (Eds.). Raven Press, New York, pp. 1-16.

Lee, C. Y., Huang, M-C., and Bonilla, C. A. (1973). Mode of action of purified basic proteins from three rattlesnake venoms on neuromuscular junctions of the chick biventer cervicis muscle. In *Animal and Plant Toxins*, E. Kaiser (Ed.). Wilhelm Goldmann, München, pp. 173-178.

Lee, C. Y., Chen, Y. M., and Karlsson, E. (1976). Post-synaptic and musculotropic effects of notexin, a pre-snyaptic neurotoxin from the venom of *Notechis scutatus scutatus* (Australian tiger snake). *Toxicon 14*:493-494.

Lee, C. Y., Ho, C. L., and Eaker, D. (1977). Cardiotoxin-like action of a basic phospholipase A isolated from *Naja nigricollis* venom. *Toxicon 15*:355-356.

Lomba, M. G. (1969). Estudos sôbre a distribuiçao e excreçao da crotoxina-131I em caes. Doctoral Thesis, State University of Campinas, Campinas, Sao Paulo, Brazil.

Lüllman-Rauch, R., and Thesleff, S. (1979). Effects of taipoxin on the ultrastructure of cholinergic axon terminals in the mouse adrenal medulla. *Neuroscience 4*:837-841.

Maeda, N., Tamiya, N., Pattabhiraman, T. R., and Russell, F. E. (1978). Some chemical properties of the venom of the rattlesnake, *Crotalus viridis helleri. Toxicon 16*:431-442.

Magazanik, L. G., and Slavnova, T. I. (1978). Effects of presynaptic polypeptide neurotoxins from tiger snake venom (Notechis-II-5 and notexin) on frog neuromuscular junction. *Physiol. Bohemoslov. 27*:421-429.

Magazanik, L. G., Gotgilf, I. M., Slavnova, T. I., Miroshnikov, A. I., and Apsalon, V. R. (1979). Effects of phospholipase A_2 from cobra and bee venom on the pre-synaptic membrane. *Toxicon 17*:477-488.

Malik, K. V., and McGiff, J. C. (1976). Cardiovascular actions of prostaglandins In *Prostaglandins: Physiological Pharmacological and Pathological Aspects*, S. M. M. Karim (Ed.). MTP Press, Lancaster, pp. 103-200.

Margolis, J., Bruce, S., Starzecki, B., Horner, G. J., and Halmagyi, D. F. J. (1965). Release of bradykinin-like substance (BKLS) in sheep by venom of *Crotalus atrox. Aust. J. Exp. Biol. Med. Sci. 43*:237-244.

Markwardt, F., Barthel, W., Glusa, E., and Hoffman, A. (1966a). Über die freisetzung biogener amine aus blutplättchen durch tierische gifte. *Naunyn Schmiedebergs Arch. Pharmak. Exp. Pathol. 252*:297-304.

Markwardt, F., Barthel, W., Glusa, E., Hoffman, A., and Walsmann, P. (1966b). Über eine aminfreisetzende komponente des *Crotalus terrificus* — giftes. *Biochem. Z. 346*:351-356.

Marlas, G. (1978). Relationship between the pharmacological post-synaptic effect of crotoxin and its phospholipasic activity. Doctoral Thesis, University of Paris.

Mebs. D. (1970). A comparative study of enzyme activities in snake venoms. *Int. J. Biochem.* 1:335-342.

Minton, S. A., (1956). Some properties of North American pit viper venoms and their correlation with phylogeny. In *Venoms,* E. E. Buckley and N. Porges (Eds.). Am. Assoc. Adv. Sci., Pub. No. 44, Washington, D.C., pp. 145-151.

Muramatsu, I., Fujiwara, M., Miura, A., Hayashi, K., and Lee, C. Y. (1980). Beta-bungarotoxin and parasympathetic nerve blocking action. *J. Pharmacol. Exp. Ther.* 213:156-160.

Nair, B. C., Nair, C., and Elliott, W. B. (1979). Isolation and partial characterization of a phospholipase A_2 from the venom of *Crotalus scutulatus salvini. Toxicon* 17:557-569.

Oldendorf, W. H. (1971). Brain uptake of radiolabelled amino acids, amines and hexoses after arterial injection. *Am. J. Physiol.* 221:1629-1639.

Oshima, G., Sato-Ohmori, T., and Suzuki, T. (1969). Proteinase, arginineester hydrolase and a kinin releasing enzyme in snake venoms. *Toxicon* 7:229-233.

Ownby, C. L., Bjarnason, J., and Tu, A. T. (1978). Hemorrhagic toxins from rattlesnake (*Crotalus atrox*) venom. *Am. J. Pathol.* 93:201-218.

Pardridge, W. M., Connor, J. D., and Crawford, I. L. (1975). Permeability changes in the blood-brain barrier: Causes and consequences. *Crit. Rev. Toxicol.* 3:159-199.

Pellegrini Filho, A., Brazil, O. V., Dias Fontana, M., and Laure, C. J. (1978). The action of crotamine on skeletal muscle: An electrophysiological study. In *Toxins: Animal, Plant and Microbial,* P. Rosenberg (Ed.). Pergamon Press, Oxford, pp. 375-382.

Possani, L. D., Sosa, B. P., Alagón, A. C., and Burchfield, P. M. (1980). The venom from the snakes *Agkistrodon bilineatus taylori* and *Crotalus durissus totonacus*: Lethality, biochemical and immunological properties. *Toxicon* 18:356-360.

Prado-Franceschi, J. (1970). Thesis, State University of Campinas, Campinas, Sao Paulo, Brazil.

Prado-Franceschi, J., Tavares, D. G., Hertel, R., and Lobo de Araújo, A. (1981). Effects of convulxin, a toxin from the rattlesnake venom, on platelets and leukocytes of anaesthetized rabbits. *Toxicon* 19:661-666.

Rocha e Silva, M., Beraldo, W. T., and Rosenfeld, G. (1949). Bradykinin, a hypotensive and smooth muscle stimulating factor released from plasma globulin by snake venoms and by trypsin. *Am. J. Physiol.* 156:261-273.

Rothschild, A. M. (1966). Mechanism of histamine release by animal venoms. *Mem. Inst. Butantan Simp. Internac.* 33:467-476.

Rothschild, A. M., and Rothschild, Z. (1979). Liberation of pharmacologically active substances by snake venoms. In *Handbook of Experimental Pharmacology,* vol. 52: *Snake Venoms,* C. Y. Lee (Ed.). Springer-Verlag, Berlin, pp. 591-628.

Rovere, A. A., Raynald, A. C., Berman, J. M., Sanroman, A., and Garcia, C. A. (1978). Precocious effect of *Crotalus durissus terrificus* venom on the kidney circulation. *Acta Physiol. Latin Am. 28*:133-140.

Ruiz, C. E., Schaeffer, R. C., Weil, M. H., and Carlson, R. W. (1980). Hemostatic changes following rattlesnake (*Crotalus viridis helleri*) venom in the dog. *J. Pharmacol. Exp. Ther. 213*:414-417.

Russell, F. E., and Long, T. E. (1961). Effects of venoms on neuromuscular transmission. In *International Symposium on Myasthenia Gravis*, H. Viets (Ed.). Blackwell Scientific Publications, Oxford, pp. 101-116.

Schaeffer, R. C., Carlson, R. W., Puri, V. K., Callahan, G., Russell, F. E., and Weil, M. H. (1978a). The effects of colloidal and crystalloidal fluids on rattlesnake venom shock in the rat. *J. Pharmacol. Exp. Ther. 206*:687-695.

Schaeffer, R. C., Carlson, R. W., Whigham, H., Weil, M. H., and Russell, F. E. (1978b). Acute hemodynamic effects of rattlesnake *Crotalus viridis helleri* venom. In *Animal, Plant and Microbial Toxins*, P. Rosenberg (Ed.). Pergamon Press, Elmsford, N.Y., pp. 383-391.

Schaeffer, R. C., Carlson, R. W., and Weil, M. H. (1979a). Effects of antivenin and corticosteroid analogs on rattlesnake venom shock in the rat. *J. Pharmacol. Exp. Ther. 211*:409-414.

Schaeffer, R. C., Pattabhiraman, T. R., Carlson, R. W., Russell, F. E., and Weil, M. H. (1979b). Cardiovascular failure produced by a peptide from the venom of the Southern Pacific rattlesnake *Crotalus viridis helleri*. *Toxicon 17*:447-453.

Seki, C., Vidal, J. C., and Barrio, A. (1980). Purification of gyroxin from a South American rattlesnake (*Crotalus durissus terrificus*) venom. *Toxicon 18*:235-247.

Strong, P. N., Heuser, J. E., and Kelly, R. B. (1977). Selective enzymatic hydrolysis of nerve terminal phospholipids by β-bungarotoxin: Biochemical and morphological studies. In *Cellular Neurobiology*, F. Fox, Z. Hall, and R. B. Kelly (Eds.). A. R. Liss, New York, pp. 227-249.

Tu, A. T. (1977). *Venoms: Chemistry and Molecular Biology*. John Wiley, New York, 560 pp.

Vargaftig, B. B. (1967). Antagonisme par les analgésiques non-narcotiques de la libération de kinines plasmatiques due au venin de *B. jararaca* et la kallikréine pancréatique. *Med. Pharmacol. Exp. 17*:517-526.

Vargaftig, B. B. (1974). Search for common mechanism underlying the various effects of putative inflammatory mediators. In *The Prostaglandins*, vol. II, P. W. Ramwell (Ed.). Plenum, New York, pp. 205-276.

Vargaftig, B. B., and Dao Hai, N. (1972). Selective inhibition by mepacrine of the release of "rabbit aorta contracting substance" evoked by the administration of bradykinin. *J. Pharm. Pharmacol. 24*, 159-161.

Vargaftig, B. B., Bhargava, N., and Bonta, I. L. (1974). Haemorrhagic and permeability increasing effects of *"Bothrops jararaca"* and other Crotalidae venoms as related to amine or kinin release. *Agents and Actions 4*:163-168.

Vargaftig, B. B., Prado-Franceschi, J., Chignard, M., Lefort, J., and Marlas, G. (1980). Activation of guinea-pig platelets induced by convulxin, a substance extracted from the venom of *Crotalus durissus cascavella*. *Eur. J. Pharmacol.* 68:451–464.

Vick, J. A. (1971). Symptomatology of experimental and clinical crotalid envenomation. In *Neuropoisons, Their Pathophysiological Actions*, L. L. Simpson (Ed.). Plenum, New York, pp. 71–86.

Vogt, W. (1978). Role of phospholipase A_2 in prostaglandin formation. In *Advances in Prostaglandin and Thromboxane Research*, vol. 3, C. Galli, G. Galli, and G. Porcellati (Eds.). Raven Press, New York, pp. 89–95.

Witham, A. C., Remington, J. W., and Lombard, E. A. (1953). Cardiovascular response to rattlesnake venoms. *Am. J. Physiol.* 173:535–541.

Wohlman, A., Syed, M., and Ronchi, M. (1968). The effect of chymotrypsin on the gastro-intestinal absorption, tissue penetration, and pharmacological activity of drugs. I. Penetration of penicillin through the blood-brain barrier and blood-retinal barriers. *Can. J. Physiol. Pharmacol.* 46:815–818.

Zeigler, A. (1978). Effects of taipoxin on cholinergic and adrenergic transmission in guinea-pig atrium. *Neuroscience* 3:469–472.

3
Pathology of Rattlesnake Envenomation

CHARLOTTE L. OWNBY
Oklahoma State University, Stillwater, Oklahoma

Introduction 163

Local Effects 164
Hemorrhage • Myonecrosis • Inflammation, Edema, and Pain

Systemic Effects 191
Cardiovascular System • Respiratory System • Lethal Effects • Urinary System • Nervous System

Conclusions 202

References 203

INTRODUCTION

Rattlesnake envenomation induces a wide range of pathologic changes in the victim. These can be divided into systemic effects and local effects. Systemic effects include changes in the cardiovascular, nervous, respiratory, and urinary systems such as hemorrhage, hypotension, shock, coagulation, hemolysis, hemoconcentration, lactacidemia, and death. Local effects include changes near the site of the bite such as hemorrhage, myonecrosis, edema, and pain. The clinical picture of rattlesnake poisoning has been presented in Chapter 1 of this book and thus will not be repeated here. Since the clinical observations of rattlesnake envenomation are often inconsistent owing to variation in the interval between poisoning and patient care and other factors, many investigators have studied the pathology of rattlesnake poisoning by carefully controlled experimental injection of venom. The purpose of this chapter is to discuss and evaluate the state of our knowledge of the pathologic changes induced by rattlesnake envenomation and experimental injection of rattlesnake venom. Since the emphasis of this book is on rattlesnake venom, discussion will center on pathologic changes induced by venom of snakes belonging to two genera: *Crotalus* and *Sistrurus*. However, when needed to properly evaluate the mechanism of venom-induced changes, information on other venoms will be included. Morphologic changes will be emphasized since Chapter 2 covers physiologic and pharmacologic changes.

LOCAL EFFECTS

The most outstanding local effects of rattlesnake poisoning are edema, hemorrhage, and myonecrosis. In cases of severe poisoning these can lead to complete destruction and subsequent dysfunction or loss of an extremity. An example of the type of local tissue damage which can occur is shown in Figure 1. In the United States the main problem in rattlesnake bite cases is local tissue damage owing to the inability of polyvalent Crotalidae antivenin to prevent these changes (Minton, 1954; Stahnke, 1966).

The histopathology of rattlesnake poisoning has been studied by many (Flexner and Noguchi, 1905; Pearce, 1909; Taube and Essex, 1937; Fidler et al., 1938, 1940; Efrati and Reif, 1953; Beamer et al., 1960; Russell, 1960; Trevino, 1968; Tu et al., 1969; Tu and Homma, 1970; Homma and Tu, 1971), but the fine structural changes are only now being extensively investigated. Figure 2 is an example of local tissue damage induced by *Crotalus atrox* viewed with the light microscope.

Hemorrhage

Perhaps the most striking local symptom of rattlesnake poisoning is hemorrhage. Of eleven rattlesnake venoms tested, only one (*Crotalus scutulatus*) did not produce hemorrhage upon subcutaneous injection (Freiderich and Tu, 1971; Tu, 1977). Other venoms tested were *Crotalus atrox, C. adamanteus, C. basiliscus, C. durissus, C. durissus terrificus, C. durissus totonacus, C. horridus horridus, C. horridus atricaudatus, C. viridis viridis*, and *Sistrurus miliarius barbouri*. Even though the presence of hemorrhage has been recorded since the earliest observations of snakebite, little was known about the precise mechanism of venom-induced hemorrhage until recently. There are at least two possible mechanisms by which loss of blood from the vascular system can occur: per diapedesis (Figure 3) or per rhexis (Figure 4). In hemorrhage per rhexis the vascular wall, especially

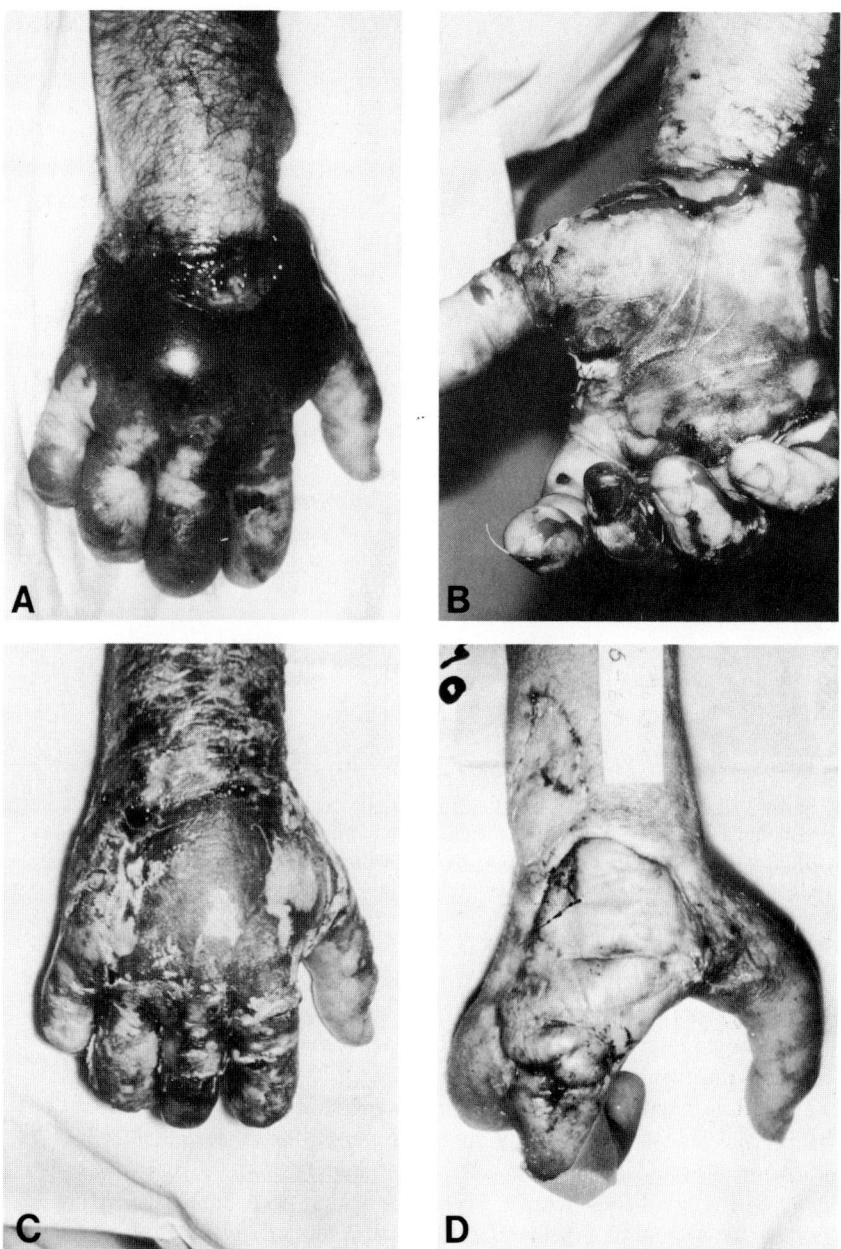

Figure 1 Example of local tissue damage resulting from envenomation by an Eastern diamondback rattlesnake (*Crotalus adamanteus*). A. Dorsal surface of hand at 4 days after bite. B. Ventral surface of hand at 15 days. C. Dorsal surface at 4 weeks. D. Dorsal surface of hand approximately 6 months after envenomation; note loss of two fingers and some skin over dorsum of hand. The patient received 13 units of antivenom intravenously. (Courtesy of William J. Bailey, M.D., P.A., F.A.C.S., Naples, Florida.)

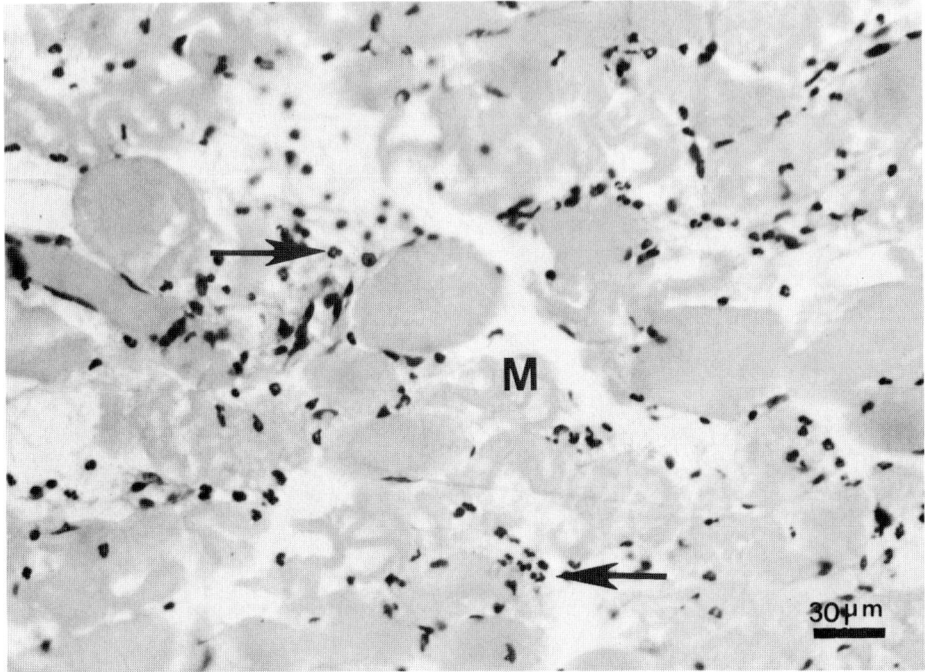

Figure 2 Light micrograph of skeletal muscle from a mouse 24 hr after the intramuscular injection of *C. atrox* venom. Myonecrosis, hemorrhage, edema, and inflammation were prominent features. Note necrotic muscle cells (M) and polymorphonuclear leukocytes (arrows).

the endothelium, is destroyed, whereas in hemorrhage per diapedesis only the intercellular junctions are modified to allow blood components to escape. In hemorrhage per diapedesis the endothelial cells remain morphologically intact.

Early light microscopic studies of rattlesnake venom-induced hemorrhage revealed disintegration of the endothelium with consequent tears in the vessel walls and subsequent hemorrhage. Flexner and Noguchi (1905) suggested that rattlesnake venom acted directly on endothelium to produce actual tears in the walls and allow escape of blood cells. The wall appeared to be dissolved at the point where blood cells escaped. These observations were confirmed by Pearce (1909) in his study of a glomerular lesion induced by *C. adamanteus* venom. Capillary walls were swollen and granular; the nuclei were swollen or pyknotic and reduced in number, suggesting endothelial destruction. A more complete histologic study was not made until 1937 when Taube and Essex investigated the microscopic changes induced by the intravenous injection of rattlesnake venom in dogs. They found disintegration of the walls of capillaries and venules

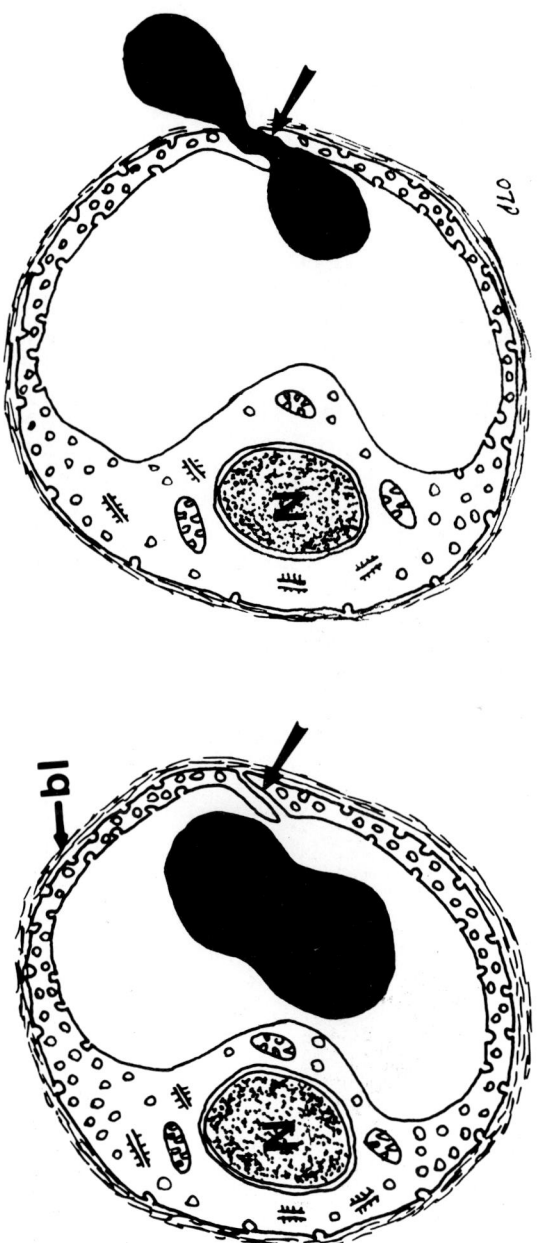

Figure 3 Diagram illustrating hemorrhage per diapedesis. Erythrocyte squeezes through intercellular junction into connective tissue surrounding capillary. Note there is no major change in the endothelial cell itself. N, nucleus of endothelial cell; bl, basal lamina; arrow, intercellular junction.

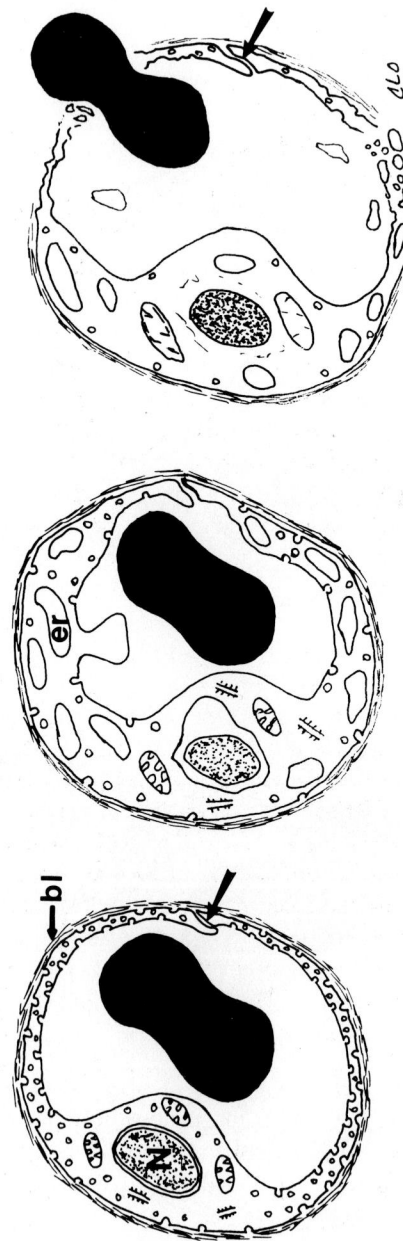

Figure 4 Diagram illustrating hemorrhage per rhexis. The endothelial cell is injured to cause dilatation of the perinuclear space and endoplasmic reticulum, blebbing of endothelium, and eventually, lysis of the plasma membrane of endothelial cells. Erythrocytes escape through gaps in the endothelial cell, while the intercellular junction remains intact. N, nucleus of endothelial cell; bl, basal lamina; arrow, intercellular junction; er, endoplasmic reticulum. See text for further discussion of hemorrhage per rhexis induced by venom and pure hemorrhagic toxins.

in many organs and proposed that "endothelial cells first contract, then swell and finally burst and dissolve, leaving gaps in the walls of blood vessels." Fidler et al. (1938, 1940) thought a subcutaneous route of venom injection would more closely simulate actual snakebite than the intravenous route used by Taube and Essex, so they studied histopathologic changes resulting from a subcutaneous injection of *C. atrox* venom into monkeys. At the site of injection the skin and subcutaneous tissue contained smaller blood vessels and capillaries which had ruptures in their walls. They concluded that *Crotalus* venom could cause varying changes in vascular walls. Low doses could cause increased permeability and diapedesis of erythrocytes, whereas higher doses might cause necrosis, intravascular thrombosis, and actual rupture of the wall.

Later light microscopic studies by Trevino (1968) reported swollen and necrotic endothelial cells in vessels after the intramuscular injection of *C. atrox* venom into pigs. In some cases there was a "frank discontinuity of the walls with consequent extravasation of blood." Homma and Tu (1971) demonstrated that *C. atrox* venom caused arterial lesions containing degenerating endothelial cells, hemorrhage, and mural thrombi. They also reported that of nine rattlesnake venoms tested, *C. adamanteus* and *C. durissus totonacus* caused the most severe hemorrhage. Owing to the relatively poor resolution of the light microscope, it was impossible to determine precisely the mechanism of hemorrhage.

The first electron microscopic study of the pathogenesis of hemorrhage induced by rattlesnake venom was done by Ownby et al. (1974) and Ownby (1975). *Crotalus atrox* venom was injected intramuscularly into Swiss-Webster white mice, and muscle samples were taken 2 min, 30 min, and 3 h after injection. Control capillaries were normal in appearance (Figure 5). The initial effect of the venom appeared to be dilatation of the perinuclear space and endoplasmic reticulum of endothelial cells (Figure 6). Many cells were swollen; blebs were observed on swollen and nonswollen cells (Figure 7). Subsequent changes included rupture of the plasma membranes of endothelial cells, extravasation of blood (Figure 8), and often, platelet aggregations (Figure 9). The authors concluded that *C. atrox* causes hemorrhage by direct lytic action on endothelial cells of capillaries, i.e., hemorrhage per rhexis. Since these studies were done with crude venom, it is possible that the lytic action was due either to components other than hemorrhagic ones or to a synergistic action with other venom components such as phospholipase A. For more details see Ownby et al. (1974).

By fractionating venom and isolating pure hemorrhagic toxins, it was possible to study the pathogenesis of hemorrhage induced by a single known toxin. Bjarnason and Tu (1978) reported the isolation of five hemorrhagic toxins from the venom of the Western diamondback rattlesnake (*C. atrox*). All five were found to be proteolytic.

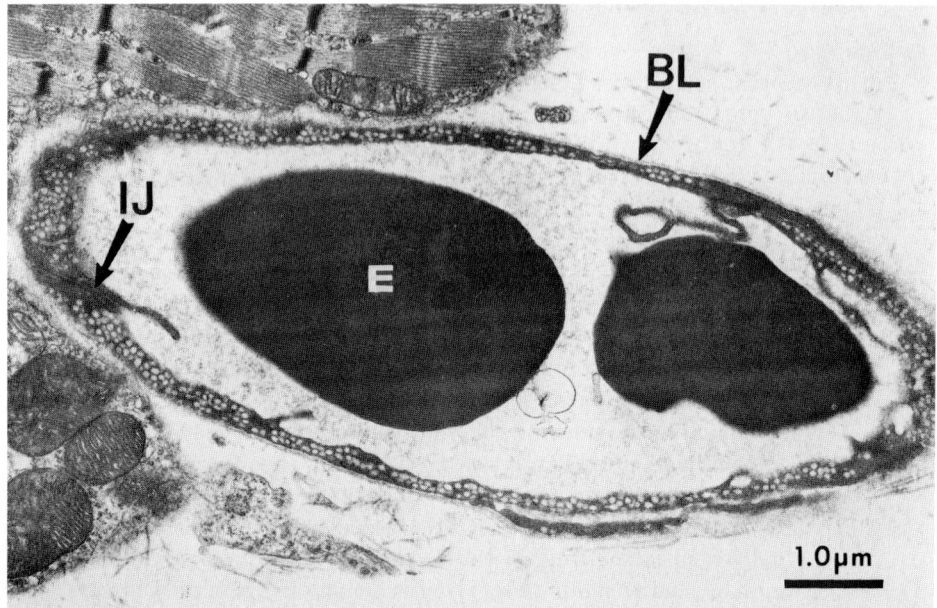

Figure 5 Electron micrograph of capillary from a mouse injected with physiologic saline (PSS) as control. Note the thin, uniform endothelium containing many pinocytotic vesicles. Also note the smooth, intact basal lamina underlying the endothelium (BL), the intercellular junction (IJ), and the erythrocytes in the lumen (E).

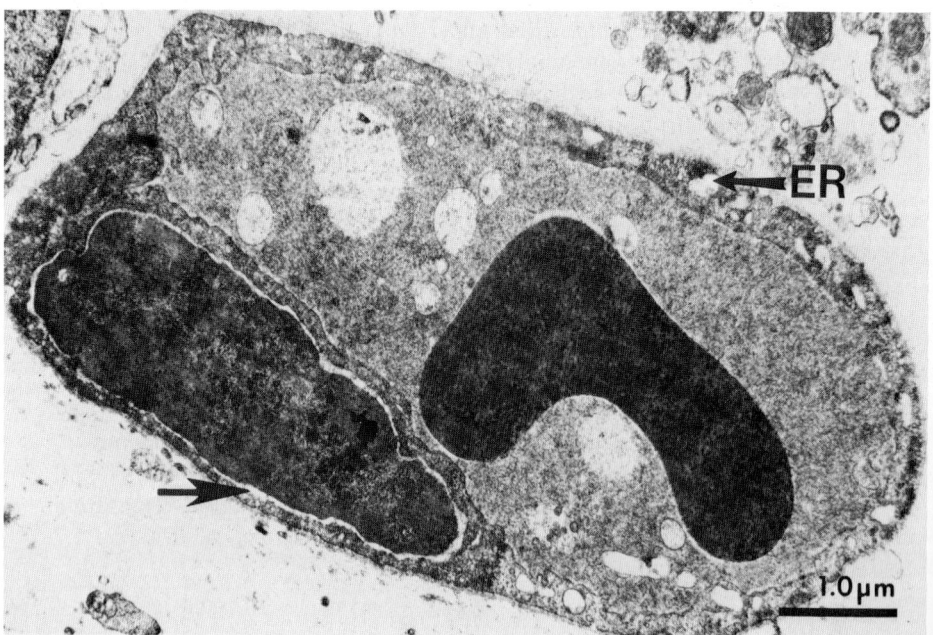

Pathology of Envenomation

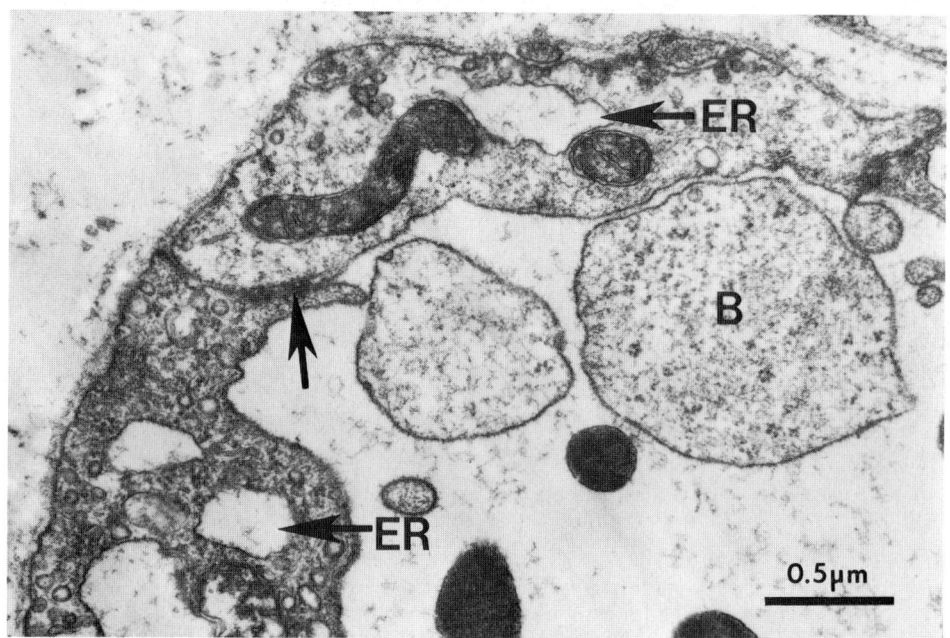

Figure 7 Electron micrograph of capillary from a mouse injected with *C. atrox* venom. Note dilatation of endoplasmic reticulum (ER) in both endothelial cells and blebbing (B) of swollen endothelial cell into the lumen. Intercellular junction (arrow) is intact. (From Ownby, Kainer, and Tu, 1974. Reprinted with permission of the American Association of Pathologists.)

Ownby et al. (1978) reported the mode of action of three of these hemorrhagic toxins (HT). HT*a* has a molecular weight of 68,000 and is acidic, HT*b* has a molecular weight of 24,000 and is basic, and HT*e* has a molecular weight of 25,700 and is acidic (isoelectric point of 5.6). HT*a* induced extensive hemorrhage within 5 min after intramuscular injection into mice. Light microscopy revealed a connective tissue filled with erythrocytes and ruptured vessels (Figures 10 and 11); electron microscopy showed that HT*a* appeared to act directly on endothelial cells, resulting in lysis of their plasma membranes and leakage of plasma and erythrocytes. The endothelium of affected vessels appeared to have become very thin prior to actual breakage (Figure 12). Gaps in the endothelium appeared to be within endothelial cells rather than between cells, as indicated by the presence of intact intercellular junctions immediately

Figure 6 Electron micrograph of capillary from a mouse injected with *C. atrox* venom. Note dilatated perinuclear space (arrow), swollen endoplasmic reticulum (ER), and presence of dense flocculent material in the lumen.

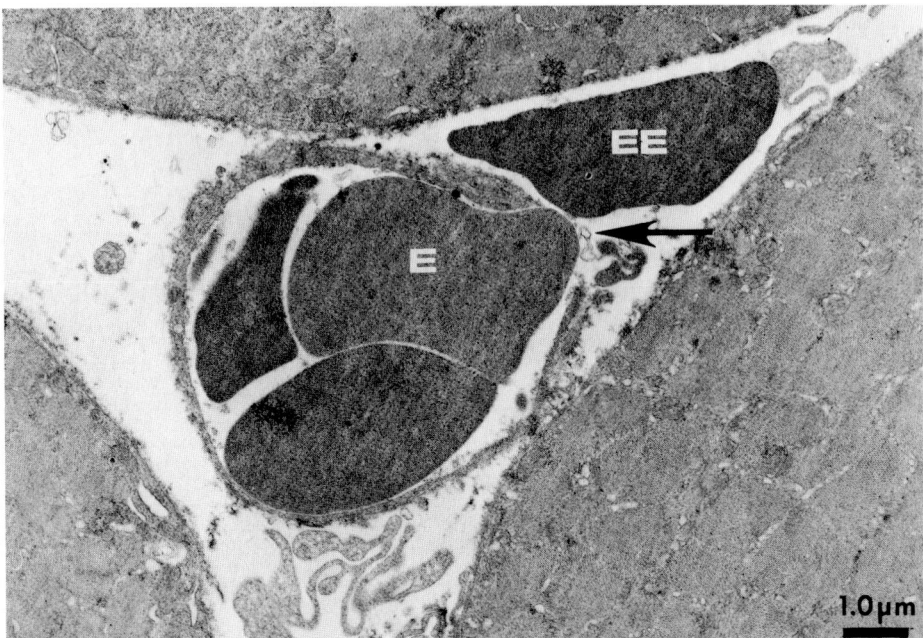

Figure 8 Electron micrograph of capillary from a mouse injected with *C. atrox* venom. The endothelium has a large gap (arrow) in it which contains a membranous structure. Note extravasated erythrocyte in connective tissue (EE) and a second in the gap (E). The endothelium is disorganized and ruptured. (From Ownby, Kainer, and Tu, 1974. Reprinted with permission of the American Association of Pathologists.)

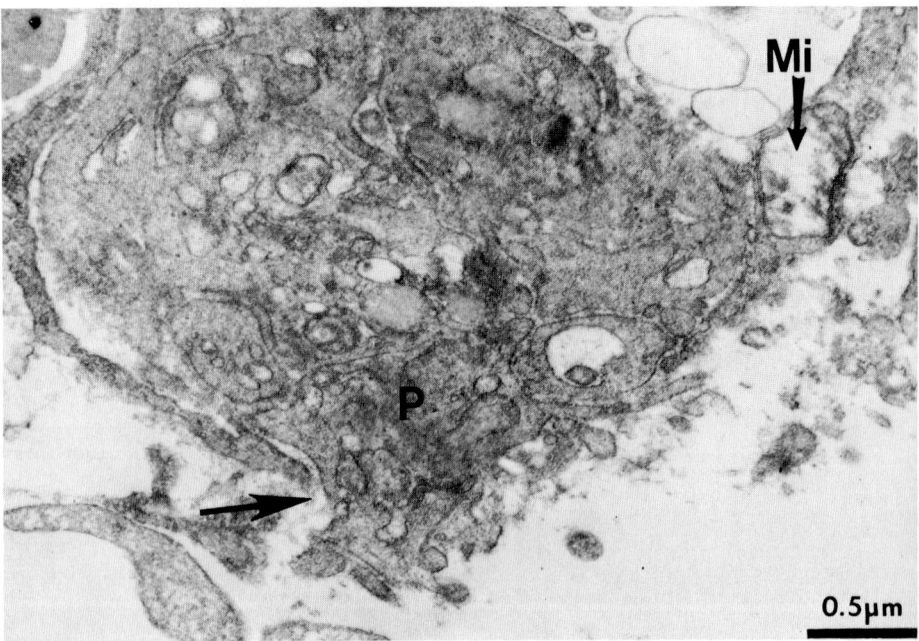

Pathology of Envenomation

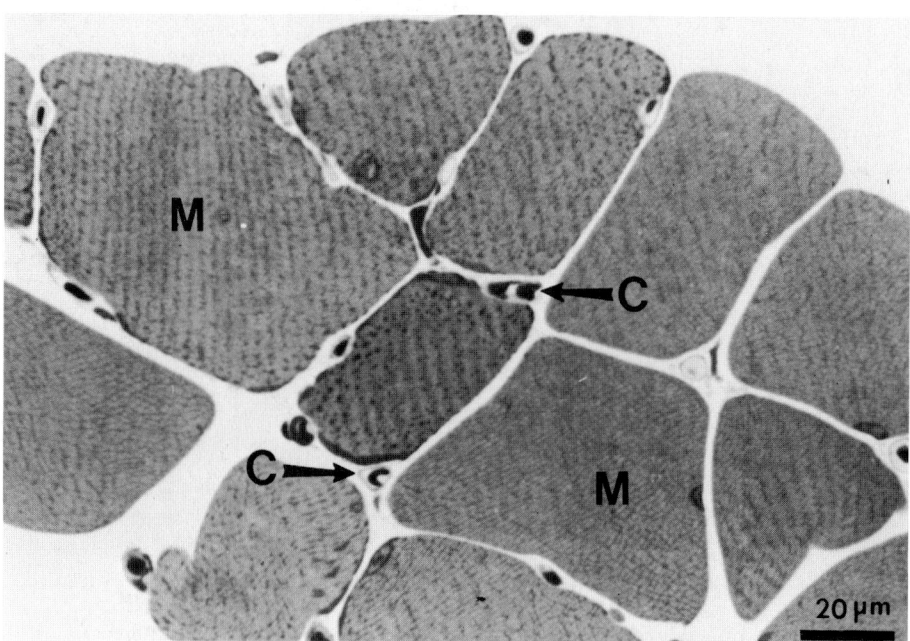

Figure 10 Light micrograph of muscle taken from a mouse injected with PSS as a control. Capillaries (C) and muscle cells (M) are intact; the connective tissue is free of erythrocytes.

adjacent to the openings (Figures 13 and 14). In some vessels the endothelium was broken down into numerous small vesicles, which resulted again in gaps in the capillary wall. Fibrin was observed around some of the damaged capillaries. A common observation was the formation of platelet aggregations in damaged vessels, especially at the site of endothelial rupture (Figure 14). Some of the platelets appeared to have undergone the platelet release action as indicated by apparent loss of granules. Even in these vessels, intact intercellular junctions were present. Collagen associated with damaged capillaries was usually disorganized, and the basal lamina was usually absent near and beneath ruptured endothelium. The connective tissue surrounding damaged vessels was filled with a flocculent material, plasma, extravasated erythrocytes, and the normal connective tissue elements.

Figure 9 Electron micrograph of capillary from a mouse injected with *C. atrox* venom. Note gap in endothelium (arrow) which is plugged by clump of platelets (P). Damage to the endothelium is indicated by swollen mitochondrion (Mi). (From Ownby, Kainer, and Tu, 1974. Reprinted with permission of the American Association of Pathologists.)

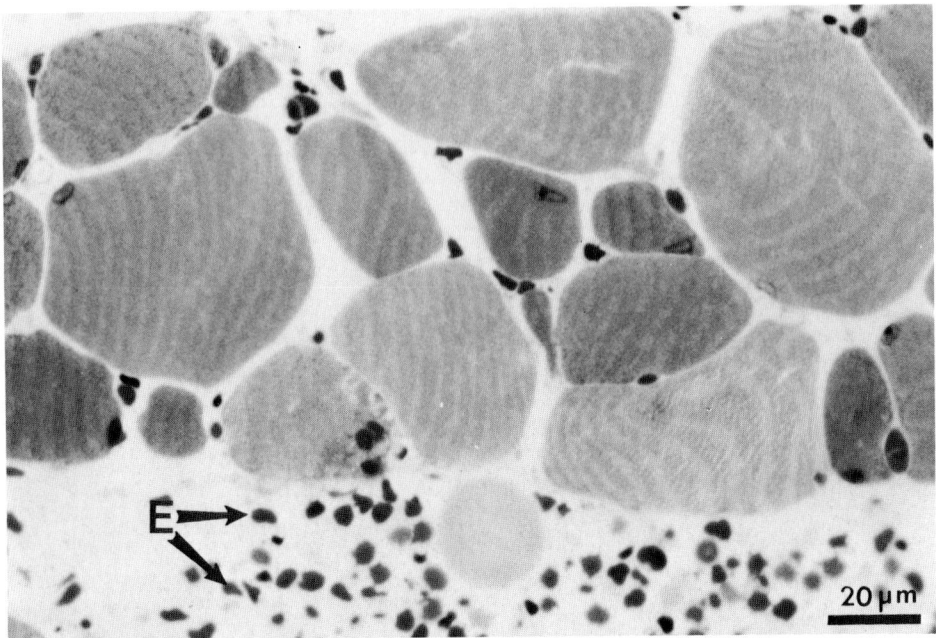

Figure 11 Light micrograph of muscle taken from a mouse injected with pure hemorrhagic toxin from *C. atrox* venom. Note presence of many erythrocytes (E) in the connective tissue indicating hemorrhage, and scarcity of intact capillaries (compare with Figure 10).

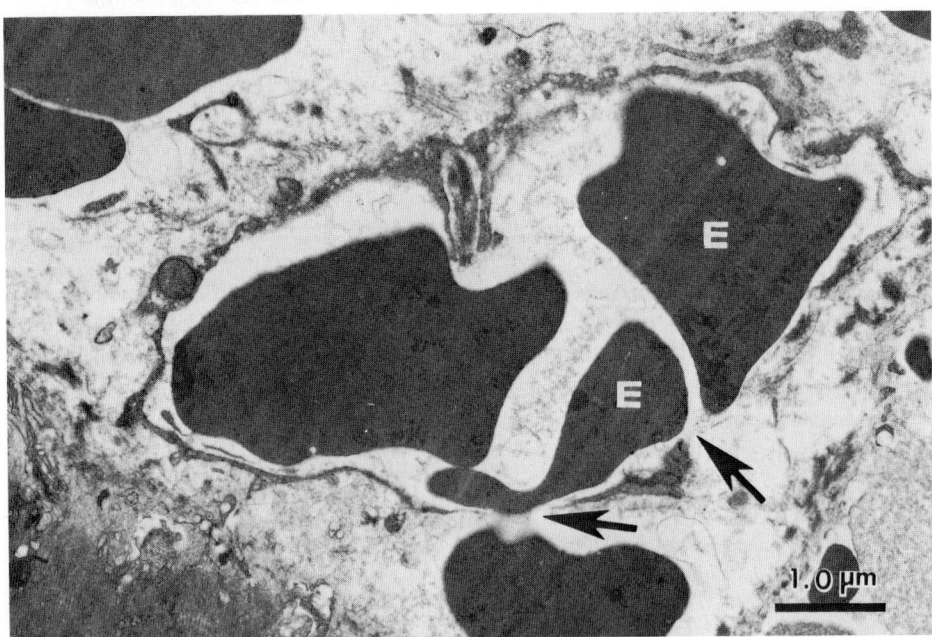

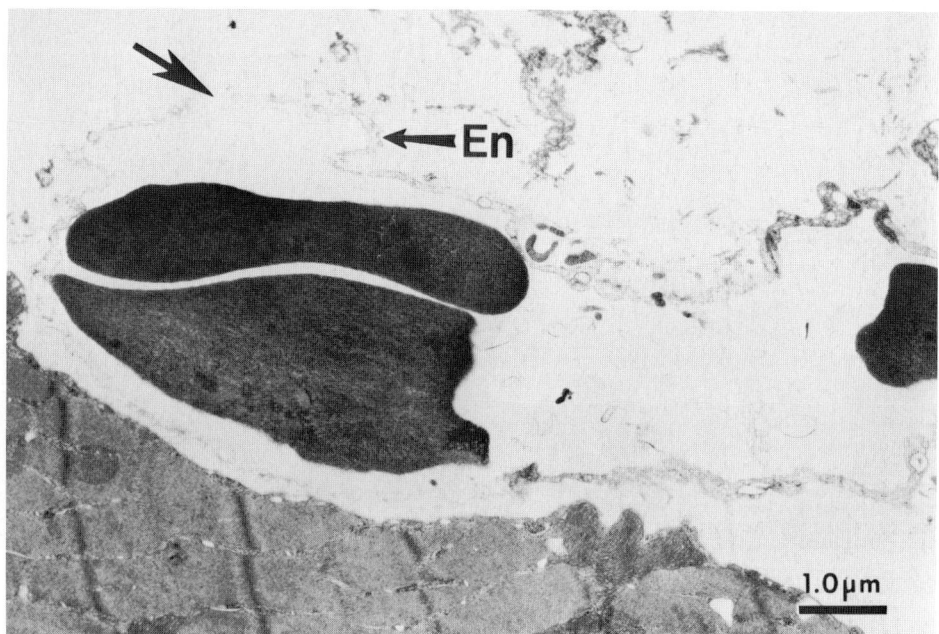

Figure 13 Electron micrograph of capillary from mouse injected with hemorrhagic toxin *a*. Note the very thin endothelium (En) and the gap in the vessel wall (arrowhead). Also note the absence of collagen and basal lamina. (From Ownby, Bjarnason, and Tu, 1978. Reprinted with permission of the American Association of Pathologists.)

Light microscopic examination of muscle injected with hemorrhagic toxin *e* revealed a connective tissue packed with erythrocytes and extruded plasma. The few intact capillaries were filled with erythrocytes and platelet aggregations. HT*e* acted rapidly to cause hemorrhage within 2 min after injection. It appeared to act on the endothelium to cause hemorrhage per rhexis. Endothelial cells became extremely thin, and erythrocytes were observed passing through the thin portions of endothelial cell cytoplasm (Figure 15). In some vessels swollen endothelial cells were seen adjacent to normal cells (Figure 16). This effect was accompanied by disruption of collagen and basal lamina, intravascular hemolysis, and slight swelling of muscle. There appeared to be some intravascular lysis

Figure 12 Electron micrograph of capillary from mouse injected with hemorrhagic toxin *a*. Note thinness of endothelium and presence of large gaps (arrows) through which erythrocytes (E) are escaping. Note also the absence of disruption of collagen and the basal lamina (compare with Figure 3). (From Ownby, Bjarnason, and Tu, 1978. Reprinted with permission of the American Association of Pathologists.)

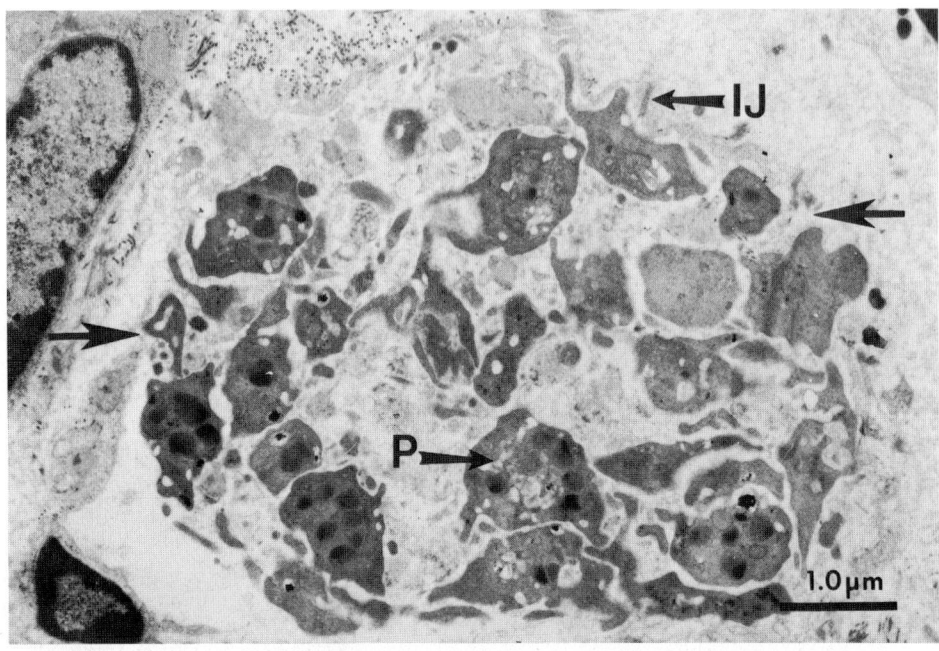

Figure 14 Electron micrograph of a capillary from mouse injected with hemorrhagic toxin *a*. The endothelium is very thin and broken in several places (arrows); the lumen is filled with an aggregation of platelets (P). Note intact intercellular junction (IJ). (From Ownby, Bjarnason, and Tu, 1978. Reprinted wtih permission of the American Association of Pathologists.)

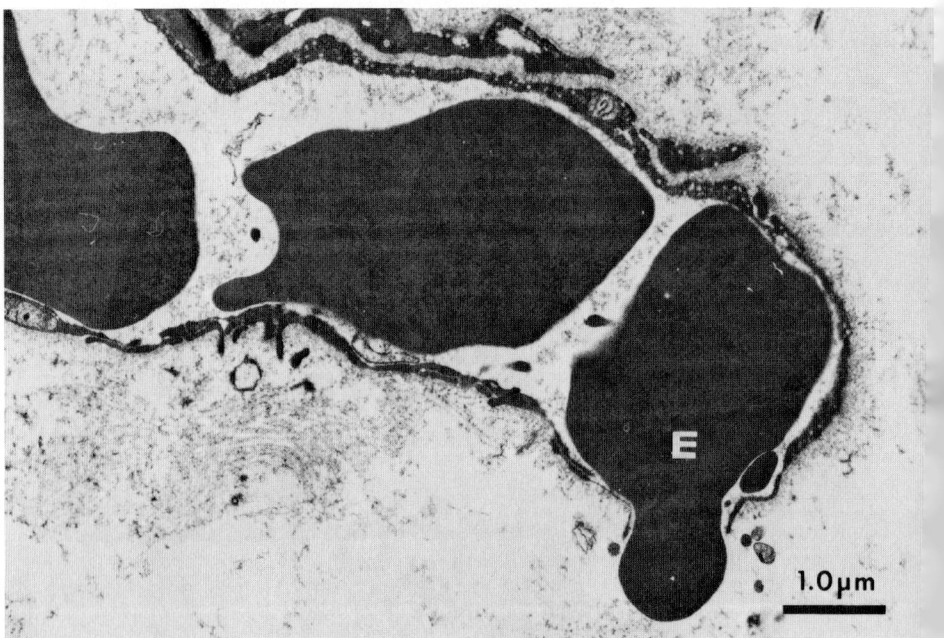

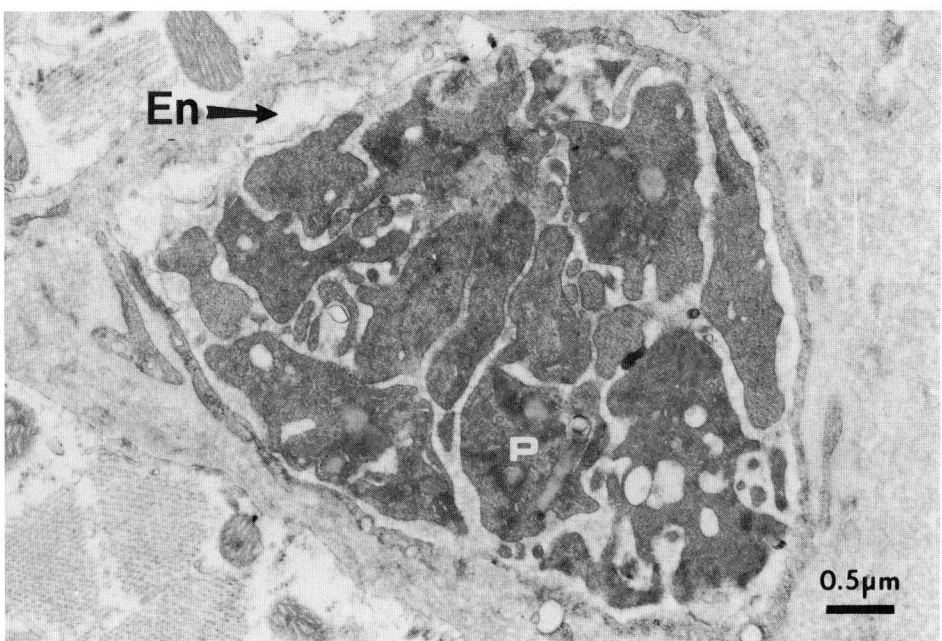

Figure 16 Electron micrograph of a capillary from mouse injected with hemorrhagic toxin *e*. Note swollen endothelial cell (En) next to nonswollen cell. Lumen is filled with platelets (P); note flocculent material in connective tissue around vessel. (From Ownby, Bjarnason, and Tu, 1978. Reprinted with permission of the American Association of Pathologists.)

of other blood cells, probably eosinophils. Many damaged vessels were plugged with platelet aggregations (Figure 16), and the surrounding area was filled with a dense flocculent material and erythrocytes.

Gross examination of muscle injected with hemorrhagic toxin *b* revealed no hemorrhage at 5 min, but there was a large hemorrhagic area over the thigh at 3 h after injection. At the light microscopic level, interstitial edema was evident, and the connective tissue was filled with extravasated erythrocytes. In some areas few intact capillaries were observed, and muscle necrosis was prominent. In these areas erythrocytes appeared to be located within damaged muscle cells. This is quite a contrast to the other toxins investigated, since both HT*a* and HT*e* are hemorrhagic but not myotoxic. The pathogenesis of hemorrhage induced by HT*b* was very similar to that induced by HT*a* except

---

Figure 15 Electron micrograph of a capillary from mouse injected with hemorrhagic toxin *e*. Note erythrocyte (E) passing through ruptured endothelium. (From Ownby, Bjarnason, and Tu, 1978. Reprinted with permission of the American Association of Pathologists.)

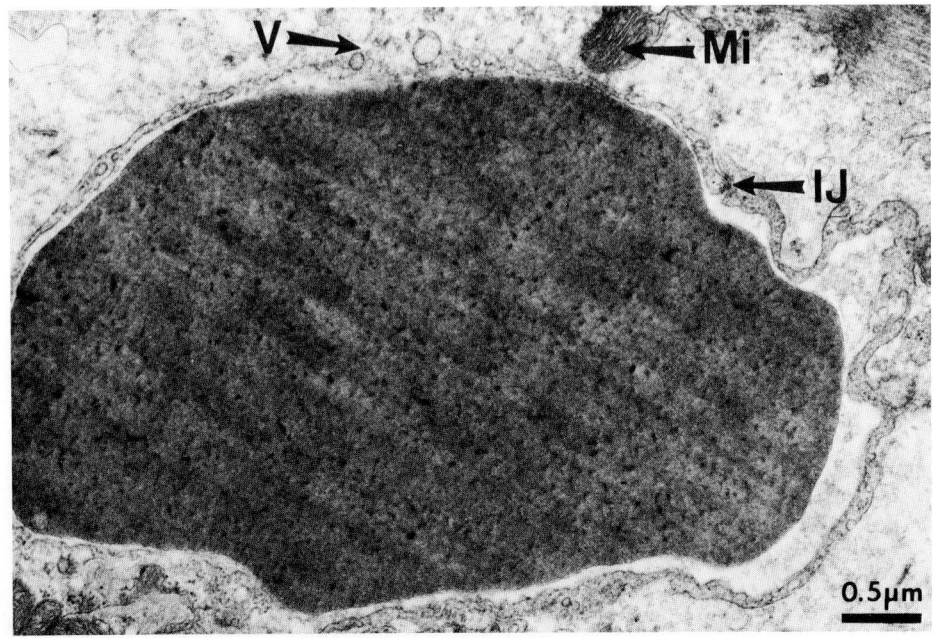

Figure 17 Electron micrograph of a capillary from mouse injected with hemorrhagic toxin *b*. Note vesiculation (V) of endothelium and presence of intact intercellular junction (IJ). Also note free mitochondrion (Mi). (From Ownby, Bjarnason, and Tu, 1978. Reprinted with permission of the American Association of Pathologists.)

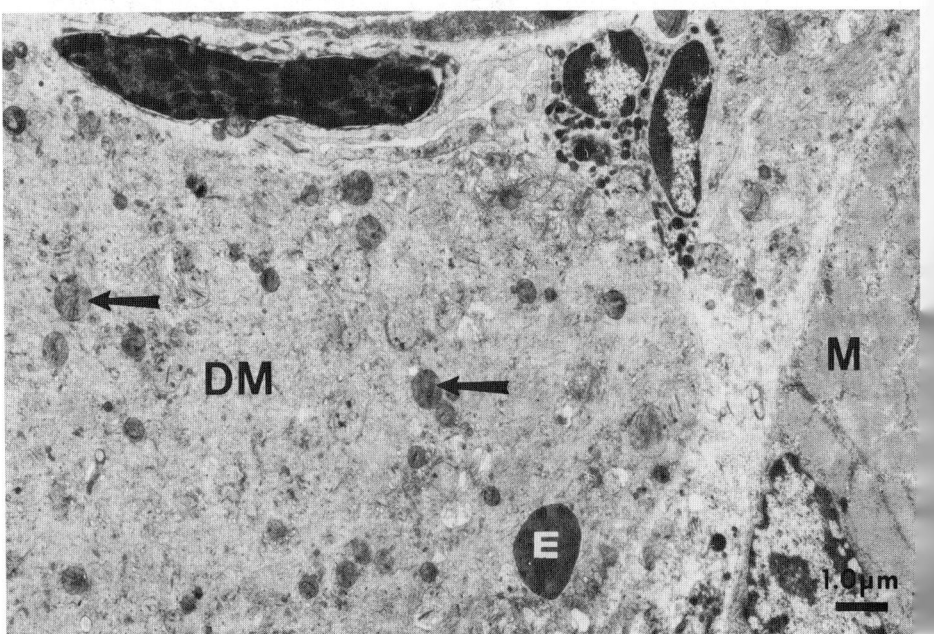

that it was much slower acting, and it also caused necrosis of muscle. HT*b* caused endothelial cells to undergo a vesiculation which resulted in gaps in the capillary wall through which blood escaped (Figure 17). Intercellular junctions adjacent to ruptured endothelium remained intact. HT*b* also caused marked necrosis of muscle as indicated by the rupture of sarcolemma, breakdown of sarcoplasmic reticulum into numerous small vesicles, disoriented myofilaments, and abnormal mitochondria (Figure 18). Erythrocytes appeared to be located inside some damaged muscle cells. There was an apparent increase in the amount of glycogen in some ruptured muscle cells. Mitochondrial changes in damaged muscle cells ranged from slight swelling to severe swelling and, most commonly, a stacking or reduplication of cristae. Numerous mitochondria of various shapes and sizes were observed in the connective tissue surrounding damaged muscle cells.

All of the hemorrhagic toxins, HT*a*, HT*b*, and HT*e*, act directly on the capillary wall, causing degeneration of endothelial cells which results in rupture of their plasma membranes, producing gaps through which blood escapes. These results confirm those reported by Ownby et al. (1974) using the crude venom of *C. atrox*.

It seems clear that, at least with *C. atrox* venom, local hemorrhage is due to a direct lytic action of hemorrhagic components on the endothelium of capillaries, i.e., hemorrhage per rhexis. However, this may not be the mechanism of hemorrhage induced by other crotaline venoms or indeed by venom of noncrotaline snakes. For example, Ohsaka et al. (1973, 1975) have shown that venom of the habu (*Trimeresurus flavoviridis*) causes hemorrhage per diapedesis. These differences could be real or due to differences in route of injection, doses of venom injected or amount of time between injection and tissue sampling.

Myonecrosis

In addition to hemorrhage, necrosis of skeletal muscle near the site of the bite is a frequent consequence of rattlesnake envenomation. It is this destruction of muscle which leads to loss of function or complete loss of extremities in severe snakebite cases.

Many investigators have reported necrosis of muscle after the experimental injection of rattlesnake venom. Most of the studies were either made grossly or histologically, and only a few have been made with the electron microscope.

Beamer et al. (1960) used the light microscope to study tissue taken from rabbits injected either intramuscularly or intracutaneously with *C. atrox* venom. They described coagulation necrosis and liquefaction of muscle at the site of

---

Figure 18 Electron micrograph of muscle from mouse injected with hemorrhagic toxin *b*. Note degenerating muscle cell (DM) and normal muscle cell (M). Also note erythrocyte (E) and abnormal mitochondria (arrows). (From Ownby, Bjarnason, and Tu, 1978. Reprinted with permission of the American Association of Pathologists.)

injection. Early changes in muscle included karyopycnosis, karyolysis, and swelling of cells, followed by increased eosinophilia and liquefaction. Trevino (1968) also described the disintegration of muscle after the injection of *C. atrox* venom in swine. Early changes included fusion of myofibrils, loss of striations, and a homogeneous eosinophilic appearance, i.e., Zenker's degeneration of muscle. Subsequent changes included vacuolation of the muscle cell owing "either to lysis and precipitation of the proteins or loss of sarcoplasm through damaged sarcolemma." Eventually, liquefaction necrosis followed with invasion of neutrophils.

Homma and Tu (1971) investigated local lesions induced by the intramuscular injection of 28 crotaline venoms (9 rattlesnake venoms) into mice. Light microscopic examination of muscle 24 h after injection revealed three different types of myonecrosis: (1) coagulative, (2) myolytic, and (3) mixed, including fibers undergoing both coagulative and myolytic necrosis. Rattlesnake venoms, *C. adamanteus, C. atrox, C. basiliscus basiliscus, C. durissus durissus, C. durissus totonacus, C. horridus atricaudatus, C. horridus horridus,* and *C. viridis viridis,* produced myonecrosis of the coagulation type, whereas *C. durissus terrificus* caused myonecrosis of the mixed type. It is worth noting that *C. durissus terrificus,* which is known to contain a neurotoxin, crotoxin, produces myolytic necrosis in addition to the coagulative type. Since Homma and Tu found that the elapid venoms all caused a myolytic type of necrosis and all other crotalid venoms caused a coagulative type, *C. durissus terrificus* seems to be a combination of both of these types of venoms. The authors point out, however, that "there was no completely pure type of muscle necrosis."

The first electron microscopic study of rattlesnake venom-induced myonecrosis was by Stringer et al. (1972). They investigated the effect of crude *C. viridis viridis* venom on mouse skeletal muscle 3 h after intramuscular injection of one-fourth LD_{50} dose of venom. Light and electron microscopic observation of injected muscle revealed that the lesions were focal, i.e., degenerating cells were interspersed among cells of normal appearance. Initial changes in degenerating cells were dilatation of the longitudinal sarcotubular system and appearance of myeloid figures in swollen sarcoplasmic reticulum (SR) vesicles. As the SR increased in size, it lost the normal organization between myofibrils, and eventually the large vacuoles of SR broke into smaller and fewer vesicles. As the SR increased in size and number, breakdown of myofibrils began with disappearance of the H band and M line. Disruption of the Z line followed, and subsequently the myofibrils lost all organization. Eventually, all that was left of the myofibrils was a coagulation of light and dark masses. T tubules were of normal morphology throughout the degenerative process, but mitochondria began to degenerate when myofibril breakdown occurred. Only during later stages of degeneration did rupture of the basal lamina and plasma membrane (sarcolemma) occur.

More recently, Ownby et al. (1976) reported an electron microscopic study of a pure myotoxic component isolated from crude *Crotalus viridis viridis*

Pathology of Envenomation

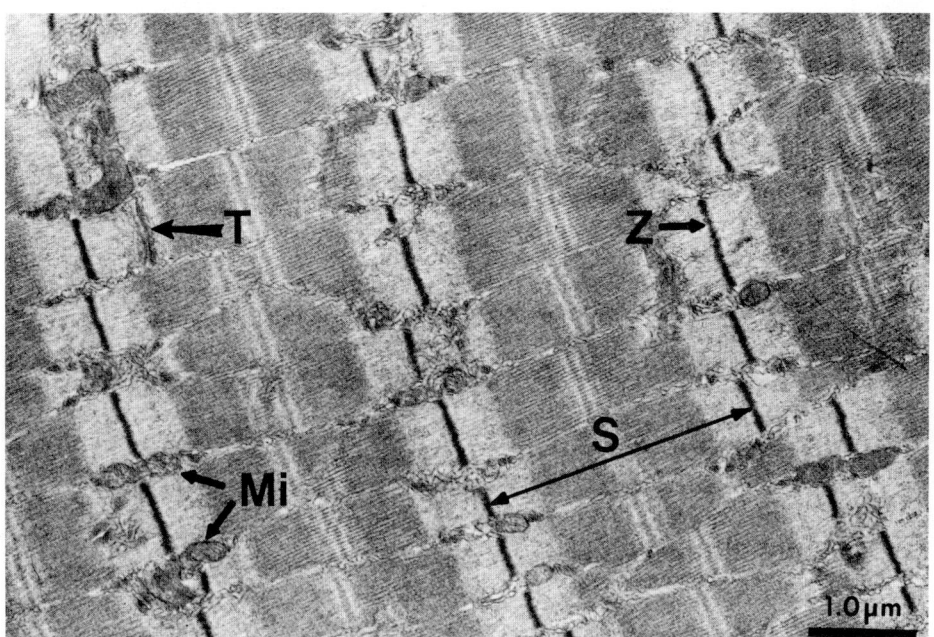

Figure 19 Electron micrograph of muscle from mouse injected with physiologic saline as a control. Note sarcomeres (S), Z lines (Z), mitochondria (Mi), and T tubule (T). Sarcoplasmic reticulum vesicles are small and difficult to see at this magnification.

venom. This component, myotoxin *a*, has been well studied (Ownby et al., 1976; Cameron and Tu, 1977, 1978; Fox et al., 1979; Bailey et al., 1979). It is a small (4400 daltons), basic (isoelectric point of 9.6) polypeptide consisting of 39 amino acid residues, including 10 lysine and 1 arginine.

Intramuscular injection of 1.5 µg/g of pure myotoxin *a* (in physiologic saline) into mice induces a series of degenerative events in muscle cells reported by Ownby et al. (1976). Light microscopic examination revealed focal lesions containing areas of partial vacuolation of muscle cells at 6, 12, and 24 h and complete vacuolation and loss of striations at 48 and 72 h. Hemorrhage, hemolysis, or damage to connective tissue cells was not observed. Electron microscopy revealed the basis of the vacuolation as dilatation of sarcoplasmic reticulum and the perinuclear space. The typically flattened cisterns of sarcoplasmic reticulum (Figure 19) were rounded and enlarged, and they increased in size as degeneration progressed (Figures 20 and 21). The perinuclear space was usually enlarged at one pole of the nucleus (Figure 22). At 24 h after injection the SR was even more enlarged and disorganized and occupied more of the cytoplasm. Many of the swollen cisterns contained membranous figures (Figures 23 and 24), indicating damage to SR membranes. At 48 h after injection the SR had broken

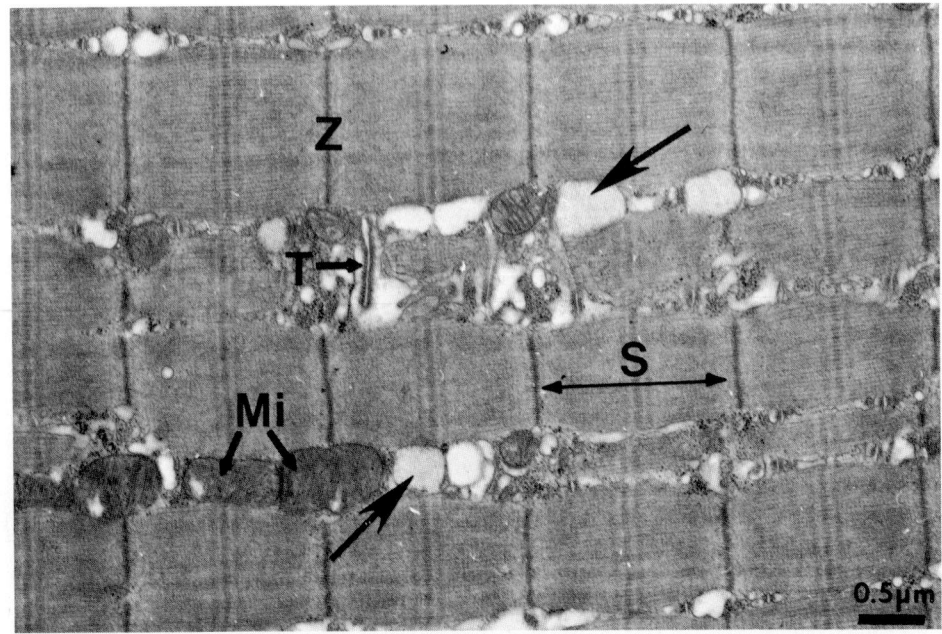

Figure 20 Electron micrograph of muscle from mouse 3 hr after injection of myotoxin *a*. Note slight dilatation of sarcoplasmic reticulum (arrows). Sarcomeres (S) and T tubules (T) are intact; note mitochondria (Mi) and Z lines (Z).

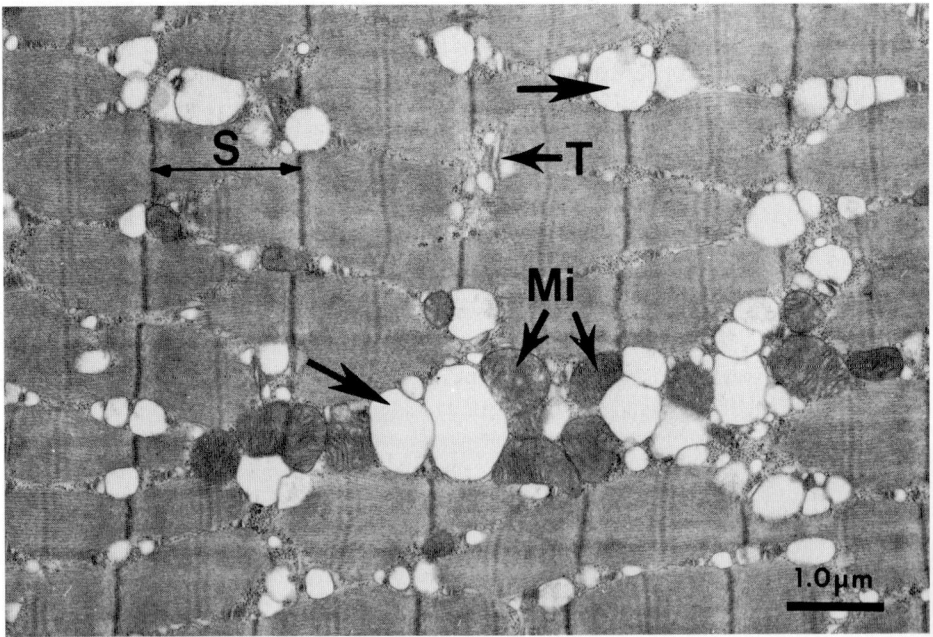

Figure 21 Electron micrograph of muscle from mouse 12 hr after injection of myotoxin *a*. Note dilatated sarcoplasmic reticulum (arrows). Sarcomeres (S), mitochondria (Mi), and T tubules (T) are all normal.

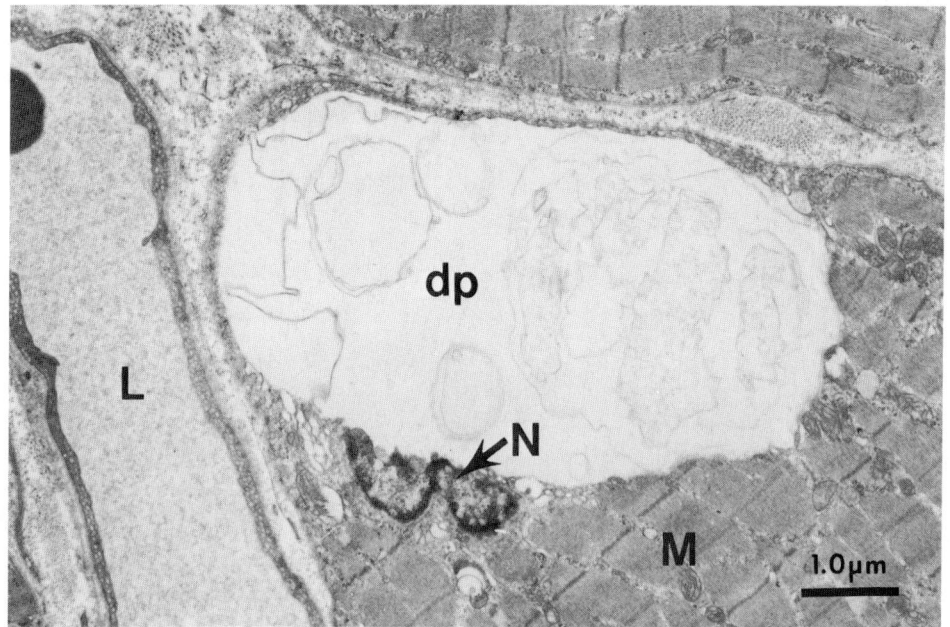

Figure 22 Electron micrograph of muscle 12 hr after injection of myotoxin *a*. Muscle cell (M), nucleus (N) of muscle cell, lumen of capillary (L). Note marked dilatation of the perinuclear space (dp) which is filled with membranous structure.

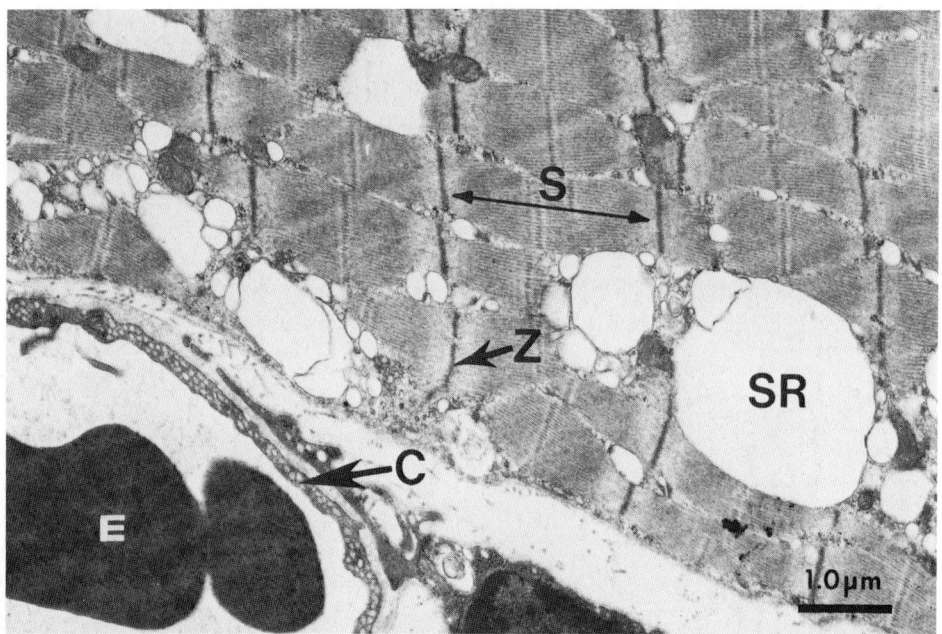

Figure 23 Electron micrograph of muscle 24 hr after injection of myotoxin *a*. Sarcoplasmic reticulum is greatly dilatated (SR), but sarcomeres (S) and Z lines (Z) are intact. Note also that capillary (C) and erythrocytes (E) are not damaged.

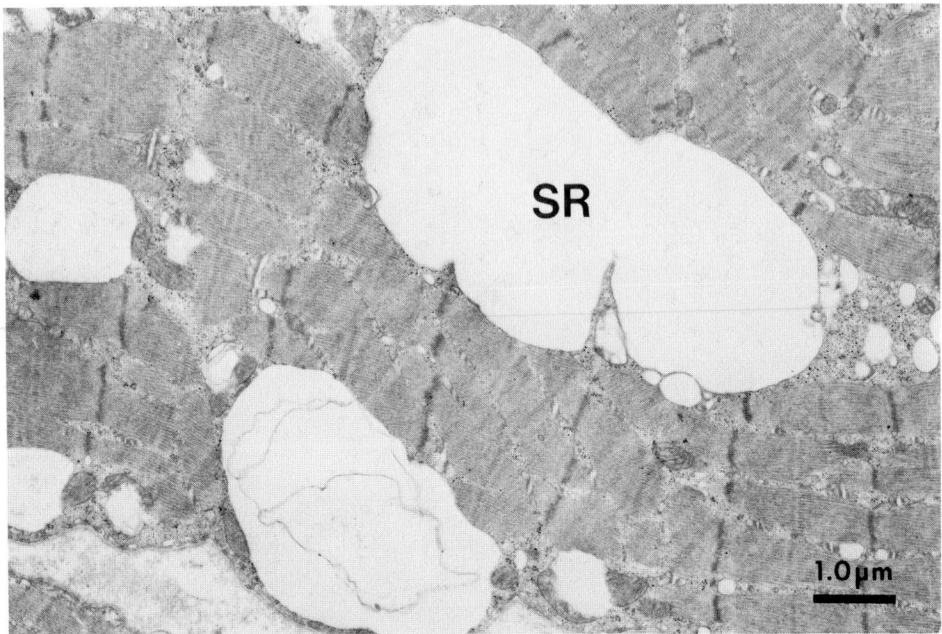

Figure 24 Electron micrograph of muscle 24 hr after injection of myotoxin *a*. Note greatly enlarged sarcoplasmic reticulum (SR); one contains a membranous structure.

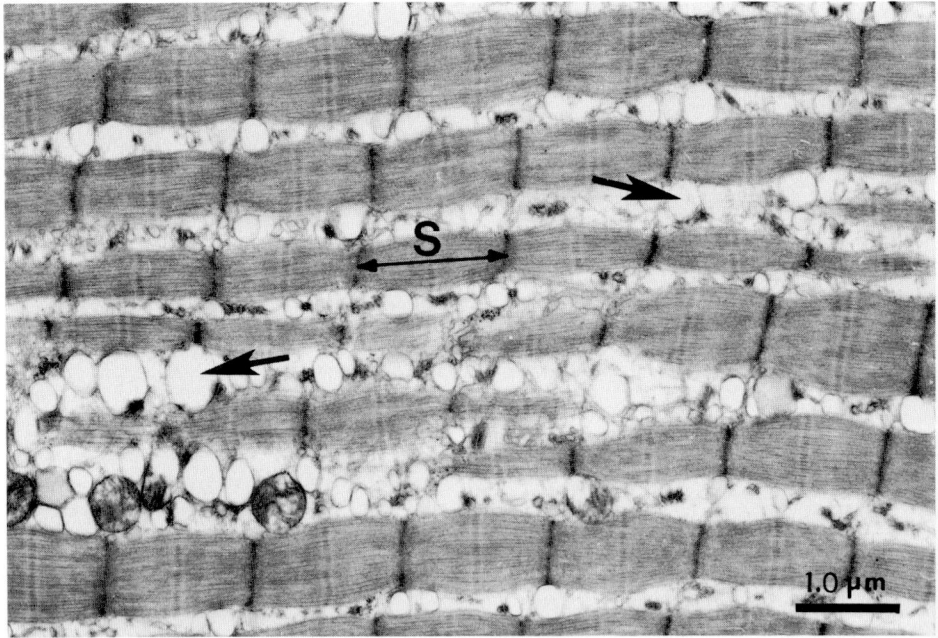

Figure 25 Electron micrograph of muscle 48 hr after injection of myotoxin *a*. Note numerous smaller dilatated cisterns of sarcoplasmic reticulum (arrows). Sarcomeres (S) are still intact.

Pathology of Envenomation

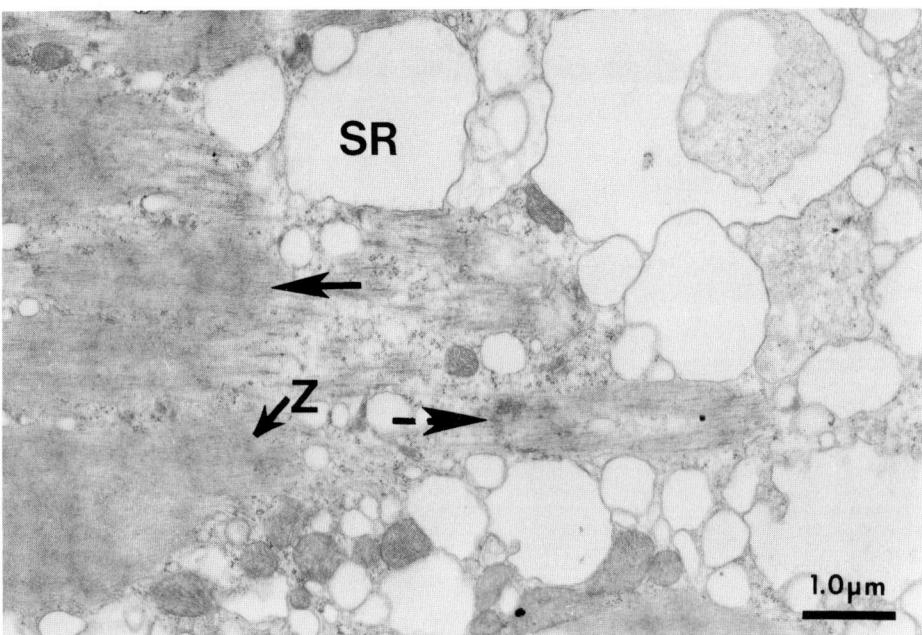

Figure 26 Electron micrograph of muscle 72 hr after injection of myotoxin *a*. Note degenerating myofibrils (arrows). Z line material (Z) is all that can be identified of sarcomeres. Dilatated sarcoplasmic reticulum (SR) is also present.

down into numerous smaller vesicles, and there was some disintegration of myofibrils (Figure 25). Also, at this time mitochondria were slightly swollen and lacked cristae but were not completely degenerated. Noticeably, however, at this time transverse tubules (T tubules) were normal in appearance. By 72 h after injection many fibers lacked the typical striated appearance. In these fibers intact sarcomeres were separated by loose actin and myosin filaments (Figure 26). Dissolution of myofilaments occurred first between Z lines; the Z line retained the longitudinal orientation of the myofilaments. Intact T tubules were present in these areas. The sarcolemma and basal lamina of necrotic muscles were still intact. The pathogenesis of myonecrosis induced by myotoxin *a* is summarized diagrammatically in Figure 27. Myonecrosis induced by myotoxin *a* was similar to that caused by crude *C. viridis viridis* venom, but not identical. Stringer et al. (1972) also observed hemorrhage, hemolysis, severe mitochondrial changes, and destruction of the external lamina and sarcolemma. None of these changes was observed following injection of myotoxin *a* alone. Therefore, it appears that myotoxin *a* acts specifically on skeletal muscle cells. Ownby et al. (1976) concluded that the primary site of action was on the sarcoplasmic

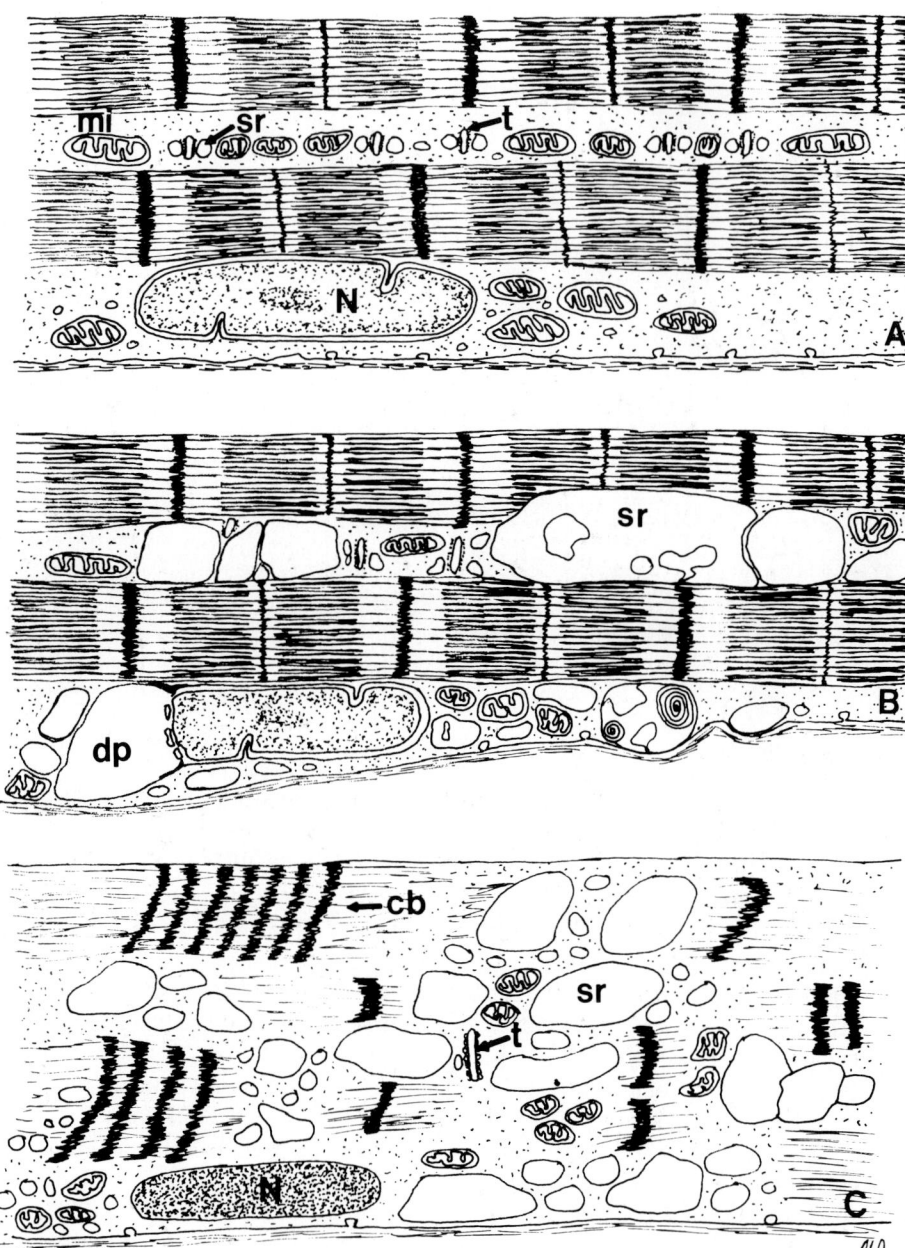

Figure 27 Diagram illustrating the pathogenesis of myonecrosis induced by myotoxin *a* from *C. viridis viridis* venom. A. Portion of skeletal muscle cell before injection of myotoxin. B. Portion of same muscle cell 24 hr after injection of myotoxin. Note swollen sarcoplasmic reticulum and perinuclear space;

reticulum either directly or indirectly through the sarcolemma, and they proposed that myotoxin *a* might affect transport of ions such as sodium by inhibiting the Na^+/K^+ ATPase. Inhibition of the Na^+/K^+ ATPase would cause an increase in the influx of Na^+ into the cell; water would follow, and the cell would begin to swell. The water would initially be taken up by the sarcoplasmic reticulum, but eventually the entire cell would swell. Continued loss of the ability of the cell to control its volume would lead to cell death.

Another small, basic muscle-damaging polypeptide isolated from rattlesnake venom is crotamine. It was isolated from *Crotalus durissus terrificus* var. *Crotaminicus* venom (Goncalves, 1956) and has been shown to cause skeletal muscle stiffness after injection into mice, rats, and dogs (Brazil et al., 1979). Cheymol et al. (1971) found that in the rat the action of crotamine was on the muscle cell membrane, causing a change in calcium or sodium ion permeability which led to contracture that in turn induced an efflux of potassium.

Recently Cameron and Tu (1978) reported that crotamine causes vacuolation of skeletal muscle cells after intramuscular injection into mice. Light microscopic examination of muscle taken 72 hr after injection showed a vacuolation effect similar to the one induced by myotoxin *a* isolated from *C. viridis viridis* venom. In addition to causing similar pathologic alterations in skeletal muscle cells, crotamine and myotoxin *a* are similar chemically (Cameron and Tu, 1978) and immunologically (Ownby et al., 1979). Comparison of the chemical structure of crotamine and myotoxin *a* (Cameron and Tu, 1978) shows that they are very similar: crotamine has 3 disulfide bridges and 42 amino acid residues, myotoxin *a* has only 2 disulfide bridges and 39 residues; they are both basic polypeptides. Ownby et al. (1979) recently showed that antiserum specific to myotoxin *a* reacts with some component in *C. durissus terrificus* venom; this precipitin line identifies completely with a line forming between the antiserum and myotoxin *a*. Unpublished results in our laboratory show that antiserum to myotoxin *a* reacts with myotoxin *a* and crotamine in gel diffusion to show complete identity. Therefore, crotamine and myotoxin *a* are very similar in their chemical formulas and effects on skeletal muscle cells and are immunologically indistinguishable. It is possible that there are other similar myotoxic components present in crotaline venoms, and these components may be responsible for the myonecrosis commonly induced by rattlesnake envenomation.

---

myofibrils are still intact. C. Portion of same cell 72 hr after injection of myotoxin. Note swollen sarcoplasmic reticulum, contraction bands and dissolution of myofibrils and myofilaments. T tubules are still intact. N, nucleus; mi, mitochondrion; sr, sarcoplasmic reticulum; t, T tubule; dp, dilatated perinuclear space; cb, contraction band. See text for discussion and electron micrographs of myotoxin-injected muscle.

Apparently there are two different myotoxins in *C. viridis viridis* venom: myotoxin *a*, a small, basic polypeptide, and viriditoxin, a larger, acidic protein. Viriditoxin, recently isolated by Fabiano and Tu (1980), is both myotoxic and hemorrhagic. Also, hemorrhagic toxin *b* isolated from *C. atrox* venom causes myonecrosis (Ownby et al., 1978). Nothing is known about the pathogenesis of myonecrosis induced by these toxins. Other rattlesnake venom components which cause myonecrosis include crotoxin, a presynaptic toxin isolated from *C. durissus terrificus* venom (see Chapter 4) and Mojave toxin isolated from *C. scutulatus* venom (see Chapter 4). Injection of crotoxin or homologous phospholipase A intramuscularly caused degeneration of muscle fibers beginning at 6 hr after injection, with complete necrosis occurring at 24 hr (Gopalakrishnakone, 1979).

The myonecrosis induced by Mojave toxin was similar to that caused by crotoxin (Gopalakrishnakone et al., 1980).

It appears then that there are different types of myotoxins in rattlesnake venoms. Some, such as myotoxin *a* and crotamine, are primarily myotoxins, having little effect on other tissues; some such as hemorrhagic toxin *b* from *C. atrox* venom and viriditoxin (*C. viridis viridis*) are equally hemorrhagic and myotoxic; and some, such as crotoxin/homologous phospholipase A and Mojave toxin, are neurotoxic and myotoxic.

It is interesting to speculate on the mechanism of myotoxicity of these different toxins. There is considerable evidence that myotoxin *a* and crotamine are chemically and immunologically very similar. Morphologic evidence indicates that their mode of action might also be similar or the same. Ownby et al. (1976) proposed that myotoxin *a* might interfere with Na^+ transport by inhibiting the Na^+/K^+ ATPase in the sarcolemma. Increased Na^+ influx would lead to increased water influx, cell swelling, and eventually cell death. This sequence of events has been shown to occur in other types of cells after ouabain treatment (Ginn et al., 1968). Crotamine has been shown to affect ionic channels, especially sodium channels. Several groups of investigators have provided experimental evidence that crotamine interacts with some molecule which regulates the permeability of the sodium channel to sodium. First, tetrodotoxin, which specifically blocks sodium channels, prevents the effects of crotamine (Cheymol et al., 1971; Pellegrini et al., 1978; Chang and Tseng, 1978). Also, a decrease in the extracellular sodium concentration prevents crotamine effects (Pellegrini et al., 1978; Chang and Tseng, 1978). Both of these facts indicate a increase in sodium permeability. In addition, Chang and Tseng (1978) reported that crotamine induced an increase in the influx of ^{24}Na; pretreatment with ouabain, an inhibitor of Na^+/K^+ ATPase, actually enhanced the effects of crotamine. Their conclusion was that crotamine binds with a molecule that restricts the permeability of the sodium channel to sodium, thereby causing increased permeability to sodium.

Although these types of experiments have not yet been done with myotoxin *a*, it is reasonable to expect their modes of action to be very similar, if not identical. These types of myotoxins may be present in venoms other than *C. durissus terrificus* and *C. viridis viridis*. In fact, Gerencser and Tu (1977) have reported changes in sodium transport across frog skin induced by *C. atrox* venom.

Less is known about the mode of action of other myotoxins such as hemorrhagic toxin *b* and viriditoxin. Since these toxins both cause hemorrhage, it is possible that their myotoxic action is actually an indirect effect owing to ischemia. This is purely speculative, and they too could affect sodium transport, but in a nonspecific way. The mechanism of myonecrosis induced by crotoxin and Mojave toxin is also unknown. However, recent electron microscopic data suggest that crotoxin (and phospholipase A) might cause necrosis of skeletal muscle via increase in intracellular calcium concentration (Gopalakrishnakone, 1979). This idea is based on the observation of calcium granules in the mitochondria of degenerating fibers following injection of crotoxin-phospholipase A. That increased levels of calcium can lead to cell necrosis has been proposed by Duncan (1978) and by Schanne et al. (1979). It has been shown that induction of liver cell death by 10 different toxins is dependent on extracellular calcium. Interaction of the toxins with the cell membrane does not require calcium, and the toxins may injure the plasma membrane by different mechanisms. But an influx of calcium across the membrane is required for necrosis to occur (Schanne et al., 1979).

Since Mojave toxin has been shown to reduce the ability of isolated sarcoplasmic reticulum vesicles to sequester calcium (Cate and Bieber, 1976), it might be speculated that it too could increase the cellular calcium concentration and lead to death.

Inflammation, Edema, and Pain

The local reaction to venom is complicated in some respects by a severe inflammatory reaction. It is not clear whether this is directly due to venom components or indirectly due to the autopharmacologic action of venom components. Regardless of its cause, it is definitely a part of the pathology of the local tissue effects of envenomation.

Inflammation is "the reaction of tissues to an irritant; a process which begins following sublethal injury to tissue and ends with complete healing" (Smith et al., 1972). Briefly, the process of inflammation involves an initial constriction of blood vessels, followed by dilation and increased permeability, increase in rate of blood flow followed by decrease in rate and stasis, margination of erythrocytes and leukocytes followed by diapedesis of erythrocytes and emigration of leukocytes, and exudation of serum.

In response to rattlesnake envenomation, the viable tissue becomes inflamed in an attempt to prevent or decrease the effects of the toxins and to promote healing. There is an increase in blood to the affected area and exudation of substances that can neutralize an irritant, such as humoral antibodies if present, and reactive cells, especially leukocytes. Thus, if the irritant cannot be destroyed it can at least be confined.

There is ample evidence for the presence of leukocytes in the local reaction to rattlesnake envenomation (Fidler et al., 1940; Beamer et al., 1960; Homma and Tu, 1971) (also see Figure 2).

In snakebite cases in which small amounts of venom are injected by the snake, the inflammatory reaction is probably successful in preventing permanent tissue damage and in effecting healing. There are reports of the absence of local tissue damage in mild poisoning cases (Russell, 1960; Russell et al., 1975; Garfin et al., 1979).

However, in severe snakebite cases, the amount of venom and the rapidity of tissue damage, especially hemorrhage, plus systemic involvement overcome the inflammatory reaction, and it is unable to prevent local tissue destruction.

The mediators of inflammation are thought to be histamine (early changes within 30-60 min of injury), serotonin, and most importantly, the kinins, bradykinin and kallidin (Smith et al., 1972). There is substantial evidence that treatment of rats with high molecular weight substances and peptides causes release of serotonin and histamine (reviewed by Ohsaka, 1979). In fact, rattlesnake venoms are known to induce the release of histamine, serotonin, bradykinin, and ATP in the tissues of the snakebite victim (Russell, 1965). Therefore, it is possible that the inflammatory reaction seen in snakebite cases is due to an indirect effect of snake venom components. The exact role of the inflammatory reaction in local tissue destruction by rattlesnake venom is not known, but it is clear that it is part of a complex response of the tissues to venom. However, antiinflammatory compounds such as acetylsalicylic acid, phenacetin, methapyrilene, and sodium salicylate will not prevent local tissue damage induced by *C. atrox* venom (Ownby, 1975; Ownby et al., 1975). Also, purified hemorrhagic and myotoxic components induce specific effects without inducing a substantial inflammatory reaction (Ownby et al., 1976; Ownby et al., 1978).

One of the most spectacular effects of rattlesnake envenomation is tremendous local swelling which may be used as a valuable diagnostic sign of actual envenomation (Russell et al., 1975; Garfin et al., 1979; Clement and Pietrusko, 1978, 1979). This swelling usually begins within 5-30 min after the bite and spreads very rapidly from the point of venom injection (Parrish et al., 1965; Clement and Pietrusko, 1978, 1979). Maximum swelling usually occurs at 24-36 hr after envenomation (Vick, 1971). Progressive swelling in the affected extremity usually indicates significant envenomation.

Experimental injection of rattlesnake venom also induces marked swelling around the site of injection. The basis for this swelling is local edema, i.e., excessive accumulation of fluid in the intercellular spaces. Such an increase in the amount of fluid causes an enlargement in the spaces between adjacent cells, fibrils, and other tissue elements. Thus the histopathologic image is one in which cells are separated by large spaces which are usually filled by an exudate. Grossly, the edematous part is swollen.

However, even though swelling is a common effect of rattlesnake envenomation, very little is known about the mechanism of venom-induced edema or about the role of edema in the local tissue destruction.

Most pathologists would classify rattlesnake venom-induced edema as either an inflammatory edema or an edema resulting from serous inflammatory exudation (Smith et al., 1972). This type of edema results from an increase in the capillary wall to large molecules such as albumin and globulins. Derangement of vasomotor innervation is known to cause this type of edema, but in snake envenomation the actual cause in unknown. In rattlesnake poisoning the pathology is complicated by the presence of hemorrhagic components as well as by components which are autopharmacologically active. Vick et al. (1963) was able to prevent edema in isolated canine lungs induced by *C. adamanteus* venom in the presence of plasma. He did this either by heating the plasma to 56°C for 30 min or by using epsilon aminocaproic acid. These results suggest that edema could be due to activation of a plasma enzyme and therefore not due directly to venom components.

It is possible that edema is part of the inflammatory reaction induced by venom-released histamine, serotonin, and bradykinin. However, in at least one venom, *Agkistrodon piscivorus,* there appears to be a component more active in causing edema than the kininogenase present in the venom (Bhargava et al., 1972).

Immediate pain around the site of the bite occurs in most rattlesnake envenomations (Vick, 1971; Russell et al., 1975; Garfin et al., 1979). Since it is known that injection of bradykinin either intradermally or intraperitoneally induces pain (Dickerson et al., 1965; Lim et al., 1967), it is likely that this is the source of pain in snakebite cases. According to Brown (1973), "the acute pain that occurs shortly after bite inflicted by crotaline snakes may be examined by release of and subsequent algesia induced by bradykinin, histamine (mast cells) and/or serotonin (platelets)."

SYSTEMIC EFFECTS

Cardiovascular System

Perhaps the most striking effects of rattlesnake envenomation are on the cardiovascular system. Although the effect of venom on this system is profound and complex, it is not completely understood. Since the subject is treated fully in Chapter 2, it will only be briefly reviewed here in order to complete the discussion of pathologic effects of rattlesnake venom.

There have been many studies on the response of the cardiovascular system to rattlesnake venom (for a review see Tu, 1977; Marsh and Whaler, 1978; Lee and Lee, 1979). It is known that acute hypotension usually follows a severe rattlesnake bite (Clement ant Pietrusko, 1978, 1979). Experimentally, rattlesnake venom causes a rapid drop in systemic arterial blood pressure and a decrease in heart rate. These effects seem to vary depending on the species, the route and amount of venom injected, and many other parameters.

A fall in systemic arterial pressure has been described after the intravenous injection of crude venom from *C. atrox* (Russell et al., 1962; Halmagyi et al., 1965; Vick et al., 1966; Vick, 1973), *C. adamanteus* (Vick et al., 1966; Phillips, 1972; Vick, 1973), *C. horridus horridus* (Witham et al., 1953; Phillips, 1972; Vick, 1973), *C. cerastes, C. adamanteus,* and *C. durissus durissus* (Vick, 1973), and *Crotalus viridis helleri* (Schaeffer et al., 1973, 1976; Carlson et al., 1975).

In general, the intravenous injection of rattlesnake venom causes a precipitous drop in systemic arterial blood pressure and total systemic resistance; cardiac output increases. This initial reaction is followed by a hypotensive state in which there is an increase in total systemic resistance, decrease in cardiac output, hypoproteinemia, and increase in packed cell volume (Witham et al., 1953; Halmagyi et al., 1965; Schaeffer et al., 1973). Increase in vascular permeability to albumin and erythrocytes can lead to a reduction in blood volume (Carlson et al., 1975; Witham et al., 1953). Low blood pressure continues, and the animal is in shock. Since hypotension and shock are discussed in Chapter 2, no further mention of them will be made here.

There have been a few reports on the effects of isolated crotaline toxins on the heart. Bonilla and Fiero (1971) isolated small, basic proteins from *C. adamanteus* and *C. horridus atricaudatus* venom which caused ischemia, myocardial failure, and eventually, death. Abel et al. (1973) studied the fine structural changes in cat myocardium induced by the intravenous injection of the basic protein of *C. adamanteus* venom. Light microscopic examination revealed extensive subendocardial and myocardial hemorrhages and edemic lungs with interstitial and alveolar hemorrhage. Special staining (basic fuchsin) showed an area of myocardial ischemia. Electron microscopic examination of cats with ischemic myocardium revealed denser and disrupted Z lines, swollen, elongated mitochondria with few or no calcium granules, and a decreased amount of glycogen and an increased amount of lipid. Intercalated disks, sarcomeres, myosin and actin myofilaments, sarcoplasmic reticulum, and transverse tubules all appeared normal. Fat emboli were also observed in coronary blood vessels. These changes are all consistent with myocardial ischemia, although they do not indicate the cause. However, because they are similar to changes induced by other pharmacologic or disease agents, Abel et al. (1973) suggested that anoxia might be the final common factor causing heart failure.

Rattlesnake envenomation causes changes in the number and behavior of platelets. LaGrange and Russell (1973) reported that in humans small amounts of rattlesnake venom caused little or no change in platelet numbers, whereas large amounts caused a rapid and marked decrease in the number of platelets. Lyons (1971) described a drop in the number of platelets after human envenomation by *Crotalus ruber ruber*. He proposed that either peripheral destruction or consumption by hemostatic processes could be responsible for the decrease. Either of these is possible since crotaline venoms do contain components capable of destroying platelets such as phospholipase A or cytotoxins (Tu and Giltner, 1974). Also, there is evidence for involvement of platelets in hemostatic mechanisms (Homma and Tu, 1971; Ownby et al. 1974). Davey and Luscher (1967) found components in *C. terrificus terrificus* venom which caused platelets to aggregate and which are different from coagulant, proteolytic, and phospholipase activities of whole venom.

An experimental study of platelet levels after intramuscular injection of *C. viridis helleri* venom into rabbits (LaGrange and Russell, 1973) showed that there was an initial decrease in the number of platelets, a later rise and overshoot, and then a final decline to near control levels.

Hasiba et al. (1975) described a human snakebite case in which there was a thrombocytopenia after envenomation by *C. horridus horridus*. The patient had both a thrombocytopenia and hypofibrinogenemia. The disorder was described as being DIC-like (disseminated intravascular coagulation). The venom was also shown to aggregate platelets in vitro.

More recently, Ruiz et al. (1980) studied the effects of *C. viridis helleri* venom on homeostasis in dogs. There was a marked decrease in the number of platelets 5 min after intravenous injection of the venom, and there was an even lower number at 30 min. Morphologic changes were observed in megakaryocytes at 30 min, and at 24 hr there were fewer than normal numbers in bone marrow smears. Speculating on the cause of the marked decline in platelet numbers, Ruiz et al. (1980) suggested that it could be the result of aggregation or of increased adhesion to vessel walls or suppression of bone marrow. Also, endothelial damage and exposure of collagen and/or the release of platelet aggregators such as ADP may contribute.

Often, nonclotting blood is found in persons bitten by rattlesnakes (Clement and Pietrusko, 1979). In fact, venom interference in the blood coagulation sequence has long been recognized. Defibrination syndromes have been described following envenomation by rattlesnakes (McCreary and Wurzel, 1959; Andrews, 1960; Weiss et al., 1969) and are thought to be due to the depletion of fibrinogen caused by the action of a thrombin-like enzyme in the venom (Rosenfeld et al., 1968). In one case envenomation by *C. adamanteus* caused an afibrinogenemia (lack of fibrinogen in the blood) 3 hr after the bite, but coagulation was normal 1 day after the bite (antivenin was given). Nonclotting blood was also found in a victim bitten by *C. horridus horridus* (Hasiba et al., 1975). The blood had a decreased amount of fibrinogen (hypofibrinogenemia), and fibrin split products were present. In vitro the venom converted fibrinogen to a clot which lysed quickly.

Simply stated, coagulation is the conversion of fibrinogen, a plasma protein, to fibrin, catalyzed by thrombin. Calcium ions and a heat labile accelerator speed up the process, and the fibrin formed is made insoluble by fibrin-stabilizing factor. Thrombin is produced by a complex series of enzymatic events in which prothrombin is changed to thrombin. The generation of thrombin may be initiated by intrinsic factors (Hageman factor, XII) independently of tissue damage or extrinsic factors (tissue thromboplastin, III) dependent upon tissue injury. Also, there are anticoagulants in normal plasma which remove thrombin upon its formation and keep the blood in a fluid state. A fibrinolytic enzyme which dissolves clots is also present in normal plasma. In fact, many factors contribute to maintaining the fluid state of the blood, including the integrity of the endothelial lining, the normal separation of blood cells and platelets from the endothelium by the laminar flow of the plasma, and the viscosity of blood and its rate of flow. Thus, many factors are involved in maintaining the proper consistency and flow of blood in the body.

There is some confusion about the nature and mechanism of the coagulation disturbance in snakebite victims. This is due in part to the complex reactions involved in vivo and in part to the discrepancies in results obtained in vitro. For example, *C. atrox* venom has been reported as being unable to clot either plasma or fibrinogen (Eagle, 1937; Didisheim and Lewis, 1956; Jiménez-Porras, 1961; Denson et al., 1972); yet others report direct clotting of fibrinogen (Oshima et al., 1969; Copley et al., 1973).

Rattlesnake venoms have been shown to contain components which cause coagulation (Rosenfeld et al., 1968). Denson et al. classified the coagulant activity of several venoms, including some rattlesnake venoms, as follows: *C. durissus durissus, C. viridis helleri,* and *C. adamanteus* all had thrombin-like activity; *C. viridis helleri* had a Factor X activator; *C. atrox* and *C. scutulatus scutulatus* had neither of these activities; and none of these venoms had a prothrombin activator.

A thrombin-like enzyme has been isolated from the venoms of *C. adamanteus* (Markland et al., 1970; Markland and Damus, 1971; Damus et al., 1972) and *C. horridus horridus* (Bonilla, 1975).

Damus et al. (1972) reported that the thrombin-like enzyme from *C. adamanteus* venom acts directly on fibrinogen to convert it to a fibrin clot. It does not activate Factor XIII (fibrin stabilizing factor), does not aggregate platelets, and does not activate components of the fibrinolytic system. It can, however, lyse fibrin by itself if present in high concentrations. Apparently what follows injection of the enzyme is a primary consumption of fibrinogen with a secondary fibrinolysis. The amount of enzyme required for fibrinolysis is at least two orders of magnitude higher than that needed for defibrination. This dose-related response might explain some of the conflicting reports in the literature.

Later studies of this thrombin-like enzyme showed that it has thrombin-like action of prothrombin, but requires calcium ions and phospholipid for activity (Pirkle et al., 1976).

A component which activates Factor X is present in *C. viridis helleri* venom (Denson et al., 1972) but not in *C. durissus durissus, C. adamanteus, C. atrox, C. scutulatus scutulatus,* or *Sistrurus miliarus barbouri* venom.

More recently, Reid and Theakston (1978) have reported interesting results relating changes in coagulation activity of *C. atrox* venom to the age of snakes. Venom from snakes aged 2-8 months clotted fibrinogen directly; venom from snakes aged 9-10 months clotted plasma but not fibrinogen; and venom from older snakes did not clot either plasma or fibrinogen. Qualitatively, the type of activity changes as the snake ages, from direct thrombin-like (first 8 months), to thromboplastin-like (8-10 months), to no procoagulant activity (from 1 year and older).

In order to determine the action of rattlesnake venom on the blood coagulation sequence, experiments should be done in which the age of the snake donating the venom and the dose of venom injected are strictly controlled. Also, experiments done on coagulation in vitro should be correlated with in vivo experiments, since the results might be different depending on the role of tissue factors. For a review of the effects of rattlesnake venom on blood coagulation, the reader is referred to Rosenfeld et al. (1968) and Tu (1977).

Hemolysis is the release of hemoglobin from red blood corpuscles into their bathing fluid. Venom hemolysis, that caused by snake venom, is due to lysis of erythrocytes by venom components either directly or indirectly. Elapid venoms contain both direct and indirect hemolytic factors, but rattlesnake venoms apparently contain only indirect hemolytic factors. Hemolysis has been reported after rattlesnake bite (Russell, 1960) but is apparently uncommon except after envenomation by *C. durissus terrificus* (Minton, 1954). Its role in the induction of death or vascular collapse is not clear. For a review of snake venom induced hemolysis, see Tu (1977).

Systemic hemorrhage is a frequent complication of severe rattlesnake envenomation (Sabback et al., 1977; Clement and Pietrusko, 1979). Although little is known about the actual organs involved or the extent of systemic hemorrhage in human snakebite cases, a substantial amount of work has been done using experimental envenomation.

Following intravenous injection or infusion of crotaline venoms into various experimental animals, hemorrhage has been observed in practically all the serous and mucous membranes and many of the parenchymatous organs (Taube and Essex, 1937). Witham et al. (1953) observed petechial (small, pinpoint) hemorrhages at the margins of the eyelids and gums in dogs after intravenous injection of *C. horridus* venom. Hemorrhage has been observed in the lung (Russell et al., 1962), diaphragm (Carlson et al., 1975; Schaeffer et al., 1976, 1978, 1979a,b), peritoneum (Schaeffer et al., 1978, 1979a,b; Ruiz et al., 1980), and intercostal muscles (Carlson et al., 1975; Schaeffer et al., 1978, 1979a,b). In addition, hemorrhage and/or the presence of blood in the lumen of the intestines has been reported (Witham et al., 1953; Carlson et al., 1975; Schaeffer et al., 1976, 1978, 1979a,b). There have been a few reports of hemorrhagic areas (petechia) in the myocardium, usually at the apex of the left ventricle in the epicardium (Schaeffer et al., 1978, 1979a,b). Ruiz et al. (1980) reported hemorrhage in the spleen, liver, and retroperitoneum, but not in the heart and lungs. The animals which had spleen and liver hemorrhages did not survive the venom injection.

Schaeffer et al. (1979b) found that the intravenous injection of a pure peptide from *C. viridis helleri* venom induced hemorrhage in the small intestine and abdominal cavity as did crude venom, but it did not cause petechial hemorrhages in the diaphragm, peritoneum, intercostal muscles, or the epicardium.

Since all these studies were done using an intravenous injection, it is possible that they may not accurately reflect the pathology of actual snakebite. In most snakebite cases, the venom is deposited either subcutaneously or intramuscularly. However, experiments using intravenous infusion might be comparable to severe rattlesnake bites.

Fidler et al. (1940) reported systemic hemorrhage after subcutaneous injection of rattlesnake venom into monkeys. Hemorrhage was observed in the left ventricle beneath the endocardium, usually in the interventricular septum. In one animal there was hemorrhage in the papillary muscles. A more extensive study was done by Beamer et al. (1960) in which subcutaneous injection of venom was compared with intramuscular injection. Essentially no difference was found, and hemorrhage was observed in the kidneys, heart muscle, and lungs and occasionally in the cerebrum, cerebellum, and brain stem.

The degree and extent of hemorrhage in organ systems may depend on the amount and route of venom injection. Probably the more severe the envenomation, the more widespread will be the hemorrhagic effect.

Respiratory System

Most physicians who have treated rattlesnake bite victims would agree that the most common cause of death is respiratory arrest. Thus, it is clear that rattlesnake venoms damage the respiratory system; but whether this is due to a direct action of venom components or due to an indirect action of venom on the cardiovascular or nervous system is not known.

At least with two rattlesnake venoms, *C. durissus terrificus* and *C. scutulatus*, the effect on respiration appears to be through the action of venom components on the nervous system, i.e., the neuromuscular junction. Thus, crotoxin and Mojave toxin, the lethal toxins from these two venoms, respectively, prevent proper stimulation of the diaphragm muscle and thus cause respiratory arrest. Crotoxin is heterogenous and contains phospholipase A_2 as the primary active component (Tu, 1977). Mojave toxin is interesting because not only does it interfere with neuromuscular transmission, but it also causes hemorrhage in the alveoli and interstitium of lungs of injected animals (Gopalakrishnakone et al., 1980). Crotoxin, however, caused none of these effects in the lungs.

The observation of pathologic changes in lungs after experimental injection of crude rattlesnake venoms has been reported by many. Most of the changes are either hemorrhagic or could be due to hemorrhage and/or anoxia. Taube and Essex (1937) described a marked diffuse congestion in the lungs of dogs injected intravenously with rattlesnake venom. If death was delayed, hemorrhage, edema, fibrin deposition, and alveolar necrosis were observed. Beamer et al. (1960) reported the presence of extensive hemorrhage and edema in the lungs of rabbits injected either intramuscularly or subcutaneously with *C. atrox* venom, whether they died or were killed 48 or 72 hr after injection. They also noted that the severity of the lesions depended on the survival time after injection and on the dose of venom received.

In a study to determine whether *C. atrox* venom acts directly on the lungs, Russell et al. (1962) infused an isolated cat lung preparation with crude venom. They found hemorrhage, edema, and alveolar disruption and concluded that blood pooled in the lungs.

In summary, pathologic studies indicate the presence of hemorrhage, edema, and alveolar necrosis; but all of these changes could be due to hemorrhagic toxins or other cytolytic toxins. Whether any are specific for lung tissue is not known, nor is the role of lung pathology in venom-induced death.

One of the most complete studies done on the effects of rattlesnake venom on the respiratory system was that of Halmagyi et al. (1965). They used a slow intravenous infusion of *C. atrox* venom into sheep and measured several physiologic parameters as well as some morphlogic ones. A large dose of venom (60-600 μg/kg) induced shock and death owing to respiratory failure in most of the animals. Hemorrhage and increased secretions were present in lung airways, but edema was not a common finding. Thromboemboli were usually present and, according to these investigators, are the cause of pulmonary hypertension. Another respiratory effect of venom was alveolar hypoventilation, apparently caused by a venom-induced decrease in the sensitivity of respiratory neurons. In summary, the experiments of Halmagyi et al. (1965) indicated "that the respiratory effect of *Crotalus* venom may be dominant in determining its physiological action."

Vick et al. (1966) and Vick (1973) reported marked effects of rattlesnake venoms on respiration. *Crotalus adamanteus* and *C. atrox* venoms both led to rapid cessation of respiration after intravenous injection into dogs. It is interesting that, although the primary mechanism of death was respiratory failure, if artificial ventilation was used the animals eventually died of cardiovascular failure.

Thus, rattlesnake venom probably does induce changes in the respiratory system, and these changes can lead to respiratory failure and death. However, the severity of the respiratory damage depends on the amount and route of venom injection as well as the kind of venom. It is possible that there are toxins which act specifically on the respiratory system, but it is also possible that respiratory effects are due to the action of other components or the combined action of several components.

Lethal Effects

Rattlesnake venoms are complex mixtures containing proteins having many different pharmacologic actions; therefore, they have very complex biologic actions. The body also responds to the venom in a complex way, making determination of the exact causes of venom-induced tissue damage and death difficult. Some believe that death is due to the direct action of venom components on respiration, whereas others believe death is due to respiratory failure brought on as a secondary effect of venom on the cardiovascular system. Still others believe death is due to the combination of venom-induced damage to cardiovascular, respiratory, and nervous systems as well as interference with hemostatic mechanisms.

Respiratory paralysis of a peripheral origin has been implicated as the cause of death of crotoxin from *C. durissus terrificus* venom (Brazil, 1972; Breithaupt, 1976). Mojave toxin from *C. scutulatus* venom probably causes death by neurotoxic blockage of motor nerve terminals which leads to respiratory arrest (Gopalakrishnakone et al., 1980). Thus, death induced by these two venoms is due to neurotoxins which prevent impulse transmission from the phrenic nerve to the diaphragm.

The lethal effect of other rattlesnake venoms is not understood. It is possible that they too contain similar neurotoxic components, but their effect is less pronounced compared with the action of other toxins, such as those acting on the cardiovascular system. The actual cause of death might depend on the rapidity of the onset of intoxication, as was the case with Mojave toxin (Gopalakrishnakone et al., 1980). With rapid onset death was due to respiratory paralysis, but with slow onset cardiovascular changes were more prominent and death appeared to be due to circulatory failure.

The mechanism of death due to rattlesnake envenomation remains to be completely determined.

Urinary System

Although there have been reports of renal failure in victims of rattlesnake enenomation (Amorim and Mello, 1954; Danzig and Abels, 1961; Amorim, 1971), there have been few detailed studies of the pathogenesis of kidney lesions induced by rattlesnake venom, either clinically or experimentally. Renal failure can be due to events occurring outside the kidney such as reduced renal blood flow, hemolysis, hemoglobinuria, and blockage of the excretory system. Cardiac failure, water and sodium depletion, and shock can all lead to a decrease in the amount of blood reaching the kidneys. Or renal failure can be due to changes originating within the kidney such as glomerulopathy, tubule disease, or renal vascular and interstitial disease.

In the case of snakebite the exact cause of renal failure is difficult to determine owing to the complexity of events induced by envenomation. The question which needs to be answered is whether or not renal failure of snakebite results directly from specific venom components or indirectly from venom-induced shock or other vascular changes, such as hemorrhage.

Perhaps the earliest investigation of the effects of rattlesnake venom on the kidney was done by Pearce (1909) in which he injected venom of *C. adamanteus* intravenously into rabbits. Using the light microscope, he identified two types of lesions in the glomeruli: one hemorrhagic, the other exudative. Both lesions were limited to the glomerulus; tubules appeared normal, and Pearce explained the lesion as being due to an "endotheliolytic body of crotalus venom."

Taube and Essex (1937) reported capillary congestion in the glomeruli of kidneys from dogs injected intravenously with rattlesnake venom (species not given). Histopathology also included swollen nuclei in parietal epithelial cells of Bowman's capsule, swollen endothelium, and degenerating tubular cells throughout the entire kidney. Studies of dogs injected with lower doses and of tissues taken 20 hr after injection of venom showed necrosis of tubules, especially in the subcapsular zone. The proximal convoluted tubules were the most degenerated. Hemorrhage was present throughout the kidney, and the collecting tubules were filled with fibrin and blood.

Beamer et al. (1960) noted lesions in the kidneys 24 hr after either intramuscular or subcutaneous injection of *C. atrox* venom. The glomerular endothelium was swollen, and there was a granular material in Bowman's space. Granular and hyalin casts were present in tubular lumens, and the lining epithelial cells were in various stages of degeneration, from slightly swollen to necrotic.

The most complete histopathologic study of snake venom-induced renal changes was by Amorim and Mello (1954), reviewed by Amorim (1971). Three *C. terrificus terrificus* snakebite cases were studied for renal lesions (Amorim and Mello, 1954). The primary lesion was found to be in the renal tubules, especially those in the intermediate or boundary zone (between cortex and medulla), and in the ascending limb of the loop of Henle and was of a degenerative parenchymatous type. In affected tubules the cells were undergoing granular and hyaline degeneration, pyknosis and karyolysis, and complete disruption. Hyaline casts containing hemoglobin were present in the lumens of damaged tubules. There was also a strong inflammatory reaction consisting of edema, neutrophils, lymphocytes, and plasma cells. The medulla lacked an inflammatory reaction, but hemoglobin casts were present.

Experimental injection of crotaline venom (Amorim, 1971) produced the same kind of kidney lesions as those found in human snakebite victims reported above. To summarize, the primary lesion was of ascending limb of the loop of Henle or the intermediate segment of the nephron; glomerular lesions were absent as were any in the convoluted tubules; however, some Bowman's spaces and tubules contained albumin. Hemoglobin cases were common in the ascending limb, distal convoluted tubules, and collecting tubules. Amorim (1971) proposed that South American crotaline venom contains a fraction toxic to the kidney which acts specifically on the intermediate segment of the renal tubule.

Hadler and Brazil (1966) studied kidney lesions in dogs caused by intravenous injection of crotoxin, a neurotoxic component of *C. durissus terrificus* venom. Early changes included congestion of glomerular capillaries, increased thickness of the basement membrane, and dying glomerular cells, whereas later changes involved the tubules more than the glomeruli. As with other crotaline venoms, damage to capillaries was severe. These changes were different from those observed after injection of crude venom. Hadler and Brazil (1966) suggest that the differences could be due to a more important role for a nephrotoxic factor, lysolecithin, formed by crotoxin in high concentration in the kidney in crotoxin intoxication, whereas crude venom is more potent in causing shock and therefore renal ischemia and tubular lesions.

More recently, Schmidt et al. (1976) reported an electron microscopic study of renal lesions induced by the experimental injection of *C. atrox* venom intravenously into mice. In tissue taken 9 hr after venom injection, lesions of the renal corpuscles and proximal convoluted tubules were observed. Visceral epithelial cells of the renal corpuscle were swollen, contained swollen endoplasmic reticulum and mitochondria, and had cytoplasmic blebs and microvillar projectins. Parietal epithelial cells were similar except they lacked microvillar projections. Areas of lysed mesangial cells and abundant collagen fibers were common. Epithelial cells of the proximal convoluted tubules contained a larger than normal number of lysosome-related structures. The authors compared these changes with those induced by sea snake venom, and they stated that damage induced by crotaline venom was more severe. However, they did not discuss the pathogenesis of rattlesnake venom-induced kidney lesions, nor did they speculate as to the presence of a "nephrotoxin" in *C. atrox* venom.

Indeed, in these studies the use of crude venom makes it difficult to postulate specific kidney lesions. Most, if not all, of the changes described could be secondary effects of shock or possibly due to hemorrhagic or other toxins. For example, if the basement membranes of glomerular capillaries were destroyed, by hemorrhagic components as some have proposed (Ohsaka et al., 1973), fibrin and fibrinogen would be extruded into urinary spaces and the mesangium. This could explain the presence of such materials and the presence of hemorrhage. Also, in diseases characterized by the lysis of erythrocytes or muscle cells (both occur after rattlesnake envenomation), large protein molecules (hemoglobin or myoglobin) pass the glomerular capillaries and may be found in other parts of the kidney. However, until more work is done with purified toxins, we cannot rule out the existence of a specific renal toxin in rattlesnake venom.

Nervous System

Although rattlesnake venoms are not generally noted for inducing disturbances in the nervous systems of snakebite victims, the first neurotoxin isolated from snake venom was from the venom of the South American rattlesnake, *Crotalus durissus terrificus* (Slotta and Fraenkel-Conrat, 1938, 1939). The toxin crotoxin has been shown to be heterogeneous, containing hemolytic, hyaluronidase, phospholipase A_2, and smooth muscle-stimulating activities as well as being neurotoxic. For a review of the chemistry and pharmacologic action of crotoxin, see Tu (1977) as well as Chapters 2 and 4 of this book. Most of the studies on crotoxin have been either biochemical or physiological, and there have been no published studies of the pathologic effects of crotoxin on nervous tissue.

Russell and Bohr (1962) studied the effects of several venoms on the central nervous system. *Crotalus adamanteus, C. atrox,* or *C. horridus* venoms in doses of 0.1 to 5.0 mg were injected into the lateral ventricles of the brain of anesthetized cats. They noted marked changes in behavior and motor and parasympathetic dysfunction; these changes were attributed to a direct action of the venoms on the central nervous system. However, the only control was the injection of Ringer-Tyrode solution. Perhaps a more appropriate control would be a solution containing an amount of protein equivalent to the amount of venom used. There was no report on the morphologic appearance of brain tissue following the injection of venom.

Vick et al. (1964) and Vick and Lipp (1970) have reported physiologic studies on the effects of rattlesnake venoms on the central nervous systems of dogs and rhesus monkeys. Crude venom was given either intramuscularly or intravenously, and cortical electrical activity (EEG) was recorded. These investigators concluded that changes in the normal EEG pattern were probably the result of cerebral anoxia and not a direct action of venom components on nervous tissue. Anoxia results from depression of respiration by venom, and as the animal becomes more anoxic, brain function becomes depressed.

Martin and Rosenberg (1968) studied the effects of *C. adamanteus* venom and phospholipase A on the squid giant nerve fibers. Electron microscopic examination revealed that, although other venoms (*Naja naja* and *Agkistrodon piscivorus piscivorus*) and phospholipase A caused an accumulation of cytoplasmic globules in the Schwann sheath, rattlesnake venom had practically no effect on the sheath.

More recently, a lethal toxin from Mojave rattlesnake (*C. scululatus*) venom has been isolated (Bieber et al., 1975) and has been shown to block the neuromuscular junction by interfering with transmitter release (Gopalakrishnakone et al., 1980). Mojave toxin is apparently similar to crotoxin (from *C. durissus terrificus* venom).

It is possible that all rattlesnake venoms contain components toxic to the nervous system. However, they may be present in lower amounts or may be less potent in some venoms than in others.

CONCLUSIONS

Rattlesnake envenomation results in both local and systemic changes. Local morphologic changes include hemorrhage, myonecrosis, and edema. These changes may be slight in a case of mild poisoning in which the body's defense system can manage them with little or no permanent alteration in tissues. On the other hand, these changes may be drastic in a case of severe poisoning in which the body's defense system cannot manage them, and the results are complete tissue destruction with loss of function and/or amputation of the affected extremity. Rattlesnake venoms appear to induce hemorrhage by destroying capillary walls, i.e., lysis of endothelial cells. Myonecrosis results from an as yet unknown action of venom components on muscle cells, but experimental evidence indicates that some of the myotoxins act by interfering with sodium transport, perhaps interacting with the sodium channel of the sarcolemma. Other myotoxins, however, may be altering the transport of calcium across muscle cell membranes.

Systemic changes occurring in rattlesnake envenomation are even less well understood. Hypotension and shock are well documented changes in the cardiovascular system, but rattlesnake venom also affects respiratory, urinary, and nervous systems, as well as hemostasis.

It seems then that the pathology of rattlesnake envenomation is a combination of changes in many different organ systems, organs, tissues, and cells. These all combine to produce the clinical symptoms of rattlesnake envenomation. Lethal effects could be due to the combination of effects on different systems or to one primary effect. Venoms from snakes of different species are certainly different, and whereas one venom is strongly hemorrhagic, another might be strongly myotoxic and weakly hemorrhagic. We should become aware of these differences and adjust our treatment procedures according to the species and perhaps even subspecies of snake as well as to the apparent severity of the bite.

REFERENCES

Abel, J. H., Nelson, A. W., and Bonilla, C. A. (1973). *Crotalus adamanteus* basic protein toxin: Electron microscopic evaluation of myocardial damage. *Toxicon 11*:59-63.

Amorim, M. F. (1971). Intermediate nephron nephrosis in human and experimental crotalic poisoning. In *Venomous Animals and Their Venoms*, vol. 2, W. Bücherl and E. Buckley (Eds.). Academic Press, New York, pp. 319-343.

Amorim, M. F., and Mello, R. F. (1954). Intermediate nephron nephrosis from snake poisoning in man. *Am. J. Pathol. 3*:479-499.

Andrews, C. E. (1960). The treatment of diamondback rattlesnake bite (*Crotalus adamanteus*). *Arch. Surg. 81*:699-705.

Bailey, G. S., Lee, J., and Tu, A. T. (1979). Conformational analysis of myotoxin *a* (muscle degenerating toxin) of prairie rattlesnake venom. *J. Biol. Chem. 254*:8922-8926.

Beamer, P. D., Boys, F. E., and Smith, H. M. (1960). Histopathology of ophidiasis in rabbits. *Exp. Med. Surg. 18*:256-265.

Bieber, A. L., Tu, T., and Tu, A. T. (1975). Studies of an acidic cardiotoxin isolated from the venom of Mojave rattlesnake (*Crotalus scutulatus*). *Biochim. Biophys. Acta 400*:178-188.

Bhargava, N., Vargaftig, B. B., de Vos, C. J., Bonta, I. L., and Tijs, T. (1972). Dissociation of oedema provoking factor of *Agkistrodon piscivorus* venom from kininogenase. In *Vasopeptides. Chemistry, Pharmacology, and Pathophysiology*, N. Back and F. Sicuteri (Eds.). Plenum, New York, pp. 141-148.

Bjarnason, J. B., and Tu, A. T. (1978). Hemorrhagic toxins from Western diamondback rattlesnake (*Crotalus atrox*) venom: Isolation and characterization of five toxins and the role of zinc in hemorrhage toxin e. *Biochemistry 17*:3395-3404.

Bonilla, C. A. (1975). Defibrinating enzyme from timber rattlesnake (*Crotalus h. horridus*) venom: A potential agent for therapeutic defibrination. I. Purification and properties. *Thromb. Res. 6*:151.

Bonilla, C. A., and Fiero, M. K. (1971). Comparative biochemistry and pharmacology of salivary gland secretions. *J. Chromatogr. 56*:253-263.

Brazil, O. V. (1972). Neurotoxins from the South American rattlesnake venom. *J. Formosan Med. Assoc. 71*:394.

Brazil, O. V., Prado-Franceschi, J., and Laure, C. J. (1979). Repetitive muscle responses induced by crotamine. *Toxicon.* 17:61-67.

Breithaupt, H. (1976). Neurotoxic and myotoxic effects of crotalus phospholipase A and its complex with crotapotin. *Naunyn-Schmiedebergs Arch. Pharmacol.* 292:271-278.

Brown, J. H. (1973). *Toxicology and Pharmacology of Venoms from Poisonous Snakes.* Ch. C Thomas, Springfield, Ill., p. 123.

Cameron, D. L., and Tu, A. T. (1977). Characterization of myotoxin a from the venom of prairie rattlesnake (*Crotalus viridis viridis*), *Biochemistry* 16:2546-2553.

Cameron, D. L., and Tu, A. T. (1978). Chemical and functional homology of myotoxin a from prairie rattlesnake venom and crotamine from South American rattlesnake venom. *Biochim. Biophys. Acta* 532:147-154.

Carlson, R. W., Schaeffer, R. C., Whigham, H., Michaels, S., Russell, F. E., and Weil, M. H. (1975). Rattlesnake venom shock in the rat: Development of a method. *Am. J. Physiol.* 229:1668-1674.

Cate, R. L., and Bieber, A. L. (1976). Effects of Mojave toxin on rat skeletal muscle sarcoplasmic reticulum. *Biochem. Biophys. Res. Comm.* 72:295-301.

Chang, C. C. and Tseng, K. H. (1978). Effect of crotamine, a toxin of South American rattlesnake venom, on the sodium channel of murine skeletal muscle. *Br. J. Pharmacol.* 63:551-559.

Cheymol, J., Gonçalves, J. M., Bourillet, F., and Roch-Arveiller, M. (1971). Action neuromusculaire comparee de la crotamine et du venin de *Crotalus durissus terrificus* var. *crotaminicus*. II. Sur preparations isolees. *Toxicon* 9:287-289.

Clement, J. F., and Pietrusko, R. G. (1978). Pit viper snakebite in the United States. *J. Fam. Pract.* 6:269-279.

Clement, J. F., and Pietrusko, R. G. (1979). Pit viper snakebite envenomation in the United States. *Clin. Toxicol.* 14:515-538.

Copley, A. L., Banerjee, S., and Levi, A. (1973). Studies of snake venoms on blood coagulation. I. The thromboserpentin (thrombin-like) enzyme in the venoms. *Thromb. Res.* 2:487-508.

Damus, P. S., Markland, F. S., Davidson, T. M., and Shanley, J. D. (1972). A purified procoagulant enzyme from the venom of the eastern diamondback rattlesnake (*Crotalus adamanteus*): In vivo and in vitro studies. *J. Lab. Clin. Med.* 79:906-923.

Danzig, L. E., and Abels, G. H. (1961). Hemodialysis of acute renal failure following rattlesnake bite, with recovery. *JAMA* 175:136-137.

Davey, M.G., and Luscher, E. F. (1967). Actions of thrombin and other coagulant and proteolytic enzymes on blood platelets. *Nature* 216:857-858.

Denson, K. W. E., Russell, F. E., Almagro, D., and Bishop, R. C. (1972). Characterization of the coagulant activity of some snake venoms. *Toxicon* 10:557-562.

Dickerson, G. D., Engle, R. J., Guzman, F., Rodgers, D. W., and Lim, R. K. S. (1965). The intraperitoneal bradykinin-evoked pain test for analgesia. *Life Sci.* 4:2063-2069.

Didisheim, P., and Lewis, J. H. (1956). Fibrinolytic and coagulant activities of certain snake venoms and proteases. *Proc. Soc. Exp. Biol. Med. 93*:10-13.

Duncan, C. J. (1978). Role of intracellular calcium in promoting muscle damage: A strategy for controlling the dystrophic condition. *Experientia 34*:1531-1535.

Eagle, H. (1937). The coagulation of blood by snake venoms and its physiologic significance. *J. Exp. Med. 65*:613-639.

Efrati, P., and Reif, L. (1953). Clinical and pathological observations on sixty-five cases of viper bite in Israel. *Am. J. Trop. Med. Hyg. 2*:1085-1108.

Fabiano, R. J., and Tu, A. T. (1980). Purification and biochemical study of viriditoxin, tissue damaging toxin, from Prairie rattlesnake venom. *Biochemistry 20*:21-27.

Fidler, H. K., Glasgow, R. D., and Carmichael, E. B. (1938). Pathologic changes produced by subcutaneous injection of rattlesnake (*Crotalus*) venom into *Macaca mulatta* monkeys. *Proc. Soc. Exp. Biol. Med. 38*:892-894.

Fidler, H. K., Glasgow, R. D., and Carmichael, E. B. (1940). Pathological changes produced by the subcutaneous injection of rattlesnake (*Crotalus*) venom into *Macaca mulatta* monkeys. *Am. J. Pathol. 16*:355-364.

Flexner, S., and Noguchi, H. (1905). On the plurality of cytolysins in snake venom. *J. Pathol. Bact. 10*:111-124.

Fox, J. W., Elzinga, M., and Tu, A. T. (1979). Amino acid sequence and disulfide bond assignment of myotoxin *a* isolated from the venom of prairie rattlesnake (*Crotalus viridis viridis*). *Biochemistry 18*:678-684.

Freiderich, C., and Tu, A. T. (1971). Role of metals in snake venoms for hemorrhagic, esterase and proteolytic activities. *Biochem. Pharmacol. 20*:1549-1556.

Garfin, S. R., Mubarak, S. J., and Davidson, T. M. (1979). Rattlesnake bites. *Clin. Orthop. 140*:50-57.

Gerencser, G. A., and Tu, A. T. (1977). Effect of *Crotalus atrox* venom on sodium transport across the frog skin. *Proc. Soc. Exp. Biol. Med. 156*:104-108.

Ginn, F. L., Shelburne, J. D., and Trump, B. F. (1968). Disorders of cell volume regulation. I. Effects of inhibition of plasma membrane adenosine triphosphatase with ouabain. *Am. J. Pathol. 53*:1041-1071.

Gonçalves, J. M. (1956). Purification and properties of crotamine. In *Venoms*, E. E. Buckley and N. Porges (Eds.). American Association for the Advancement of Science, Washington, D.C., pp. 261-274.

Gopalakrishnakone, P. (1979). Morphological studies on the effects of a phospholipase A_2 complex (crotoxin) from *Crotalus durissus terrificus* venom on muscle, nerve and neuromuscular junction in the mouse. Ph.D. Thesis, University of London, pp. 148, 263.

Gopalakrishnakone, P., Hawgood, B. J., Holbrooke, S. E., Marsh, N. A., Santana De Sa, S., and Tu, A. T. (1980). Sites of action of Mojave toxin isolated from the venom of the Mojave rattlesnake. *Br. J. Pharmacol. 69*:421-431.

Hadler, W. A., and Brazil, O. V. (1966). Pharmacology of crotoxin. IV. Nephrotoxicity. *Mem. Inst. Butantan 33*:1001-1008.

Halmagyi, D. F. J., Starzecki, B., and Horner, G. J. (1965). Mechanism and pharmacology of shock due to rattlesnake venom in sheep. *J. Appl. Physiol. 20*: 709-718.

Hasiba, U., Rosenbach, L. M., Rockwell, D., and Lewis, J. H. (1975). DIC-like syndrome after envenomation by the snake, *Crotalus horridus horridus. N. Eng. J. Med. 292*:505-507.

Homma, M., and Tu, A. T. (1971). Morphology of local tissue damage in experimental snake envenomation. *Br. J. Exp. Pathol. 52*:538-542.

Jiménez-Porras, J. M. (1961). Biochemical studies on venom of the rattlesnake, *Crotalus atrox atrox, J. Exp. Zool. 148*:251-258.

LaGrange, R. G., and Russell, F. E. (1973). Platelet studies in rabbits following *Crotalus* poisoning. In *Toxins of Animal and Plant Origin,* vol. 3, A. de Vries and E. Kochva (Eds.). Gordon and Breach, New York, pp. 1033-1038.

Lee, C. Y., and Lee, S. Y. (1979). Cardiovascular effects of snake venoms. In *Handbook of Experimental Pharmacology,* vol. 52, *Snake Venoms,* C. Y. Lee (Ed.). Springer-Verlag, Berlin, pp. 547-590.

Lim, R. K. S., Miller, D. G., Guzman, F., Rodgers, D. W., Rodgers, R. W., Wang, S. K., Chao, P. Y., and Shih, T. Y. (1967). Pain and analgesia evaluated by the intraperitoneal bradykinin-evoked pain method in man. *Clin. Pharmacol. Therap. 8*:521-542.

Lyons, W. J. (1971). Profound thrombocytopenia associated with *Crotalus ruber ruber* envenomation: A clinical case. *Toxicon 9*:237-240.

McCreary, T., and Wurzel, H. (1959). Poisonous snake bites. *JAMA 170*:268-272.

Markland, F. S., and Damus, P. S. (1971). Purification and properties of a thrombin-like enzyme from the venom of *Crotalus adamanteus* (Eastern diamondback rattlesnake). *J. Biol. Chem. 246*:6460-6473.

Markland, F. S., Damus, P. S., Davidson, T. M., and Shanley, J. D. (1970). Thrombic enzyme from *Crotalus adamanteus. Lancet 1*:1398.

Marsh, N., and Whaler, B. (1978). The effects of snake venoms on the cardiovascular and haemostatic mechanisms. *Int. J. Biochem. 9*:217-220.

Martin, R., and Rosenberg, P. (1968). Fine structural alterations associated with venom action on squid giant nerve fibers. *J. Cell. Biol. 36*:341-353.

Minton, S. A. (1954). *Venom Diseases,* Chas C Thomas, Springfield, Ill., p. 162.

Ohsaka, A. (1979). Hemorrhagic, necrotizing and edema-forming effects of snake venoms. In *Handbook of Experimental Pharmacology,* vol. 52, *Snake Venoms,* G. V. R. Born, A. Farah, H. Herken, and A. D. Welch (Eds.). Springer-Verlag, Berlin, pp. 480-546.

Ohsaka, A., Just, M., and Habermann, E. (1973). Action of snake venom hemorrhagic principles on isolated glomerular basement membrane. In *Animal and Plant Toxins,* E. Kaiser (Ed.). Goldmann, Munich, Federal Republic of Germany, pp. 93-97.

Ohsaka, A., Suzuki, K., and Ohashi, M. (1975). The spurting of erythrocytes through junctions of the vascular endothelium treated with snake venom. *Microvasc. Res.* 10:208-213.

Oshima, G., Sato-Ohmori, T., and Suzuki, T. (1969). Proteinase, arginine ester hydrolase and a kinin releasing enzyme in snake venoms. *Toxicon* 7:229-233.

Ownby, C. L. (1975). *Pathogenesis and Chemical Treatment of Hemorrhage Induced by Rattlesnake Venom.* Ph.D. Dissertation, Colorado State University, pp. 67-70, 82-101.

Ownby, C. L., Kainer, R. A., and Tu, A. T. (1974). Pathogenesis of hemorrhage induced by rattlesnake venom. *Am. J. Pathol.* 76:401-408.

Ownby, C. L., Tu, A. T., and Kainer, R. A. (1975). Effect of diethylenetriaminepentaacetic acid and procaine on hemorrhage induced by rattlesnake venom. *J. Clin. Pharmacol.* 15:419-426.

Ownby, C. L., Cameron, D., and Tu, A. T. (1976). Isolation of myotoxic component from rattlesnake (*Crotalus viridis viridis*) venom. *Am. J. Pathol.* 85:149-158.

Ownby, C. L., Bjarnason, J., and Tu, A. T. (1978). Hemorrhagic toxins from rattlesnake (*Crotalus atrox*) venom. *Am. J. Pathol.* 93:201-210.

Ownby, C. L., Woods, W. M., and Odell, G. V. (1979). Antiserum to myotoxin from prairie rattlesnake (*Crotalus viridis viridis*) venom. *Toxicon* 17:373-380.

Parrish, H. M., Silberg, S. L., and Groldner, J. C. (1965). Snakebite: A pediatric problem. *Clin. Pediatr.* 4:237-241.

Pearce, R. M. (1909). An experimental glomerular lesion caused by venom (*Crotalus adamanteus*). *J. Exp. Med.* 11:532-540.

Pellegrini-Filho, A., Brazil, O. V., Fontana, M. D., and Laure, C. J. (1978). The action of crotamine on skeletal muscle: An electrophysiological study. In *Toxins: Animal, Plant and Microbial.* Proceedings of the 5th International Symposium on Animal, Plant and Microbial Toxins, Rosenberg, P., (Ed.). Pergamon Press, London, pp. 375-382.

Phillips, S. J. (1972). The effect of snake and bee venoms on cardiovascular hemodynamics and function. In *Toxins of Animal and Plant Origin*, vol. 2, A. de Vries and E. Kochva (Eds.). Gordon and Beach, New York, pp. 683-701.

Pirkle, H., Markland, F. S., and Theodor, I. (1976). Thrombin-like enzymes of snake venoms: Actions on prothrombin. *Thromb. Res.* 8:619-627.

Reid, H. A., and Theakston, R. D. G. (1978). Changes in coagulation effects by venoms of *Crotalus atrox* as snakes age. *Am. J. Trop. Med. Hyg.* 27:1053-1057.

Rosenfeld, G., Nahas, L., and Kelen, E. M. A. (1968). Coagulant, proteolytic, and hemolytic properties of some snake venoms. In *Venomous Animals and Their Venoms*, vol. 1, W. Bücherl, E. E. Buckley, and V. Deulofeu (Eds.). Academic Press, New York, pp. 229-273.

Ruiz, C. E., Schaeffer, R. C., Weil, M. H., and Carlson, R. W. (1980). Hemostatic changes following rattlesnake (*Crotalus viridis helleri*) venom in the dog. *J. Pharmacol. Exp. Ther.* 213:414-417.

Russell, F. E. (1960). Rattlesnake bites in southern California. *Am. J. Med. Sci.* 239:1-9.

Russell, F. E. (1965). Bradykininogen levels following *Crotalus* envenomation. *Toxicon* 2:277-279.

Russell, F. E., and Bohr, V. C. (1962). Intraventricular injection of venom. *Toxicol. Appl. Pharmacol.* 4:165-173.

Russell, F. E., Buess, F. W., and Strassberg, J. (1962). Cardiovascular response to *Crotalus* venom. *Toxicon* 1:5-18.

Russell, F. E., Carlson, R. W., Wainschel, J., and Osborne, A. H. (1975). Snake venom poisoning in the United States: Experience with 550 cases. *JAMA* 233:341-344.

Sabback, M. S., Cunningham, E. R., and Fitts, C. T. (1977). A study of the treatment of pit viper envenomation in 45 patients. *J. Trauma* 17:569-573.

Schaeffer, R. C., Carlson, R. W., Whigham, H., Russell, F. E., and Weil, M. H. (1973). Some hemodynamic effects of rattlesnake (*Crotalus viridis helleri*) venom. *Proc. West. Pharmacol. Soc.* 16:58-62.

Schaeffer, R. C., Carlson, R. W., Whigham, H., Weil, M. H., and Russell, F. E. (1976). Acute hemodynamic effects of rattlesnake, *Crotalus viridis helleri*, venom. In *Toxins: Animal, Plant and Microbial*, P. Rosenberg (Ed.). Pergamon Press, Elmsford, N.Y., pp. 383-391.

Schaeffer, R. C., Carlson, R. W., Puri, V. K., Callahan, G., Russell, F. E., and Weil, M. H. (1978). The effects of colloidal and crystalloidal fluids on rattlesnake venom shock in the rat. *J. Pharmacol. Exp. Ther.* 206:687-695.

Schaeffer, R. C., Carlson, R. W., and Weil, M. H. (1979a). The effects of antivenin and corticosteroid analogs on rattlesnake venom shock in the rat. *J. Pharmacol. Exp. Ther.* 211:409-414.

Schaeffer, R. C., Pattabhiraman, T. R., Carlson, R. W., Russell, F. E., and Weil, M. H. (1979b). Cardiovascular failure produced by a peptide from the venom of the southern pacific rattlesnake, *Crotalus viridis helleri. Toxicon* 17:447-453.

Schmidt, M. E., Abdelbaki, Y. Z., and Tu, A. T. (1976). Nephrotoxic action of rattlesnake and sea snake venoms: An electron-microscopic study. *J. Pathol.* 118:75-81.

Slotta, K. H., and Fraenkel-Conrat, H. (1938). Two active proteins from rattlesnake venoms. *Nature* 142:213.

Slotta, K. H., and Fraenkel-Conrat, H. (1939). Crotoxin. *Nature* 144:290-291.

Schanne, F. A. X., Kane, A. B., Young, E. E., and Farber, J. L. (1979). Calcium dependence of toxic cell death: A final common pathway. *Science* 206:700-702.

Smith, H. A., Jones, T. C., and Hunt, R. D. (1972). *Veterinary Pathology*, 4th ed. Lea and Febiger, Philadelphia, pp. 145-154.

Stahnke, H. L. (1966). *The Treatment of Venomous Bites and Stings*. Bureau of Publications, Arizona State University, Tempe.

Stringer, J. M., Kainer, R. A., and Tu, A. T. (1972). Myonecrosis induced by rattlesnake venom. *Am. J. Pathol.* 67:127-134.

Taube, H. N., and Essex, H. E. (1937). Pathologic changes in the tissues of the dog following injections of rattlesnake venom. *Arch. Pathol.* 24:43-51.

Trevino, G. S. (1968). Pathologic effects of the venoms of *Crotalus atrox* and *Naja naja* in pigs. Ph.D. Thesis. Michigan State University, East Lansing.

Tu, A. T. (1977). *Venoms: Chemistry and Molecular Biology*. John Wiley and Sons, New York, chapters 14, 21, and 23.

Tu, A. T., and Giltner, J. B. (1974). Cytotoxic effects of snake venoms on KB and Yoshida sarcoma cells. *Res. Commun. Chem. Pathol. Pharmacol.* 9:783-786.

Tu, A. T., and Homma, M. (1970). Toxicologic study of snake venoms from Costa Rica. *Toxicol. App. Pharmacol.* 16:73-78.

Tu, A. T., Homma, M., and Hong, B. (1969). Hemorrhagic, myonecrotic, thrombotic and proteolytic activities of viper venoms. *Toxicon* 6:175-178.

Vick, J. A. (1971). Symptomatology of experimental and clinical crotalid envenomation. In *Neuropoisons. Their Pathophysiological Actions*, vol. 1, L. L. Simpson (Ed.). Plenum, New York, pp. 71-86.

Vick, J. A. (1973). Effect of actual snakebite and venom injection on vital physiological function. In *Toxins of Animal and Plant Origin*, vol. 3, A. de Vries and E. Kochva (Eds.). Gordon and Breach, New York, pp. 1001-1011.

Vick, J. A., and Lipp, J. (1970). Effect of cobra and rattlesnake venoms on the central nervous system of the primate. *Toxicon* 8:33-39.

Vick, J. A., Blanchard, R. J., and Perry, J. F. (1963). Effects of epsilon amino caproic acid on pulmonary vascular changes produced by snake venom. *Proc. Soc. Exp. Biol.* 113:841-844.

Vick, J. A., Ciuchta, H. P., and Polley, E. H. (1964). Effect of snake venom and endotoxin on cortical electrical activity. *Nature* 103:1387-1388.

Vick, J. A., Ciuchta, H. P., and Manthei, J. H. (1966). Pathophysiological studies of ten snake venoms. In *Animal Toxins*, F. E. Russell and P. R. Saunders (Eds.). Pergamon Press, Elmsford, N.Y. pp. 269-282.

Weiss, H. J., Allan, S., Davidson, E., and Kochva, S. (1969). Afibrinogenemia in man following the bite of a rattlesnake (*Crotalus adamanteus*). *Am. J. Med.* 47:625-634.

Witham, A. C., Remington, J. W., and Lombard, E. A. (1953). Cardiovascular response to rattlesnake venom. *Am. J. Physiol.* 173:535-541.

4
Presynaptic Toxins from Rattlesnake Venoms

ROBERT A. HENDON
Colorado State University, Fort Collins, Colorado

ALLAN L. BIEBER
Arizona State University, Tempe, Arizona

Introduction   211

Crotoxin   213
Brief Historical Perspective • Isolation and Characterization • Neurotoxicity and Phospholipase A_2 Activity • Subunit Interaction

Mojave Toxin   231
Brief Historical Perspective • Isolation • Biochemical Studies • Immunological Studies

An Antigenically Related Toxin   240

Summary Comparison   240

References   242

INTRODUCTION

The isolation and characterization of crotoxin by Dr. Karl Slotta and Dr. Heinz Fraenkel-Conrat in 1938 marked the beginning of serious biochemical research in a new area, the field of snake venoms. The unique nature of crotoxin, being a neurotoxin, caused it to be the subject of extensive investigation in the ensuing years. *Crotalus durissus terrificus,* the rattlesnake from which crotoxin was originally isolated, was thought to be the only crotalid containing a neurotoxic lethal component. Recently, another crotalid has been shown to possess a neurotoxin that has many similarities to crotoxin. These two toxins, crotoxin and Mojave toxin, are the subject of the following chapter.

Crotoxin is a neurotoxin composed of an acidic and basic subunit associating by ionic interaction and has definitely been shown to have a deleterious effect on neuromuscular transmission at the presynaptic terminal. It is also alleged, with more evidence accumulating, that the toxin disrupts chemical transmission at the postsynaptic terminal. The toxin is found as a large percentage of the crude venom from *Crotalus durissus* subspecies *durissus* and *terrificus:* the former occurs in Central and South America, and the latter is predominantly found in South America (See Glenn and Straight, Chapter 1 of this volume).

Mojave toxin is also a subunit neurotoxin which, like crotoxin, has been shown to affect neuromuscular transmission at the presynaptic terminal. Unlike crotoxin, Mojave toxin exerts a detrimental effect on the cardiovascular system in vivo (Beiber et al., 1975; Gopalakrishnakone et al., 1980). Mojave toxin seems to occur in specific areas with the range of *Crotalus scutulatus* (See Glenn and Straight, Chapter 1 of this volume). In the northern portion of its range, the presence of Mojave toxin in the crude venom is insignificant; however, the individuals inhabiting the southern portion possess an extremely potent venom owing to the high content of this toxin. Thus, two distinct populations of *Crotalus scutulatus* occur based on electrophoretic and biological toxicity comparisons. Whether these populations are also geographically distinct is unclear at this time.

Crotoxin and Mojave toxin require the presence of both acidic and basic subunits to express full neurotoxicity, and both possess phospholipase A_2 activity associated only with the basic subunit. Both toxins putatively act by exhausting the quanta of acetylcholine contained in the presynaptic vesicles. Neither the mechanism nor necessity of interaction for the subunits is well understood; but, it is believed that the complex dissociates at or near the target site.

Neither *Crotalus durissus* spp. nor *Crotalus scutulatus* evoke the symptomology or etiology of the majority of crotalids after envenomation: their bites more closely resemble those of the Elapidae. The anomaly of these venoms as compared with the remainder of the Crotalidae has spurred more intensive scrutiny of other crotalid venoms. Bieber et al. (1975) suggested that all crotalid venoms possess Mojave toxin in varying concentrations, and because of this difference in concentration, the powerful neurotoxic component goes undetected. This is intriguingly possible, and some species of *Crotalus*, namely *Crotalus viridis concolor, Crotalus viridis helleri, Crotalus vegrandis*, and *Crotalus basiliscus*, are receiving more intensive observation because of the severity of their bites. Hawgood (Chapter 2, p. 135) clearly shows Mojave toxin and/or a crotoxin-like substance present in the venom of several crotalids. This indicates that these compounds may be ubiquitous for venoms in this genus.

CROTOXIN

Crotoxin, for many years, was considered to be a unique neurotoxin, found only in the neotropical rattlesnake *Crotalus durissus* (subspecies *durissus* and *terrificus*), because none of the North American species evoked symptoms suggesting neurotoxic activity. This portion of the chapter is concerned with the discovery, isolation, and subsequent 40 years of biochemical inquiry into the molecular basis of activity and function of this protein.

Brief Historical Perspective

Investigative research in the area of snake venoms has expanded considerably since the advent of interest in this complex field. No work, essay, or treatise, however, especially on crotoxin, would be complete without tribute being paid to Dr. Karl Slotta and Dr. Heinz Fraenkel-Conrat. A Brazilian crotalid, *Crotalus durissus terrificus* (name changed from *Crotalus terrificus terrificus*), was known to inflict a severe bite; in many cases the consequences were fatal. Envenomation by this snake was observed to be different from most rattlesnakes by little swelling and the relative absence of pain, with the etiology progressing to extreme respiratory difficulties; all suggested some form of neurotoxin.

The Instituto Butantan at São Paulo, Brazil, is a snake venom institute possessing large quantities of crude venom and has the primary function of antivenom production, presumably for envenomated farmers of the area. Dr. Karl Slotta, a sterol chemist who had successfully competed with Dr. C. Butenandt in progesterone chemistry elucidation, went to São Paulo in 1935 to assume the directorship of the newly established Chemical Institute (see the Foreword). The general thinking at this time was that snake poison, like toad poison, must be a sterol. However, it was quickly found not to be a sterol but a protein. Dr. Heinz Fraenkel-Conrat, a recently specialized protein chemist under Dr. Max Bergman, Rockefeller University, and brother-in-law to Dr. Slotta, arrived in São Paulo for a summer vacation and remained for 1 year working at the institute. At this time the director of Butantan was a progressive administrator and hired several German scientists which had been proclaimed redundant by the Nazi party. He also enabled them to bring the latest chemical equipment of the day with them. The result of this was an institute that favored pure research, a condition that prevailed until political change replaced the director, with the subsequent departure of most of the scientists. However, during this period, in 1938, Dr. Karl Slotta and Dr. Heinz Fraenkel-Conrat achieved the isolation and crystallization of the lethal toxin from *Crotalus durissus terrificus*. Thus a field of research was begun that now encompasses the world.

Today, as a result, many pure toxins have been isolated, characterized, and extensively investigated. The first step in venom research is to repeat with other venoms what Slotta and Fraenkel-Conrat did originally with the venom from

Crotalus durissus terrificus: isolate, and characterize the major lethal component from the venom in question. The venoms of snakes were and remain the most frequently and extensively studied; however, venom research now includes arachnids (spiders and mites), apids (bees), scorpions, wasps, ants, and many marine organisms (jellyfish, sea anemones, starfish, and pufferfish). These men also reoriented the field somewhat in that their mode of investigation incorporated both biology and chemistry, as they were endeavoring to investigate the isolated lethal component of the venom rather than the crude mixture in toto. No only did Dr. Slotta and Dr. Franekel-Conrat become pioneers in the opening of a new field of research, but both have continued in their investigations and significant contributions to this field throughout the years.

Venom research has undergone another expansion in recent years by encompassing the area of neuroelectrophysiology. But, no matter what new areas are incorporated in the future, the original efforts of Slotta and Fraenkel-Conrat will stand as the first experiment. For nothing can be accomplished without first isolating and characterizing the lethal components. To these two men, this chapter is dedicated.

Isolation and Characterization

The original isolation of crotoxin by Slotta and Fraenkel-Conrat (1938) was through utilization of heat coagulation, followed by isoelectric precipitation. The resulting material was referred to as amorphous crotoxin, and it was found, by this method, that 42% of the dry content of crude venom could be isolated (Slotta, 1955). An alternate method, ammonium sulfate saturation using three steps, was found to produce a total of 37%. Crystallization of crotoxin was achieved by treating a solution of amorphous crotoxin in 1% acetic acid at 55°C, followed by pyridine titration to pH 4.4 (Slotta, 1955). This solution, cooled slowly, would result in the crystallization of crotoxin after approximately 30 min. Following 5 recrystallizations, 74 mg of crotoxin was obtained from 140 mg amorphous crotoxin (Slotta, 1955). The protein was also found to need ionic strength for solubility as only 2.3 mg was soluble in 4% NaCl (Slotta, 1955).

The molecular weight of the new toxin was investigated by Gralen and Svedberg (1938) using sedimentation and diffusion experiments in the ultracentrifuge. The result was that crotoxin behaved as a homogeneous molecule with a computed molecular weight of 30,000.

Following this, Li and Fraenkel-Conrat (1942) pursued the homogeneity of crystalline crotoxin by Tiselius electrophoretic experiments. No evidence of heterogeneity was found, and the isoelectric point was determined to be 4.71. Thus, early on crotoxin was considered a homogeneous molecular species with a definite acidic nature.

Table 1 Amino Acid Composition of Crotoxin and Crotoxin Subunits

Amino acid	Crotoxin	Crotoxin	Crotoxin	Crotoxin A subunit	Crotapotin	Crotoxin B subunit	Crotoxin B from sequence	Crotoxin phospholipase A
Lysine	12	11	13	2	2	9	10	11
Histidine	4	3	3	1	1	2	2	2
Arginine	14	10	14	2	2	8	9	12
Aspartic acid	26	19	23	10	12	9	11	11
Threonine	12	10	12	4	4	6	7	8
Serine	14	11	13	5	6	6	5	7
Glutamic acid	27	20	24	13	14	8	9	10
Proline	13	10	10	5	5	5	4	5
Glycine	25	19	23	9	10	10	11	13
Alanine	15	11	13	5	6	6	6	7
Valine	4	3	3	1	1	2	2	2
Methionine	4	2	3	1	1	2	2	2
Isoleucine	7	7	7	3	2	4	5	5
Leucine	8	7	8	1	1	6	7	7
Tyrosine	15	11	15	3	3	9	11	12
Phenylalanine	17	8	10	3	3	6	6	7
Half-cystine	30	21	30	10–11	14	10–11	14	16
Tryptophan	6	3	4	1	1	2	3	3
	253	186	228	79–80	88	109–110	123	140
	Fischer and Dorfel (1954)	Hendon and Fraenkel-Conrat (1971)	Breithaupt et al. (1974)	Hendon and Fraenkel-Conrat (1971)	Breithaupt et al. (1974)	Hendon and Fraenkel-Conrat (1971)	Fraenkel-Conrat et al. (1980)	Breithaupt et al. (1974)

Amino acid composition was analyzed by one- and two-dimensional paper partition chromatography (Slotta and Primosigh, 1951). According to these data the amino acid composition of crotoxin includes arginine, lysine, aspartic and glutamic acids, phenylalanine, tryptophan, tyrosine, glycine, alanine, proline, serine, threonine, cystine, and methionine. Using special techniques of one-dimensional chromatography, it was also demonstrated that crotoxin contained valine, histidine, leucine, and isoleucine. It was known that sulfur occurred in crotoxin as a normal disulfide bridge rather than as thiolactone (Slotta and Fraenkel-Conrat, 1937, 1938). From amino nitrogen determinations it was felt that more aspartic and glutamic acid residues were present than were found.

A complete quantitative amino acid analysis was performed by Fischer and Dörfel in 1954 using a one-dimensional chromatographic technique followed by colorimetric determination of the individual amino acids as copper salts after ninhydrin development. Table 1 compares this analysis with the more recent analyses of Hendon and Fraenkel-Conrat (1971), Breithaupt et al. (1971), and Fraenkel-Conrat et al. (1980).

Investigations on crotoxin continued in 1950 with the finding by H. and J. Fraenkel-Conrat that chemical modification serves to totally deactivate the crotoxin molecule. Thus, crotoxin appears extremely sensitive to almost any type of chemical interference, with resultant abolishment of toxicity. However, it was also found that crotoxin was relatively temperature insensitive up to 70-75°C. This is unique in that temperatures that normally denature most biologically active proteins are actually utilized in the isolation scheme of crotoxin.

During the early 1950s questions began to emerge concerning the homogeneity of the crotoxin molecule. Ghosh and De, as early as 1939, had attempted to separate the possible complex by salt precipitation. However, the results were inconclusive concerning the homogeneity of the molecule.

Slotta and Fraenkel-Conrat (1939) had proposed the possibility that crotoxin could be a mixture of two proteins. Then, in 1953, Slotta proposed that uniform and pure protein molecules can be composed of subunits possessing the same or different biological activities. This suggestion, which was later found to be remarkably accurate, was that crotoxin could represent an easily crystallizable saltlike compound between an acidic protein (which was considered a hemolysin) and several basic polypeptide groups which were responsible for the toxic effect. Neumann and Habermann (1955) also suspected that the crotoxin molecule was, in actuality, composed of more than one distinctly dissimilar molecular species. Data to support this came in 1954 and again in 1956 from experiments of Fraenkel-Conrat and Singer, Neumann and Habermann (1955), and Habermann (1957). In both cases fluorodinitrobenzene was used to block amino groups, with the resultant separation of two distinct fractions, one precipitating

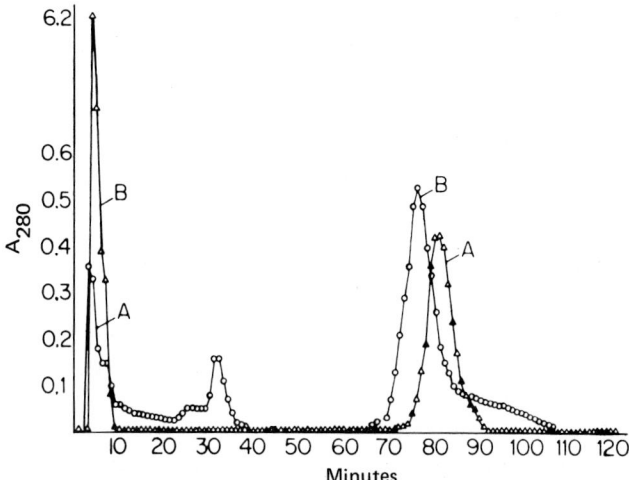

Figure 1 Chromatographic separation of crotoxin components. △–△ DEAE cellulose (urea, pH 6.0) chromatography of 30 mg of crotoxin: 0.9 × 30 cm column; 20 ml/hr flow rate; 2 ml fraction; 400 ml, 0 to 0.5 M NaCl linear gradient. O–O Carboxymethyl cellulose chromatography in ammonium formate buffer, pH 4.0 (0.1 M): 30 mg crotoxin: 0.9 × 30 cm column; 20 ml/hr flow rate; 4 ml fractions; 600 ml, 0 to 2 M NaCl linear gradient. (From Hendon and Fraenkel-Conrat, 1971.)

while the other remained soluble. Amino acid compositions revealed the insoluble fraction to be a basic protein, rich in arginine and lysine and low in dicarboxylic acids. Conversely, the water soluble fraction was acidic and contained little lysine and arginine. This fraction was shown to have serine as the N terminal and aspartic acid as the C terminal, followed by threonine, serine, and phynylalanine. Molecular weights were determined as 12,000 (water soluble) and 18,000 (water insoluble). However, work was not further pursued because the severity of the fluorodinitrobenzene (FDNB) reaction resulted in two biologically inactive molecules.

Attempts at further isolation and characterization of the "now supposed" crotoxin complex slowed owing to the difficulty of separation of the two components by chemical means without loss of biological activity. Crotoxin awaited the development of refined physiochemical methods, specifically column chromatographic techniques, and more gentle chemical modification methods.

In 1971 two laboratories working independently of one another succeeded in separating crotoxin into its individual subunits. Hendon and Fraenkel-Conrat (1971) achieved the separation using fluorodinitrobenzene resulting in the same biologically inactive water soluble and insoluble fractions, with additional confirmation utilizing maleic anhydride to block amino groups. Further evidence was achieved by employing a gentle reversible block on the amino groups with methyl maleic anhydride; thus, chemical modification with resultant separation and recovery of biologically active subunits was shown. Separation was also achieved by the physicochemical methods of column chromatography using carboxymethyl cellulose at low pH and DEAE-cellulose in 6 M urea (Figure 1). It was found that crotoxin was, indeed, a complex consisting of two totally dissimilar proteins, and that reversible denaturation did not affect the activity of the molecular association. Habermann and Rübsamen (1971) had reported on the separation of crotoxin into multiple components, although no definitive data were shown. Breithaupt et al. (1971) succeeded in separation of the complex to individual subunits using column chromatographic techniques with recovery of biologically active subunits. With the latest data it was shown that crotoxin actually consisted of the strongly basic phospholipase A with relatively low toxicity and an acidic protein that was biologically inactive.

Both laboratories found the basic protein possessing the phospholipase A and indirect hemolytic activity of the complex, and upon recombination that toxicity was restored to the same level as that of native crotoxin (unfractionated crotoxin).

Unfortunately, nomenclatural dissimilarities appeared in the literature which have led to considerable confusion. The following is a table for clarification for the discussion in the remainder of this chapter.

Hendon and Fraenkel-Conrat	Breithaupt and Habermann
Crotoxin	Crotoxin
Crotoxin A, CA, acidic subunit	Crotapotin
Crotoxin B, CB, basic subunit	Phospholipase A
	Crotactin (believed to be a mixture of crotapotin and PLA and since discarded from use)

From this point on we will refer to the acidic subunit as crotoxin A and the basic subunit as crotoxin B.

These findings represented the end of considering crotoxin a homogeneous, monodisperse protein species. Crotoxin, in this early sense, became and is a misnomer and now represents a molecular complex consisting of two markedly distinct and dissimilar proteins interacting to express the full biological activity long associated with crystalline crotoxin. The investigation immediately turned to physical characterization of the individual subunits. Hendon and Fraenkel-Conrat (1971) found the ratio of acidic to basic amino acids, disregarding the presence of amide groups, to be 5.0. Crotoxin A contains one tryptophan per 8400 daltons and B contains two valines, methionines, histidines, and tryptophans per 13,000 daltons.

The crotoxin complex subjected to the accepted treatment in preparation for SDS gels (heating in 1% SDS, 1% mercaptoethanol and electrophoresed in 0.2% SDS) showed one strong band corresponding to a protein of molecular weight 13,000, with analyses of the separated crotoxin B fractions giving similar results. It was found that crotoxin A, subjected to mercaptoethanol, did not appear on polyacrylamide gels. However, if the mercaptoethanol was omitted, a diffuse band was observed. It was suggested that crotoxin A was composed of short chains held by disulfides, and in the presence of mercaptoethanol, the disulfides were sheared, with the chains becoming too small for detectable staining or migrating completely through the gel. This was confirmed by Horst et al. (1972) and Breithaupt et al. (1974) with the finding that crotoxin A consists of three separate peptide chains held together by disulfide bridges (Table 2).

Molecular weights were assigned on the basis of SDS electrophoretic data corroborated by gel filtration data on G-75 Sephadex in guanidine-HCl (Horst et al., 1972). Gel filtration also provided additional evidence for the existence of disulfide bridges connecting individual crotoxin A chains. Experiments with crotoxin A revealed a protein of 8000-10,000 molecular weight only when mercaptoethanol was omitted. These experiments assigned molecular weights of 9000 to crotoxin A and 13,000 to crotoxin B. Isoelectric focusing gels gave isoelectric points of 3.7 and 8.6, respectively, for the two subunits. Breithaupt et al. (1974) reported pI values of 3.2 and 9.7 with molecular weights of 8900 and 14,500, respectively, for the individual subunits.

In 1980 the sequence of crotoxin B was completed and reported by Fraenkel-Conrat et al. (Figure 2). From the sequence it was found that previously reported amino acid analysis values (Hendon and Fraenkel-Conrat, 1971) were low by approximately 11% (see Table 1). It was also suggested that the middle portion of the sequence, representing the most variable portion of the molecule and, presumably, the surface of the subunit, could be the binding domain for crotoxin B to crotoxin A.

The original molecular weight of 30,000 assigned to the crotoxin complex has yet to be found. Breithaupt et al. (1971, 1974), with a molecular weight range of 23,000-24,000, and Hendon and Fraenkel-Conrat (1971), with a 21,000-22,000 range, could not be explained away. Attempts to explain the difference on the basis of crotoxin being a glycoprotein proved fruitless as experimental tests for a carbohydrate moiety were negative. The discrepancy

Table 2 Amino Acid Composition of Crotoxin A (Crotapotin) Chains A, B, and C

Amino acid	Chain A	Chain B	Chain C	Crotapotin average	Crotoxin acidic subunit
Lysine	1	1	0	2	2
Histidine	1	0	0	1	1
Arginine	1	1	0	2	2
Aspartic acid	5	6	1	12	10
Threonine	2	2	0	4	4
Serine	3	1	2	6	5
Glutamic acid	3	6	5	14	13
Proline	2	1	2	5	5
Glycine	6	3	1	10	9
Alanine	3	3	0	6	5
Valine	0	1	0	1	1
Methionine	0	1	0	1	1
Isoleucine	0	2	0	2	3
Leucine	1	0	0	1	1
Tyrosine	3	0	0	3	3
Phenylalanine	1	1	1	3	3
Half-cystine	1	5	2	14	10–11
Tryptophan	7	0	0	1	1
	40	34	14	88	79–80
	Breithaupt et al. (1974)	Breithaupt et al. (1974)	Breithaupt et al. (1974)	Breithaupt et al. (1974)	Hendon and Fraenkel-Conrat (1971)

between these molecular weights and Svedberg's original finding remain a mystery.

At the present time, crotoxin is considered to be a molecular complex of 22,000–24,000 daltons consisting of two distinctly dissimilar protein subunits, both of which are essential for the full expression of biological activity and which associate by chemical interactions other than covalent bonding. One

Presynaptic Toxins from Venoms 221

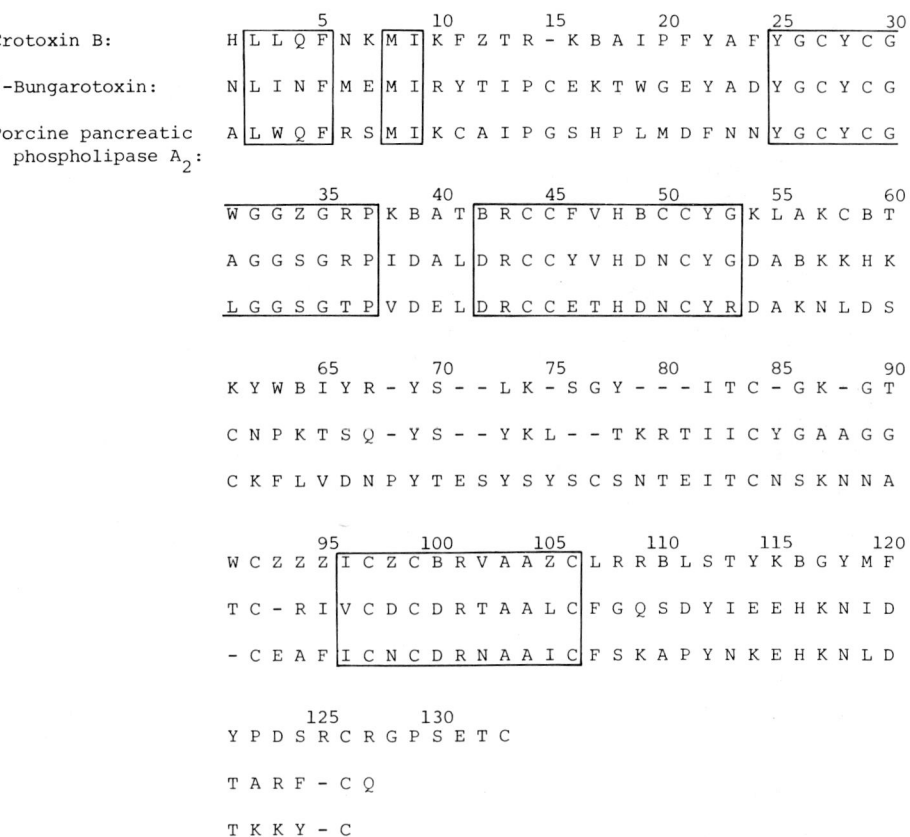

Figure 2 Comparison of amino acid sequences of three phospholipase A_2 active proteins. (From Fraenkel-Conrat et al., 1980.)

subunit, crotoxin A, possesses an isoelectric point representing a highly acidic protein and is composed of at least two and probably three polypeptide chains connected by disulfide bonds. The remaining subunit, crotoxin B, is one polypeptide chain intramolecularly cross-linked by disulfide bridges and is a highly basic protein. The mode and necessity of interaction for these two subunits will be discussed in the section Subunit Interaction.

Neurotoxicity and Phospholipase A_2 Activity

The severe toxicity of *Crotalus durissus terrificus* venom was the major reason for instigating research in this area in the late 1930s. When crotoxin was isolated and found to produce all the symptoms incurred by the victim, it consequently represented the major lethal component of the venom. Table 3 illustrates the toxicity of *Crotalus durissus terrificus,* crotoxin, and the crotoxin subunits as determined by numerous investigators.

Not only is the bite of *Crotalus durissus terrificus* considered severe, it is unique among pit vipers and especially among rattlesnakes. The neurotoxic action is peripheral in origin and is very similar to that elicited by curare (Tu, 1977). However, another crotalid, *Crotalus scutulatus,* has been shown to possess a similar toxin which is the subject of the second half of this chapter. The majority of rattlesnake bites result in extensive tissue damage, swelling, and considerable pain, with the possibility of permanent impairment of locomotor function. Neither the neotropical rattlesnake nor the Mojave rattlesnake induce any of these symptoms. Conversely, there is little swelling or pain and practically no tissue damage, yet the distinct possibility of death looms large because of its neurotoxic action. This neurotoxicity separates *Crotalus durissus terrificus* and *Crotalus scutulatus* from the remainder of the crotalids.

The biological activity of crotoxin is severely diminished or totally inhibited with almost any type of chemical modification (Fraenkel-Conrat and Fraenkel-Conrat, 1950) and serves to distinguish crotoxin from many biologically active proteins (Slotta, 1956). Insulin, for example, needs phenolic, carboxyl, and disulfide groups to be effective; however, amino, hydroxyl, amide, and guanidyl groups do not seem essential. Esterification of carboxyl groups or extensive acetylation of amino groups results in extensive detoxification of crotoxin. Inactivation is also accomplished by sulfation of aliphatic hydroxyls, iodination of the phenolic groups, or coupling of the latter with imidazole groupings. Total inactivation results from formaldehyde treatment, especially in alkaline solutions. These data are taken as indicative proof that some or all indole, guanidyl, and amide groups are essential for the activity of crotoxin (Fraenkel-Conrat and Fraenkel-Conrat, 1950).

During the 1960s, specifically 1966, a series of papers appeared by Brazil investigating the pharmacologic action of crotoxin. The findings indicated species specificity to crotoxin sensitivity, with pigeons being the most sensitive and rats the least. Flaccid paralysis was discerned in all animal species, and it was found that artificial respiration could maintain the animals after cessation of spontaneous respiration, thus indicating the toxin's precise effect on the respiratory center. Dogs exhibited defecation, vomiting, salivation, albuminurea, hemoglobinuria, oliguria, and to some degree, opacity and exfoliation of the cornea. The second paper of the series concerned the neuromuscular effects and

Table 3 Toxicity Comparisons for Crotoxin and Crotoxin Subunits

	LD$_{50}$ (μg/g)	Route of injection
Crotalus durissus terrificus crude venom		
Schöttler (1951)	0.60	i.v.
Brazil (1972)	0.169	i.v.
Friederich and Tu (1971)	0.35	i.v.
(a)	0.30	i.p.
Hendon and Fraenkel-Conrat (1971)	0.12	i.p.
Crotoxin		
Schöttler (unpublished data)	0.27	i.v.
Rübsamen et al. (1971)	0.108	i.v.
Brazil (1972)	0.082	i.v.
Rübsamen et al. (1971)	0.50	s.c.
Hendon and Fraenkel-Conrat (1971)	0.06	i.p.
Crotapotin, crotoxin A		
Rübsamen et al. (1971)	>50	i.v.
Hendon and Fraenkel-Conrat (1971)	>50	i.p.
Crotoxin B, phospholipase A		
Rübsamen et al. (1971)	0.54	i.v.
Hendon and Fraenkel-Conrat (1971)	0.25	i.p.

Source: U.S. Amphibious Forces, U.S. Department of the Navy, Bureau of Medicine and surgery (1968). *Poisonous Snakes of the World: A Manual for Use*. NAV MED P-5099, U.S. Government Printing Office.

concluded that blockage of neuromuscular transmission was probably the sole cause of crotoxin-induced paralysis. This blockade was found to be a nondepolarizing type, with a decrease in sensitivity at the end plate to the depolarizing action of acetylcholine as the major factor involved. It is believed that the affinity of crotoxin for the receptor is a relatively strong one.

Hadler and Brazil (1966) was the first to initiate the use of radiolabeled crotoxin which could be utilized as a highly sensitive probe; this was illustrated later by Jeng et al. (1978) in studies concerning subunit association within the complex. Brazil showed that crotoxin could be iodinated and, most importantly, that the procedure did not alter the biological activities associated with the molecule.

As research progressed from *Crotalus durissus terrificus* crude venom to an intensive study of crotoxin, the homogeneity of the toxin came under increased scrutiny. Several investigators had proposed that crotoxin was not homogeneous, although the idea was not fact, and the possible multiple components were not available. Thus, at this time, a summary of known facts about crotoxin would be: it is a powerful neurotoxin with associated indirect hemolytic activity; yet, the hemolytic potential is not believed to play a significant role in the lethal activity of the toxin. However, in light of the finding that crotoxin is a complex composed of dissimilar proteins, the investigation probed the individual subunits in attempts to determine which possessed the lethal effect and which possessed the hemolytic effect. The investigation pursued the idea of two activities (neurotoxic and hemolytic); therefore, each subunit must possess one or the other.

Hendon and Fraenkel-Conrat (1971) and Rübsamen et al. (1971) found that separation of the complex resulted in total abolishment of the neurotoxic activity, while crotoxin B retained the full hemolytic potential of the native complex and crotoxin A seemed devoid of any biological activity. Later investigations confirmed that the A subunit alone possessed no enzymatic or toxic activity. The B subunit, however, became the subject of some scrutiny concerning its indirect hemolytic potential. Rübsamen et al. (1971) and Habermann et al. (1972) reported that the phospholipolytic ability of crotoxin B was inhibited upon reintroduction of the A subunit. Hendon and Fraenkel-Conrat (1971) did not observe this effect; however, they did find that the crotoxin B hemolytic ability was far more pronounced than that of purified phospholipase A_2 from *Crotalus adamanteus*. In experiments comparing the two against mammalian erythrocytes, crotoxin B was five times more effective. Conversely, in comparisons of the two against a specific phosphatidylcholine substrate, crotoxin B was five times less effective than pure phospholipase A_2 from *C. adamanteus* (Hendon and Fraenkel-Conrat, 1971). Thus, it appears that the crotoxin B indirect hemolytic potential is phospholipase-like, yet is not equivalent to a true phospholipase A_2.

These findings confirmed previous observations of Slotta (1953) that the hemolytic potential was not significant in the lethal action of the venom. Table 4 shows the comparison of phospholipase A_2 and the indirect hemolytic and lethal potential of crotoxin B. It can be seen that the B subunit alone is dramatically lacking in toxicity, with purified phospholipase A_2 being totally nontoxic. These findings case doubt on the age-old notion that indirect hemolytic activity must proceed by the following mechanism:

Phospholipase A_2 + phosphatidylcholine = lysophosphatidylcholine
Lysophosphatidylcholine + erythrocytes = hemolysis

This general mechanism, requiring phospholipase A_2, no longer provides a satisfactory explanation per se. Crotoxin B, by the fact that it is extremely weak in phospholipase A_2 hydrolyzing ability, yet very strong in hemolytic

Table 4 Biological Activities of Crotoxin and Crotoxin Subunits

	Toxicity	Hemolytic activity (direct)	Hemolytic activity (indirect)	Phospholipase A_2	Presynaptic transmission blocking ability (prejunctional)[a]	Synaptic transmission blocking ability (postjunctional)[b]
Crotoxin	++++	(−)	++++	++++	++++	++++
Crotoxin A	(−)	(−)	(−)	(−)	(−)	(−)
Crotoxin B	(++)	(−)	++++	+++	++++	++++
PLA_2 (*C. adamanteus*)	(−)	(−)	++	+++++	(−)	(−)
	Hendon and Fraenkel-Conrat (1971); Rübsamen et al. (1971); Breithaupt et al. (1974)	Hendon and Fraenkel-Conrat (1971)	Hendon and Fraenkel-Conrat (1971); Jeng et al. (1978)	Hendon and Fraenkel-Conrat (1971); Breithaupt (1976a,b)	Brazil and Excell (1971); Brazil et al. (1973); Hawgood and Smith (1976) Chang et al. (1977a,b); Hanley (1978); Bon et al. (1979); Gopalakrishnakone and Hawgood (1979); Hawgood and Santana de Sa (1979)	Hanley (1978); Hawgood (1979); Bon et al. (1979)

[a] Prejunctional blocking ability denotes synaptic transmission interference by disturbance of acetylcholine containing presynaptic vesicles.
[b] Postjunctional blocking ability denotes synaptic transmission interference by any means of the postjunctional membrane. It does not necessarily involve the cholinergic receptor.

potential, alludes to other factors as yet uncovered that are involved in the mechanism.

Later work by Jeng et al. (1978) showed that with in vivo conditions crotoxin and crotoxin B were more phospholipolytic than purified phospholipase A against both synthetic phosphatidylcholine and egg yolk by adding calcium and sodium deoxycholate to the suspension. However, with assays under physiological conditions in which purified phospholipase A_2, crotoxin, and crotoxin B were allowed to digest the substrate, and the lysophosphatidylcholine was isolated and tested hemolytically, it was found that phospholipase A_2 was effective. These experiments concluded that both crotoxin B and pure phospholipase A_2 could hydrolyze a phospholipid substrate without the release, to the medium, of a titratable proton.

These data lend credence to the idea that crotoxin B is a phospholipase-like agent, but the effects, hemolytically, are more severe than actual phospholipolytic ability warrants. Possibly, crotoxin B, being basic and possessing a strong positive charge at physiological pH, attaches to the erythrocyte membrane, thereby attacking not only the phosphatidylcholine in the suspension but also the phospholipid in the membrane. This would explain, to some extent, the discrepancy between phospholipolytic and hemolytic ability. Yet, this does not explain the equivalent hemolytic potential of the complex, since the complex has a pI of 4.7 and would not carry a strong positive charge at physiological pH.

Studies with ^{125}I-labeled crotoxin B and A, the complex and purified phospholipase A_2 from *C. adamanteus*, illuminated this area to some degree. Jeng et al. (1978) found that phospholipase A_2 was not significantly bound to either erythrocyte (RBC) membranes or to RBC ghosts. Conversely, it was observed that both crotoxin and crotoxin B were richly bound to these membranes. Thus, it was initially confirmed that both crotoxin and crotoxin B possessed considerable affinity for membranes. These studies also revealed that crotoxin dissociated upon binding, i.e., using ^{125}I-labeled crotoxin A and unlabeled crotoxin B for the reconstituted complex, little or no activity was found associated with the membranes. However, with ^{125}I-labeled crotoxin B and unlabeled crotoxin A, significant levels of activity were found, indicating only the B subunit binds to the membrane. For this to be the case, dissociation must occur. Thus, it appears that crotoxin and crotoxin B are membrane targeted as would be expected for a neurotoxin. Phospholipase A_2 is more of an exoenzyme; i.e., it functions outside the cells in the physiological medium, and herein lies the difference, at least partially explaining the discrepancy between the phospholipolytic and hemolytic abilities of the two proteins.

Focusing now on the lethal activity of crotoxin, both Hendon and Fraenkel-Conrat (1971) and Rübsamen et al. (1971) found that the complex would dissociate under rather mild conditions of column chromatography as well as using 6-8 M urea. Thus, the interaction of the subunits occurs by a mechanism other than covalent bonding. It was also found that neither subunit possessed lethal toxicity approaching that of the complex; crotoxin A was nontoxic, whereas crotoxin B was barely toxic. If toxicity was not associated with either subunit, then this capacity must be a product of the association of the two resulting in the proposal of synergistic interaction. Experiments confirmed that the two subunits could be reconstituted and, upon reconstitution, full neurotoxicity was restored to the complex. Thus, neurotoxicity was not a function of one of the subunits but of both. Somehow the interaction of crotoxin A and crotoxin B provides a necessity for the complex which allows the expression of neurotoxicity. The manner in which this occurs is not fully understood as yet; however, a more detailed discussion follows in the section Subunit Interaction.

Investigation also revealed that the optimum recombination ratio is, essentially, 1 mol crotoxin A and 1 mol crotoxin B. More specifically, the ratio is 0.9 mol A to 1.1 mol B. From a biological standpoint, only crotoxin B possesses significant activity (Table 4). Crotoxin A, according to the tests that have been utilized, is devoid of detectable activity. Yet, crotoxin A is necessary for the complex to express full neurotoxicity, and, in this respect, must be considered biologically essential.

Subunit Interaction

The well established fact that crotoxin is a complex composed of subunits, neither of which possess lethal potential but which when acting synergistically represent a very powerful neurotoxin, caused the unique interaction of the individual subunits to become increasingly significant.

Paradies and Breithaupt (1975) found that the complex and crotoxin A could be described as spherical in shape, whereas crotoxin B is highly asymmetric. Also, crotoxin B was found to undergo a reversible equilibrium between a dimer and tetramer, an equilibrium governed by concentration, pH, and ionic strength. Thus, the conclusions indicate that the crotoxin complex consists of two physically and chemically different molecules: one an oblate ellipsoid and the other highly asymmetric.

Jeng and Fraenkel-Conrat (1976) found that volvatoxin, a mushroom toxin consisting of two proteins, could be separated into its constituents, and that the smaller component, mol. wt. 25,000, was able to enhance the low toxicity of crotoxin B. Of interest is that the combining ratio is 1:1. This does represent a specific interaction insofar as crotoxin B is not similarly activated, nor complexed, by other acidic proteins (Horst et al., 1972).

Hendon and Fraenkel-Conrat (1976), in studies on the role of complex formation in neurotoxicity, observed indications suggesting that the crotoxin complex did not remain associated at the target for the expression of neurotoxicity. Experiments designed to prevent complex formation included injections of the two components by separate routes and time intervals: separate intraperitoneal (i.p.) injections at various time intervals and the use of 6-8 M urea to prevent complex formation. In all cases lethality was observed, although the possibility of complex formation occurring intraperitoneally could not be ruled out.

To counteract this possibility a time to death assay was employed. Eterovic et al. (1974) suggested that α-type bungarotoxins, regarded as typical postsynaptic toxins, killed mice much more rapidly than β-type (presynaptic neurotoxins), although the LD_{50} of presynaptic toxins are usually lower than postsynaptic toxins. The crotoxin complex was found to be intermediate, i.e., 20-30 min, and resembled the β-type in that death could occur over a 24 hr period. α-type always kills within 5 hr. This test was used as a standard against which separate injections of subunits A and B in 6 M urea were compared. The time to death using this procedure was comparable to that of native crotoxin, i.e., 27 min.

Thus, indications existed that the maintenance of the crotoxin complex at the target was not essential for neurotoxicity, although the data were not conclusive as it is well established that both subunits are required for neurotoxicity.

As work continued on subunit interaction, Hawgood and Smith (1976) and Hanley (1978) uncovered another previously unknown biological function for crotoxin B. In addition to phospholipase A_2 and hemolytic activity, synaptic transmission blocking ability was found to reside on the B subunit. Once again the A subunit was necessary for the expression of full lethality, whereas in its separated state it was found to possess no other identifiable function.

Hanley (1979), using fluorescence and circular dichroism, found that model spectra of the complex generated from expected contributions of subunits A and B agreed experimentally only at pH 2, where crotoxin is fully dissociated. Within the pH range of 4-10 crotoxin spectra could not be simulated by subunit contributions. Data also indicated that tryptophan residues were masked by complex formation, and that association of the subunits into the complex increased the proportion of ordered structure, especially β-sheets. Unfortunately, conformational changes resulting from subunit interaction could not be distinguished. Flexibility of the subunits was indicated from their ability to undergo reversible state changes by exposure to the denaturing action of guanidine HCl and acid pH. Complex formation, as expected, provided increased stability, and the individual subunits were found to be considerably resistant to guanidine-HCl denaturation.

Hanley concluded that the reversible states are related to the association-dissociation equilibrium of the complex, and that conformational interconversion is probably a necessity for crotoxin to act on different targets.

The documented physical information confirmed the chemical and physicochemical information in establishing that the crotoxin complex was just that, a complex of two physically and chemically dissimilar proteins, associating by an interaction other than covalent bonding. Upon separation, the acidic subunit was found lacking in any detectable biological activity, whereas the basic subunit possessed the enzymatic (phospholipase A_2), indirect hemolytic, and synaptic transmission blocking activities. Indications existed suggesting that the complex, at the target site, dissociates.

Jeng et al. (1978) and Bon et al. (1979) proposed that crotoxin A may function as a "chaperone," improving the targeting ability of the active B subunit by limiting nonspecific absorption. The suggestion was strengthened with their finding, using erythrocytes and electroplaques, that crotoxin A is released after the active subunit B has bound to the membrane.

Hendon and Tu (1979) added more support to the chaperone concept by converting the dissociable crotoxin complex to a nondissociable complex. If the complex must dissociate for lethal activity to be expressed, then inhibition of this dissociation would, consequently, eliminate the neurotoxic potential. Using dimethyl suberimidate, a bifunctional cross-linking agent, individual subunits were bound to one another irreversibly, with the significant effect that this agent allows cross-linking with no change in molecular charge. The existence of an inter-subunit cross-linked complex was shown by molecular sieve chromatography, isoelectric precipitation, and capillary isotachophoresis. Once it was known that the irreversibly cross-linked complex was achieved, biological testing began.

Intraperitoneal injection of the unpurified reaction mixture indicated a loss of neurotoxicity of approximately 80%. On purification, i.e., removal of unreacted molecules, neurotoxicity was no longer observed. Since chemical modification is well established as an inhibitor of crotoxin activity, this possible inactivation was addressed by assays for phospholipase A_2 activity.

It is well known that many presynaptic neurotoxins have associated phospholipase A_2 activity, and that this activity plays a significant role in the disruption of synaptic transmission by depletion of acetylcholine vesicles in the presynaptic terminals. Thus, phospholipase A_2 activity must be considered an integral part of presynaptic neurotoxic activity. Crotoxin has been established as a presynaptic neurotoxin (Brazil and Excell, 1971; Brazil et al., 1973; Hawgood and Smith, 1976; Chang et al., 1977a and b; Hanley, 1978; Gopalakrishnakone and Hawgood, 1980; Hawgood and Santana de Sa, 1979; Bon et al., 1979). Since crotoxin is a presynaptic neurotoxin, and phospholipase A_2 activity is associated with the complex, then a measure of the phospholipase activity could serve as an indicator of neurotoxic activity.

Assays of the cross-linked complex for phospholipase A_2 activity showed the cross-linked complex possessing equivalent phospholipase activity as the native

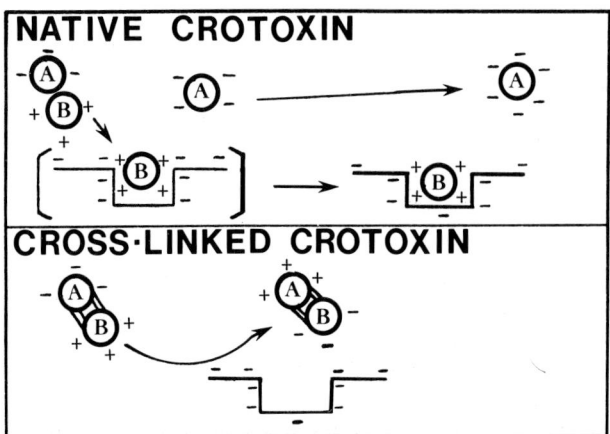

Figure 3 Proposed model for the mechanism of neurotoxic action by crotoxin. Native crotoxin: moves through physiological medium as negatively charged complex; dissociates at or near the presynaptic membrane target site; subunit B binds to the presynaptic receptor site; subunit A is released. Cross-linked crotoxin: covalently linked complex is unable to dissociate near the presynaptic membrane target. Consequently, no binding at the receptor site occurs. Nonneurotoxic. (From Hendon and Tu, 1979.)

complex. This suggests that the cross-linked crotoxin, albeit indirectly, still retains the potential for neurotoxicity. However, since the complex was found to be nonneurotoxic, this loss must be accounted for by some mechanism other than the physical presence of cross-linker.

These findings, although not definitive proof, do add support and credence to the proposed chaperone concept for crotoxin A. The proposed function of crotoxin A would be to decrease the nonspecific binding of crotoxin B by charge neutralization or masking until the target site is reached. This is somewhat supported by the findings of low toxicity for the basic subunit, i.e., requiring a much greater dose level for toxicity to be expressed. Thus, without crotoxin A the basic subunit becomes, essentially, deleted before reaching the target site. This would necessitate higher levels to overcome the deleting effect of nonspecific binding, allowing some molecules to arrive at the target. A model was proposed to explain the possible action of the complex and is shown in Figure 3.

Essentially, the association of the subunits provides a negatively charged molecular complex which would be repelled by the negatively charged membranes of the physiological system. At or near target, crotoxin A, by some mechanism that remains unclear, although steric factors are most probably involved, begins to dissociate. This would unmask crotoxin B, which, as dissociation continued, would become more progressively positively charged. By electrostatic interaction crotoxin B would be attracted to the membrane receptor site, and crotoxin A would be released to the surrounding physiological medium. Irreversible cross-linking would serve to inhibit neurotoxicity in this proposed mechanism by (1) hampering recognition of the binding site stearically and/or electrostatically and (2) preventing the dissociation necessary for the electrostatic interaction of crotoxin B with the receptor site, the result being, as was found, total abolishment of neurotoxicity.

This, essentially, is the status of crotoxin at the present time. *Crotalus durissus terrificus* from 1938 to 1975 enjoyed the notoriety of being the only known member of the Crotalidae to possess a neurotoxic venom. However, in the mid-1970s, investigators became suspicious of other crotalids, especially those known to inflict a dangerous bite. These suspicions led to investigations of the venom of *Crotalus scutulatus* (Bieber and Tu, 1974, Bieber et al. 1975; Hendon, 1975), with the finding that the Mojave rattlesnake also possesses a neurotoxic component. More extensive studies have confirmed these findings and revealed a presynaptic, acidic, subunit toxin bearing many similarities to crotoxin. Crotoxin no longer enjoys its former dubious distinction, and it is believed by these authors that as research proceeds more rattlesnakes will be added to the lists of neurotoxic venoms. However, let us now turn to a more extensive discussion of Mojave toxin from the venom of *Crotalus scutulatus*.

MOJAVE TOXIN

Brief Historical Perspective

The Mojave rattlesnake, *Crotalus scutulatus scutulatus*, was first described by Kennicott (1861). Venom from this rattlesnake species has a lower LD_{50} value than most of the other rattlesnake species found in North America (Russell, 1967). Bieber et al. (1975) reported an LD_{50} value of 0.18 mg/kg in white mice injected intravenously (i.v.) with Mojave venom obtained from the Miami Serpentarium. Glenn and Straight (1977) used venoms from individual specimens (California) and demonstrated average LD_{50} values in white mice of 0.24, 0.7, and 0.15 mg/kg for interperitoneal, intramuscular (i.m.), and intravenous injection routes, respectively. More recent studies by Glenn and Straight (1978) showed that venoms from *C. scutulatus scutulatus* specimens collected in southwestern Arizona and southern California (venom A) were more potent than venom obtained from specimens collected in central and northwestern Arizona (venom B). Two specimens collected in Utah provided venoms with low LD_{50} values and were designated as having venom A. One sample of venom from *Crotalus scutulatus salvini* presented an LD_{50} value of 0.18 mg/kg. The summary of results presented in Table 5 illustrates the differences that were found by Glenn and Straight (1978). The data clearly indicate a geographic variation

Table 5 Summary of LD$_{50}$ Values for Crude Venoms from *Crotalus scutulatus*

Crude venom	Number of individual specimens tested	i.p. LD$_{50}$ (mg/kg)	LD$_{50}$ range (mg/kg)	Collection region
C. scutulatus scutulatus				
venom A	28	0.24	0.13-0.54	southern California southwest Arizona
venom B	8	2.80	2.3-3.9	northwest Arizona central Arizona
venom A	2	0.11	0.09-0.12	Utah
C. scutulatus salvini	1	0.18		

Source: Reprinted with permission from J. L. Glenn and R. C. Straight (1978). Mojave rattlesnake *Crotalus scutulatus* venom: Variation in toxicity with geographical origin. *Toxicon 16*:81-84, Copyright 1978, Pergamon Press, Ltd.

in the lethality of venom samples obtained from the Mojave rattlesnake. The distinctive nature of venom A and venom B was substantiated by testing the efficacies of commercial antivenin and of plasma from *C. atrox* in neutralizing individual venom samples from Mojave rattlesnake specimens. Both preparations were more effective against venom B than against venom A.

The clinical symptoms elicited after envenomation by Mojave rattlesnake usually present a lower degree of pain, edema, and local tissue damage than most other rattlesnake envenomations. However, severe neurologic effects are known to result from Mojave envenomation (Russell, 1969; Russell et al. 1975). Because Mojave venom is more lethal and presents clinical symptoms different from those usually observed after rattlesnake envenomation, interest in this venom has been high in recent years. It seemed to be a likely source from which unique, potent, physiological substances might be isolated.

Isolation

Mojave Toxin

Little information about the biochemical nature of the active components in Mojave venom was available when Bieber and Tu (1974) reported the isolation of a potent protein toxin from the venom. Since the first report several fractionation schemes from different laboratories have been described in the literature, and these are discussed below.

Bieber et al. (1975) presented the details of the procedures used to isolate the most lethal protein component from Mojave venom. The purification scheme involved ion exchange chromatography on DEAE Sephadex and preparative isoelectric focusing. Cate and Bieber (1978) refined the purification method to provide the quantities of highly purified toxin necessary to undertake detailed structural studies. The purified toxin was treated with 6 M urea, and the resultant solution was subjected to DEAE chromatography in the presence of 6 M urea. Distinct, separable, highly purified subunits were obtained from the toxin by this procedure.

Hendon (1975) fractionated venom from Mojave rattlesnake by two methods. Elution from DEAE cellulose with a salt gradient gave five peaks, of which only one was lethal to mice. Purity of the lethal fraction was not assessed. A second method involved isoelectric precipitation, followed by chromatography of the isoelectric precipitate on DEAE cellulose in the presence of 6 M urea. The latter method yielded two chromatographic fractions. Neither fraction was toxic to white mice.

Pattabhiraman and Russell (1975) also fractionated venom from *C. scutulatus scutulatus*. Gel filtration on Sephadex G-100 yielded several peaks, of which two (B and C) were more lethal than the crude venom. Peak B was subjected to chromatography on DEAE Sephadex. Of the numerous peaks obtained from the DEAE column, only one, subpeak K, had a lethality greater than the crude venom. Data pertaining to the purity of the lethal fractions were not presented.

Nair et al. (1979) isolated a phospholipase A_2 in pure form from the venom of *C. scutulatus salvini*, the other known subspecies of *Crotalus scutulatus*, by a combination of gel filtration and ion-exchange chromatography. Data pertaining to lethalities of the crude venom and the purified phospholipase A_2 were not presented.

Subunits

The lethal protein isolated by Bieber et al. (1975) was shown to be an acidic protein by sucrose gradient preparative isoelectric focusing. A molecular weight of approximately 22,000 was reported on the basis of gel filtration studies. The reduced toxin yielded a single band that corresponded to a molecular weight of approximately 12,000 after electrophoresis in the presence of SDS. These data suggest that the toxin consists of subunits. Subsequent detailed studies by Cate and Bieber (1978) established that the native toxin consists of one acidic subunit (pI 3.6) and one basic subunit (pI 9.6). The subunits were resolved by chromatography on DEAE Sephadex in the presence of 6 M urea. The amino acid compositions and molecular weights of the toxin and its subunits are presented

Table 6 Amino Acid Composition of Purified Proteins from *Crotalus scutulatus* Venoms

Amino acid	Native[a] toxin	Basic subunit[a]	Acidic subunit[a]	Recombinant toxin[a]	Minor isotoxin[b]	*C. scutulatus salvini* phospholipase[c]
Lys	13	10	2	13	11	12
His	3	2	1	3	3	4
Arg	11	9	2	11	10	27
Asx	23	10	13	23	21	26
Thr	12	8	5	12	11	11
Ser	11	7	4	10	10	13
Glx	22	9	13	22	19	28
Pro	14	7	6	13	12	16
Gly	24	12	12	24	21	23
Ala	13	7	6	13	11	14
Half-cys	22	12	12	24	17	23
Val	4	2	2	4	5	6
Met	3	2	1	3	3	2
Ile	8	5	3	8	8	12
Leu	8	6	1	8	8	13
Tyr	16	12	4	16	13	15
Phe	8	6	2	8	8	8
Trp	3	2	1	3	—	—
Total residues	218	128	90	218	191	253
M_r	24,310	14,673	9,593	24,333	21,094	

[a]Cate and Bieber, 1978.
[b]Cate, 1977.
[c]Nair et al., 1979.

in Table 6. The results clearly demonstrate that the Mojave toxin is composed of two distinct kinds of subunits that are combined in a one to one ratio in the native toxin. Cate and Bieber (1978) also presented data that showed that reduction of the acidic subunit gave rise to smaller peptides, one with a molecular weight of approximately 1400 and two with molecular weights around 3900. These conclusions were derived from gel filtration studies of the reduced, alkylated, acidic subunit in the presence of 6 M guanidinium hydrochloride.

Electrophoresis and electrofocusing of the isolated lethal fractions in polyacrylamide gels showed that a major toxin and a "minor isotoxin" were present (Cate and Bieber, 1978; Cate, 1977). The results showed that the major toxin had an isoelectric point of 5.5 in this system, whereas the "minor isotoxin" had an isoelectric point of 5.7. The composition of the toxins also differed as shown in Table 6. The amino acid composition of the minor component is similar to that of Mojave toxin. However, it contains fewer amino acid residues. It is possible that this derives, in part, from loss of part of the acidic subunit by reduction. This could also account for the higher isoelectric point of the minor component, since loss of an acidic fragment would be involved. More detailed experimental work will be required to establish the relationships between these two toxic components.

Some spectral characteristics of Mojave toxin have been examined. Tu et al. (1976) obtained laser Raman spectra on solid samples and aqueous solutions of Mojave toxin that were isolated by the procedure described by Bieber et al. (1975). On the basis of the observed Raman frequencies and intensities, the authors concluded that the toxin has predominantly α-helical secondary structure and that the tyrosine residues are exposed to solvent (Figure 4). The spectra also indicated the presence of disulfide bridges and the absence of free sulfhydryl groups. The circular dichroism spectra of the native toxin, recombinant toxin, acidic subunit, and basic subunit have been obtained (Cate, 1977; Cate and Bieber, 1978). The results demonstrated that substantial differences exist between the spectra of the native toxin and the individual subunits in the long wavelength region of the ultraviolet spectra. However, recombination of the acidic and basic subunits leads to restoration of the native structure. This conclusion is supported by toxicity studies that will be mentioned later. The information currently available suggests that substantial structural changes may occur during the subunit association to form recombinant toxin.

Pattabhiraman et al. (1978) examined the lethal component in peak C mentioned earlier in the section Mojave Toxin. Their electrophoresis results showed the component to be a highly basic peptide. Amino acid analysis data were presented that indicated that 75-78 residues were present in the peptide. The molecular weight based on gel filtration studies was reported to be approximately 9000.

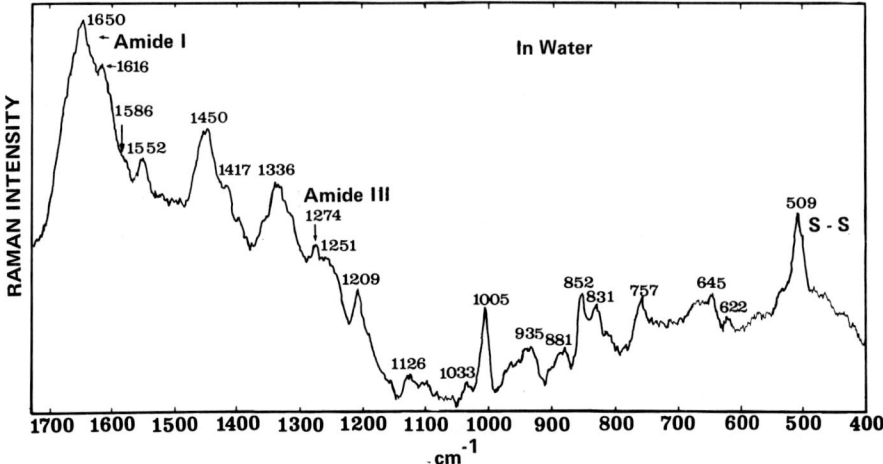

Figure 4 Raman spectrum at 32°C of Mojave toxin isolated from *C. scutulatus* venom (400–1700 cm^{-1}). Aqueous solution at pH 6.5, 74 mg toxin per milliliter. Conditions: excitation wavelength 514.5 nm; radiant power 300 mW; spectral slit width 10 cm^{-1}; scan speed 25 cm^{-1} min^{-1}; rise time 10 sec; amplification 1. (From Tu et al., 1976.)

The only other highly purified protein from *Crotalus scutulatus* venom to be characterized to any degree is the phospholipase A_2 from *C. scutulatus salvini* (Nair et al. 1979). The protein was judged to be homogenous by electrophoresis studies. The reported amino acid composition is presented in Table 6. The molecular weight derived from gel filtration studies was estimated to be 32,000, a value in good agreement with the molecular weight of 30,000 obtained by sedimentation equilibrium measurements. Electrophoresis of reduced phospholipase in the presence of SDS on acrylamide gels gave a single protein band that corresponded to a molecular weight of approximately 14,000. End group analysis indicated the presence of two amino terminal residues, lysine and serine. These data indicate the phospholipase A_2 from *C. scutulatus salvini* is a dimer composed of two nonidentical subunits, each with a molecular weight of approximately 15,000. The physical properties and structure of this phospholipase A_2 appear to correspond to those observed for purified phospholipases from *C. adamanteus* venom (Wells and Hanahan, 1969) and *C. atrox* venom (Hachimori et al., 1971).

Biochemical Studies

Very little information is available concerning the biochemical aspects of the Mojave venom. Esterolytic, proteolytic, and phospholytic activities have been measured with crude venom from *C. scutulatus scutulatus* (Kocholaty et al.

1971). Compared with other *Crotalidae* venoms, phospholytic activity was quite high, whereas the esterolytic and proteolytic activities were found at moderate levels.

The lethal toxin from the venom of *C. scutulatus scutulatus* is responsible for at least part of the phospholipase activity that the venom contains (Cate and Bieber, 1976; Cate, 1977; Cate and Bieber, 1978). The available data clearly demonstrate that the major lethal component of Mojave venom is associated with a phospholipase activity (Cate, 1977; Cate and Bieber, 1978), and that the phospholipase activity, in turn, is associated with the basic subunit of the toxin. The basic subunit is toxic (LD_{50}, 0.48 mg/kg) but much less so than the native toxin (LD_{50}, 0.056 mg/kg). The acidic subunit was totally devoid of enzymatic activity and did not kill any mice at any dose given up to 10 mg/kg. The acidic subunit, however, did reduce both the phospholipase activity and the LD_{50} of the basic subunit. Thus, the data showed that a reduction of phospholipase activity resulted in increased toxicity, suggesting that the high toxicity of the native toxin is not solely a function of the phospholipase activity.

The effects that the major toxin from Mojave venom produced on sarcoplasmic reticulum vesicles from rat skeletal muscle were determined (Cate and Bieber, 1976). The ability of sarcoplasmic reticulum to sequester Ca^{2+} was impaired, and subsequent release of Ca^{2+} from the sarcoplasmic reticulum was enhanced. These effects were dependent on the concentration of toxin and upon the length of time that the vesicles were exposed to the toxin. An increase in sacroplasmic reticulum ATPase was observed when toxin was present. This response was also dependent on toxin concentration and preincubation time. The basic subunit of Mojave toxin was even more effective than the native toxin in perturbing the Ca^{2+} sequestrations and ATPase of sarcoplasmic reticulum. The acidic subunit, on the other hand, did not significantly affect the Ca^{2+} uptake or ATPase activity of the sarcoplasmic reticulum, even at high levels. In fact, addition of acidic subunit prior to addition of basic subunit reduced the effectiveness of the basic subunit. The recombinant toxin, derived from isolated acidic and basic subunits, produced a response very similar to that observed with native Mojave toxin. It appears that the observed effects may be due to the phospholipase activity of the basic subunit leading the leaky membranes. How this relates to the toxicity of the toxin is not clear.

Of some interest is the fact that the basic subunit provided a time course for phospholipid hydrolysis that differed markedly from that observed with native or recombinant toxins. In the later instances, a substantial lag period was observed in the hydrolysis process. This lag was even more pronounced when acidic and basic subunits were added simultaneously as separate entities. It is possible that this reflects some sort of association-dissociation phenomena. It is usual to see finite amounts of basic or acid material during electrophoresis and electrofocusing studies of the highly purified toxin, evidence that association-dissociation can occur fairly easily under a variety of conditions. Detailed studies of the phospholipase activity and the structure of toxin will be needed to understand these results.

The most definitive study of the physiological effects of Mojave toxin is that of Gopalakrishnakone et al. (1980), and the results are discussed in detail by Hawgood in Chapter 2 of this volume. Of pertinence is the finding of petechial lesions specific to the lung tissue in mice that were injected with Mojave toxin. The authors suggest that these lesions may be related to prostaglandin levels which conceivably could be altered by the phospholipase-mediated release of biosynthetic precursors to prostaglandins.

Immunological Studies

The efficacies of the Wyeth antivenin and of heated plasma from *C. atrox* as agents for neutralization of Mojave venom have been studied (Straight et al., 1976; Glenn and Straight, 1977, 1978). With respect to *C. scutulatus scutulatus* venom, the neutralization capacity was far less for the more lethal variety of Mojave venom (venom A) than for the less lethal venom (venom B) or for *C. durissus terrificus*. The utility of these materials as neutralization agents for *C. viridis concolor* was greater than for Mojave venom (venom A) but less than that for *C. durissus terrificus*. Gopalakrishnakone et al. (1979) reported that rabbit antiserum raised against the crotoxin complex caused immunoprecipitation when tested against Mojave toxin. A single precipitin line was formed during gel immunodiffusion studies in which both crotoxin and Mojave toxin were tested as antigens. These data suggested that, in addition to the aforementioned chemical and structural similarities, antigenic similarities between the toxins existed. One milliliter of the antiserum was able to neutralize an amount of Mojave toxin equivalent to 80 LD_{50} doses in white mice. The crotoxin antiserum also showed moderate antigenic cross-reactivity with venom from *Crotalus horridus atricaudatus* and weak antigenic cross-reactivity with venom from *C. basiliscus*. Antigenic cross-reactivity was not demonstrable with venoms from *Crotalus durissus totonacus, Crotalus horridus, Crotalus viridis viridis, Crotalus atrox, Agkistrodon rhodostoma, Bothrops jararaca,* or *Trimeresurus graminius*. Cross-reactivity was not apparent with purified phospholipases from venoms of *Enhydrina schistosa, Naja nigricollis, Apis mellifera,* or porcine pancreas. Enzyme-linked immunosorbent assay (ELISA) was used to demonstrate that antibodies had been generated to both crotapotin and phospholipase components of crotoxin. Research in the laboratory of Bieber (unpublished results) has generated rabbit antiserum against the basic subunit from Mojave toxin. The antiserum gave rise to precipitin bands when tested against individual venom

Table 7 Cross-Reactivity of Crotalid Venom with Mojave Toxin Basic Phospholipase Antibodies

Venom	LD_{50} ($\mu g/g$)	Antigenic cross-reactivity
Crotalus scutulatus scutulatus (Mojave)	0.18 i.v.	+
C. viridis concolor (midget faded)	0.25 i.p.	+
C. durissus terrificus (South American)	0.35 i.v.	+
C. horridus atricaudatus (canebrake)	0.65 i.p.	+
C. durissus durissus (Central American)	0.67 i.p.	+
C. viridis helleri (southern Pacific)	1.29 i.v.	−
C. viridis viridis (prairie)	1.61 i.v.	−
C. adamanteus (Eastern diamond)	2.2 i.v.	−
C. viridis lutosus (Great Basin)	2.2 i.p.	−
C. viridis nuntius (Arizona prairie)	2.2 i.p.	−
C. viridis cerberus (Arizona black)	2.5 i.p.	−
C. horridus horridus (timber)	2.6 i.v.	−
C. basiliscus basiliscus (Mexican west coast)	2.8 s.c.	−
C. viridis oreganus (northern Pacific)	3.2 i.p.	−
C. atrox (Western diamond)	3.6 i.v.	−
C. molossus molossus (Northern black-tailed)	−	−

samples from several rattlesnake species. However, a greater number of rattlesnake venoms did not cause immunoprecipitation with the antiserum. The results of this preliminary study are shown in Table 7. It is of interest to note that the more lethal venoms, i.e., those with LD_{50} values less than 0.7 mg/kg, all showed cross-reactivity with the antiserum raised with Mojave basic subunit, whereas none of the other rattlesnake venoms resulted in antigenic cross-reactivity. Rabbit antiserum raised against the purified acidic phospholipase from *C. scutulatus salvini* cross-reacted with phospholipases in venoms from *C. atrox*, *C. adamanteus*, and *C. horridus horridus* but did not cross-react with phospholipases in venoms from *C. scutulatus scutulatus*, *C. durissus terrificus*, or *C. basiliscus* (Nair, et al. 1979). Thus, antigenic relationships between *C. scutulatus scutulatus* and *C. scutulatus salvini* are not evident from these studies. However, it should be borne in mind that the purified phospholipase from *C. scutulatus salvini* is an acidic phospholipase (Nair, et al. 1979) comparable to those from *C. adamanteus* (Wells and Hanahan, 1969) and *C. atrox* (Hachimori, et al., 1971), whereas the major phospholipase activity from Mojave venom is associated with a basic protein (Cate and Bieber, 1978) that is part of the major neurotoxin component in this venom.

AN ANTIGENICALLY RELATED TOXIN

Preliminary results on the fractionation of the venom from *C. viridis concolor* have been obtained in the laboratory of Bieber (Pool and Bieber, 1980). The crude venom contained phospholipase, phosphodiesterase, L-amino acid oxidase, and arginine ester hydrolase activities. Phosphomonoesterase activity was not present in measurable amounts. Fractionation on DEAE Sephadex was carried out, and 8 pooled fractions were generated from the column eluant. Fractions 1 and 7 were lethal to white mice. Fraction 1 (LD_{50}, 1.5 to 3.0 mg/kg) was further purified and was shown to contain 2 highly basic (pI values > 9), low molecular weight (4000-5000) polypeptides. Fraction 7 ($LD_{50} < 0.1$ mg/kg) was an acidic protein fraction that was subjected to further fractionation. The resultant material, on the basis of electrophoresis in polyacrylamide gels, was shown to contain 1 major protein component along with 1 other minor band. Both components had pI values less than 5.0 when the material was subjected to isoelectric focusing in polyacrylamide gels. This highly lethal, acidic, protein fraction is associated with phospholipase activity. It is also antigenically cross-reactive with the antiserum that was raised against the basic phospholytic subunit from Mojave toxin (see the section Immunological Studies). It would not be surprising if this highly toxic material is eventually shown to be a neurotoxin since it is an acidic protein associated with phospholipase activity that cross-reacts antigenically with antibodies raised against a component of an established neurotoxin from a rattlesnake venom.

SUMMARY COMPARISON

The literature cited in this chapter gives a clear indication that the venoms from two rattlesnake species contain considerable quantities of closely related but different proteins that act as extremely potent presynaptic neurotoxins. The two for which the most data are available are crotoxin and Mojave toxin. Both of these proteins have isoelectric points in the region of 4.5 to 5.5. Each is composed of two distinct, separable subunits. The larger subunit, in each case, is a basic protein (pI > 9) that has phospholipase activity associated with it. The acidic subunits have isoelectric points that are less than 4. The acidic subunit from each toxin yields three smaller polypeptides when subjected to reductive alkylation. The acidic subunit of Mojave toxin suppresses the phospholipase activity of the basic subunit. Conflicting reports exist on whether a similar

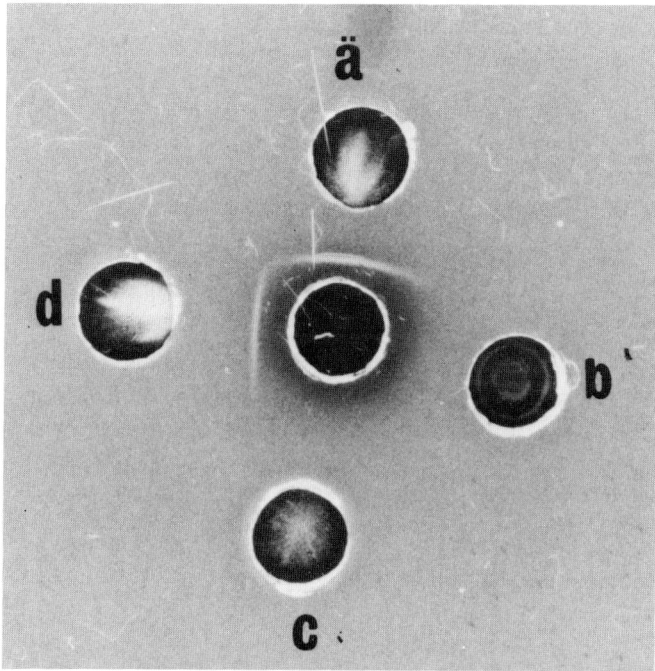

Figure 5 Ouchterlony reaction of rabbit antisera for phospholipase A-crotapotin complex (PLA-CTP) (center well). Reaction wells as follows: a, crotoxin (500 µg/ml); b, *Bitis nasicornis* venom (1 mg/ml); c, *Bitis gabonica* venom (1 mg/ml); d, Mojave toxin (500 µg/ml). (Courtesy of A. T. Tu, Colorado State University, Fort Collins, Colorado, and B. J. Hawgood, Queen Elizabeth College, University of London.)

result applies to the crotoxin complex. The basic subunits from both toxins are moderately toxic and, in each case, essentially full toxicity is restored by forming a complex with the respective acidic component. Immunological results from several laboratories have demonstrated that antisera raised against either toxin caused immunoprecipitation with the other toxin and with the respective crude venoms, i.e., antisera to crotoxin cross-reacts with Mojave toxin and vice versa (Figure 5). In addition to the above-mentioned properties, the two toxins elicit similar biological and physiological responses in test animals. Both toxins produce symptoms in test animals that are consistent with a presynaptic mode of action, and both produce demonstrable myonecrotic activity when injected into muscle. Thus, crotoxin and Mojave toxin are similar complex phospholytic toxins that appear to have multiple sites of action in test animals.

ACKNOWLEDGMENTS

The authors would like to express great appreciation to Dr. Karl Slotta and Dr. Heinz Fraenkel-Conrat for their personal commentary concerning the early years of venom research at the Instituto Butantan, São Paulo, Brazil, and for their constructive guidance and inspiration of our research activities and development in this field. We would also like to thank Dr. Hiro Ishizaki and Ms. Denise Stedman for their most valuable efforts in reviewing and commenting on the manuscript. We would like to thank Ms. Stedman for her aid in the literature search portion of this work.

This work was supported, in part, by A. T. Tu's NIH Grant 5 RO1 GM 24718 in support of R. Hendon's present research activity.

This work was supported in part by NIH Grant 1 RO1 GM 24566 to Allan L. Bieber. Allan L. Bieber also gratefully acknowledges the generosity of Dr. R. L. Straight and Mr. J. L. Glenn, who supplied many of the individual venom samples used in our immunological studies and the *C. viridis concolor* venom used in the studies described.

REFERENCES

Bieber, A. L., and Tu, A. T. (1974). Purification from an acidic toxin protein from the venom of the Mojave rattlesnake. *Fed. Proc. 33*:1564.

Beiber, A. L., Tu, T., and Tu, A. T. (1975). Studies of an acidic cardiotoxin isolated from the venom of Mojave rattlesnake (*Crotalus scutulatus*). *Biochem. Biophys. Acta 400*:178-188.

Bon, C., Changeux, J. P., Jeng, T.-W., and Fraenkel-Conrat, H. (1979). Postsynaptic effects of crotoxin and of its isolated subunits. *Eur. J. Biochem. 99*: 471-481.

Brazil, O. V. (1966a). Pharmacology of crystalline crotoxin. I. Toxicity. *Mem. Inst. Butantan. Simp. Internac. 33*(3):973-980.

Brazil, O. V. (1966b). Pharmacology of crystalline crotoxin. II. Neuromuscular blocking action. *Mem. Inst. Butantan. Simp. Internac. 33*(3):981-992.

Brazil, O. V. (1972). Neurotoxins from the South American rattlesnake venom. *J. Formosan Med. Assoc. 71*:394.

Brazil, O. V., and Excell, B. J. (1971). Action of crotoxin and crotactin from the venom of *Crotalus durissus terrificus* (South American rattlesnake) on the frog neuromuscular function. *J. Physiol. 212*:34 pp.

Brazil, O. V., Excell, B. J., and Santana de Sa, S. (1973). The importance of phospholipase A in the action of the crotoxin complex at the frog neuromuscular function. *J. Physiol. 234*:63 pp.

Breithaupt, H. (1976a). Enzymatic characteristics of *Crotalus* phospholipase A_2 and the crotoxin complex. *Toxicon 14*:221-233.

Breithaupt, H. (1976b). Neurotoxic and myotoxic effects of *Crotalus* phospholipase A and its complex with crotapotin. *Naunyn Schmiedebergs Arch. Pharmacol. 292*:271-278.

Breithaupt, H., Rübsamen, K., Walsh, P., and Habermann, E. (1971). In vivo interactions between phospholipase A and a potentiator isolated from so-called crotoxin. *Naunyn Schmiedebergs Arch. Pharmacol. 269*:403.

Breithaupt, H., Rübsamen, K., and Habermann, E. (1974). Biochemistry and pharmacology of the crotoxin complex. *Eur. J. Biochem. 49*:333-345.

Castilonia, R. R., Pattabhiraman, T. R., and Russell, F. E. (1979). Neuromuscular effects of fractions of *Crotalus scutulatus scutulatus* venom. *Proc. West. Pharmacol. Soc. 22*:205-208.

Cate, R. L. (1977). Studies of an acidic, lethal protein from *Crotalus scutulatus scutulatus* venom. Ph.D. dissertation, Arizona State University, 199 pp.

Cate, R. L., and Bieber, A. L. (1976). Effects of Mojave toxin on rat skeletal muscle sarcoplasmic reticulum. *Biochem. Biophys. Res. Commun. 72:*295-301.

Cate, R. L., and Bieber, A. L. (1978). Purification and characterization of Mojave (*Crotalus scutulatus scutulatus*) toxin and its subunits. *Arch. Biochem. Biophys. 189*:397-408.

Chang, C. Chiung, Lee, J. Dong, Eaker, D., and Fohlman, J. (1977a). The presynaptic neuromuscular blocking action of taipoxin. A comparison with β-bungarotoxin and crotoxin. *Toxicon 15*:571-576.

Chang, C. C., Su, M. J., Lee, J. D., and Eaker, D. (1977b). Effects of Sr^{2+} and Mg^{2+} on the phospholipase A and the presynaptic neuromuscular blocking actions of β-bungarotoxin, crotoxin and taipoxin. *Naunyn Schmiedebergs Arch. Pharmacol. 299*:155-161.

Eterovic, F. A., Hebert, M. S., Hanley, M. R., and Bennett, E. L. (1974). The lethality and spectroscopic properties of toxins from Bungarus multicinctus (Blyth) venom. *Toxicon 12*:1.

Fisher, F. G., and Dörfel, H. (1954). Die Aminosäuren-Zusammensetzung von Crotoxin. *Z. Physiol. Chem. 297*:278-282.

Fraenkel-Conrat, J., and Fraenkel-Conrat, H. (1950). Inactivation of crotoxin by group specific reagents. *Biochim. Biophys. Acta 5*:98-104.

Fraenkel-Conrat, H., and Singer, B. (1954). Recent chemical studies on crotoxin in venoms, E. Buckley and N. Pauges, (Eds.). First International Conference on Venoms at Annual Meeting of A.A.A.S., Pub. No. 44, p. 259.

Fraenkel-Conrat, H., and Singer, B. (1956). Fractionation and composition of crotoxin. *Arch. Biochem. Biophys. 60*:64-73.

Fraenkel-Conrat, H., Jeng, T.-W., and Hsiang, M. (1980). In *Natural Toxins: Biological Activities of Crotoxin and Amino Acid Sequence of Crotoxin B,* D. Eaker and T. Wadström (Eds.). Pergamon Press, Oxford, pp. 561, 563.

Friederich, C., and Tu, A. T. (1971). Role of metals in snake venoms for hemorrhagic, esterase and proteolytic activities. *Biochem. Pharmacol. 20*:1549-1556.

Ghosh, B. N., and De, S. S. (1939). Proteins of rattlesnake venom. *Nature 143*, 380-385.

Glenn, J. L., and Straight, R. (1977). The midget faded rattlesnake (*Crotalus viridis concolor*) venom: Lethal toxicity and individual variability. *Toxicon* 15:129-133.

Glenn, J. L., and Straight, R. (1978). Mojave rattlesnake *Crotalus scutulatus scutulatus* venom: Variation in toxicity with geographical origin. *Toxicon* 16:81-84.

Gopalakrishnakone, P., and Hawgood, B. J. (1979). Morphological changes in murine nerve, neuromuscular junction and skeletal muscle induced by the crotoxin complex. *J. Physiol.* 291:5-6.

Gopalakrishnakone, P., Hawgood, B. J., Holbrooke, S. E., Marsh, N. A., Santana de Sa, S., and Tu, A. T. (1979). An investigation of the specificity of antibodies raised to the crotoxin complex (from the venom of *C. d. terrificus*) using the methods of immunodiffusion and the enzyme-linked immunosorbant assay (ELISA). *Toxicon 17* (Suppl. 1):57.

Gopalakrishnakone, P., Hawgood, B. J., Holbrooke, S. E., Marsh, N. A., Santana de Sa, S., and Tu, A. T. (1980). Sites of action of Mojave toxin isolated from the venom of the Mojave rattlesnake. *Br. J. Pharmacol.* 69:421-431.

Gralen, N., and Svedberg, T. (1938). The molecular weight of crotoxin. *Biochem. J.* 32:1375-1377.

Habermann, E. (1957). Gewinnung und Eisenschaften von crotactin und Toxin III "aus dem Gift der brasilianischen" Klapperschlange. *Biochem. Z.* 329:405-415.

Habermann, E., and Rübsamen, K. (1971). Biochemical and pharmacological analysis of the so-called crotoxin. In *Toxins of Animal and Plant Origin*, vol. 1, A. de Vries and E. Kochva (Eds.). Gordon and Breach, London, pp. 333-341.

Habermann, E., Nalech, P., and Breithaupt, H. (1972). II. Biochemistry and pharmacology of the crotoxin complex. *Naunyn Schmiedebergs Arch. Pharmacol.* 273:313-330.

Hadler, W. A., and Brazil, O. V. (1966). Pharmacology of crystalline crotoxin. IV. Nephrotoxicity. *Mem. Inst. Butantan* 33(3):1001-1008.

Hanley, M. R. (1978). Crotoxin effects in *Torpedo californica* cholinergic excitable vesicles and the role of the phospholipase A activity. *Biochem. Biophys. Res. Commun.* 82(1):392-401.

Hanley, M. R. (1979). Conformation of the neurotoxin crotoxin complex and its subunits. *Biochemistry* 18(9):1681-1688.

Hawgood, B. J., and Santana de Sa, S. (1979). Changes in spontaneous and evoked release of transmitter induced by the crotoxin complex and its component phospholipase A_2 at the frog neuromuscular junction. *Neuroscience* 4:293-303.

Hawgood, B. J., and Smith, J. (1976). The presynaptic action of crotoxin of the murine neuromuscular junction. *J. Physiol.* 266:91-92.

Hendon, R. A. (1975). Preliminary studies on the neurotoxin in the venom of *Crotalus scutulatus* (Mojave rattlesnake). *Toxicon* 13:477-482.

Hendon, R. A., and Fraenkel-Conrat, H. (1976). The role of complex formation in the neurotoxicity of crotoxin components A and B. *Toxicon 14*:283-289.
Hendon, R. A., and Tu, A. T. (1979). The role of crotoxin subunits in tropical rattlesnake neurotoxic action. *Biochim. Biophys. Acta 278*:243-252.
Hendon, R., Roy, D., and Fraenkel-Conrat, H. (1970). Animal and plant toxins, second international symposium, *Toxicon 8*:135.
Hachimori, Y., Wells, M. A., and Hanahan, D. J. (1971). Observations on the phospholipase A of *Crotalus atrox*. Molecular weight and other properties. *Biochemistry 10*:4084-4089.
Horst, J., Hendon, R. A., and Fraenkel-Conrat, H. (1972). The active components of crotoxin. *Biochem. Biophys. Res. Commun. 46*(3):1042-1047.
Jeng, T.-W., and Fraenkel-Conrat, H. (1978). Chemical modification of histidine and lysine residues of crotoxin. *FEBS Lett. 87*(2):291-296.
Jeng, T.-W., Hendon, R. A., and Fraenkel-Conrat, H. (1978). Search for relationships among the hemolytic, phospholytic and neurotoxic activities of snake venoms. *Proc. Natl. Acad. Sci. U.S.A. 75*(2):600-604.
Kennicott, T. (1861). *Caudisona scutulata*. *Proc. Acad. Sci. Philadelphia 13*:206-208.
Kocholaty, W. F., Ledford, E. B., Daly, J. G., and Billings, T. A. (1971). Toxicity and some enzymatic properties and activities in the venoms of *Crotalidae*, *Elapidae* and *Viperidae*. *Toxicon 9*:131-138.
Li, C. H., and Fraenkel-Conrat, H. (1942). Electrophoresis of crotoxin. *J. Am. Chem. Soc. 64*:1586-1588.
Nair, B. C., Nair, C., and Elliott, W. B. (1979). Isolation and partial characterization of a phospholipase A from the venom of *Crotalus scutulatus salvini*. *Toxicon 17*:557-569.
Neumann, N. P., and Habermann, E. (1955). Uber Crotactin, das Haupttoxin des Giftes der Brasilianischen Klapperschlange (*Crotalus terrificus terrificus*). *Biochem. Z. 327*:170-185.
Paradies, H. H., and Breithaupt, H. (1975). On the subunit structure of crotoxin: Hydrodynamic and shape properties of crotoxin, phospholipase A and crotapotin. *Biochem. Biophys. Res. Commun. 66*(2):496-504.
Pattabhiraman, T. R., and Russell, F. E. (1975). Isolation and purification of the toxic fractions of Mojave rattlesnake venom. *Toxicon 13*:291-294.
Pattabhiraman, T. R., Russell, F. E., and Whigham, H. (1978). Some chemical and physiopharmacological properties of fractions from venoms of *Crotalus viridis helleri and Crotalus scutulatus scutulatus*. In *Toxins: Animal, Plant and Microbial*. P. Rosenberg (Ed.). Pergamon Press, Oxford, pp. 211-222.
Pool, W. R., and Bieber, A. L. (1980). Characterization of some components of venom from *Crotalus viridis concolor*. *Fed. Proc. 39*:1646.
Rübsamen, K., Breithaupt, H., and Habermann, E. (1971). Biochemistry and pharmacology of the crotoxin complex. *Naunyn Schmiedebergs Arch. Pharmak. 270*:274-288.
Russell, F. E. (1967). Pharmacology of animal venoms. *Clin. Pharmacol. Therap. 8*:849-873.

Russell, F. E. (1969). Clinical aspects of snake venom poisoning in North America. *Toxicon 7*:33-37.

Russell, F. E., Carlson, R. W., Wainschel, J. and Osborne, A. H. (1975). Snake venom poisoning in the United States. *JAMA 233*:341-344.

Schöttler, W. A. (1951). Toxicity of the principal snake venoms of Brazil. *J. Trop. Med. 31*:489-499.

Slotta, K. H. (1953). Zur Chemie der Schlangengifte. *Experientia 9*:81-120.

Slotta, K. H. (1955). Chemistry and biochemistry of snake venoms. *Fortschr. Chem. Org. Naturst. 12*:406-465.

Slotta, K. H. (1956). Further experiments on crotoxin in venoms, E. E. Buckley and N. Dauges (Eds.). First International Conference on Venoms at Annual Meeting of A.A.A.S., Pub. No. 44, pp. 253-258.

Slotta, K. H., and Fraenkel-Conrat, H. (1937). Estudos Quimicas Sobre os venenos ofidices. Sobre a forma de Ligacao do enxofre. *Mem. Inst. Butantan 11*:121.

Slotta, K H., and Fraenkel-Conrat, H. (1939). Crotoxin *Nature 144*:290-291.

Slotta, K. H., and Fraenkel-Conrat, H. (1938a). II. Mitt. Uber die Bindungsart des Schnefels. *Ber. Dtsch. Chem. Ges. 71*, 264.

Slotta, K. H., and Fraenkel-Conrat, H. (1938b). Schlangengifte. III. Mitt. Reiningung und Kristallisation des Klapperschlangengiftes. *Ber. Dtsch. Chem. Ges. 71*:1076-1081.

Slotta, K. H., and Primosigh, J. (1951). Estudas Quimicas Sobre as venenas ofidicos. 6. Composicao da Crotoxina. *Mem. Inst. Butantan. 23*, 51 [English translation by E. R. Hope. Defense Scient. Inform. Service, Canada (1953)]

Slotta, K. H., and Primosigh, J. (1951). Amino acid composition of crotoxin. *Nature 168*:696-697.

Straight, R., Glenn, J. L., and Snyder, C. C. (1976). Antivenom activity of rattlesnake blood plasma. *Nature 261*:259-260.

Tu, A. T. (1977). *Venoms: Chemistry and Molecular Biology.* John Wiley, New York, pp. 217.

Tu, A. T., Prescott, B., Chou, C. H., and Thomas, G. J., Jr. (1976). Structural properties of Mojave toxin of *Crotalus scutulatus* (Mojave rattlesnake) determined by laser Raman spectroscopy. *Biochem. Biophys. Res. Commun. 68*: 1139-1145.

Wells, M. A., and Hanahan, D. J. (1969). Studies of phospholipase A. I. Isolation and characterization of two enzymes from *Crotalus adamanteus* venom. *Biochemistry 8*:414-424.

5
Chemistry of Rattlesnake Venoms

ANTHONY T. TU
Colorado State University, Fort Collins, Colorado

Introduction   247

Composition of Rattlesnake Venoms (Relatively Nontoxic Components)   251
Inorganic Constituents   •   Organic Constituents   •   Enzymes

Toxic Actions of Rattlesnake Venoms   270
Local Tissue Damage   •   Lethal Toxins   •   Cardiovascular Effect   •
Hemolysis   •   Other Toxins   •   Mechanism of Overall Toxic Action

Nonlethal Actions of Rattlesnake Venoms   284
Nerve Growth Factor   •   Blood Coagulation   •   Platelet Aggregation   •
Antivenom Activity of Blood   •   Anticomplement Action   •   Autopharmacologic Action

References   299

INTRODUCTION

Rattlesnakes are strictly New World snakes belonging to the family Crotalidae. Their venoms are considerably different from those of the Elapidae and Hydrophiidae (sea snakes). Rattlesnake venoms show partial cross-reaction immunologically with those of Asian pit vipers, indicating some similarity between the venoms of Asiatic and American crotalids (Tu et al., 1980) (Figure 1). Rattlesnake venoms usually contain a large number of protein components having a variety of pharmacologic and enzymatic activities. Although the venoms of Elapidae and Hydrophiidae contain strongly lethal neurotoxins, rattlesnake

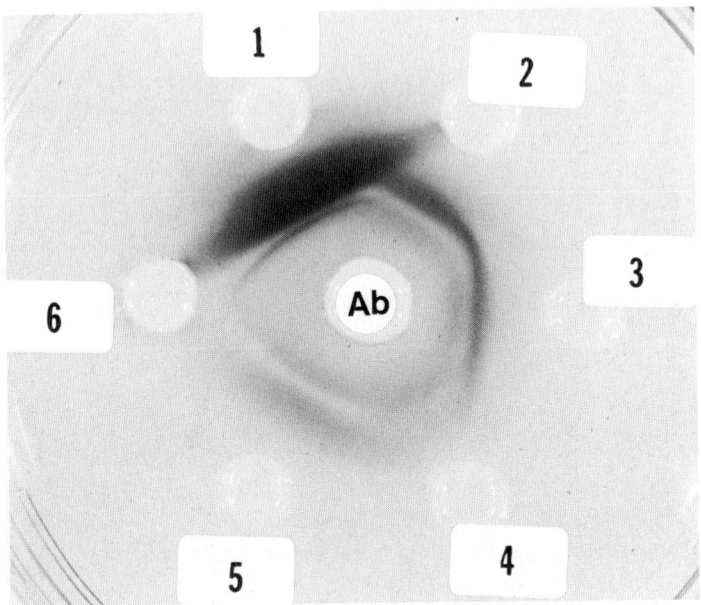

Figure 1 Immunodiffusion pattern indicating that rattlesnake venom responds to antibody of habu venom (*Trimeresurus flavoviridis*), suggesting similarity between venoms of Asian and American pit vipers. 1. *T. flavoviridis* venom from Japan. 2. *Crotalus atrox* venom from the United States. 3. *C. horridus* venom from the United States. 4. *Agkistrodon contortrix laticinctus* venom from the United States. 5. *C. durissus terrificus* venom from South America. 6. *Micrurus fulvius* (coral snake) from the United States.

venoms contain only small amounts of neurotoxins, with, of course, some notable exceptions: the neotropical rattlesnake (*Crotalus durissus terrificus*) and the Mojave rattlesnake (*Crotalus scutulatus*) venoms contain potent presynaptic toxins (Chang and Lee, 1977; Gopalakrishnakone et al., 1980). Because of the unique properties of the toxins from these two rattlesnake venoms, both are discussed in detail in a separate chapter of this book (see Chapter 4).

Clinically, most envenomation by rattlesnakes does not exhibit neurotoxic symptoms; however, by fractionation methods small amounts of neurotoxin can be isolated from rattlesnake venoms. This suggests that rattlesnake venoms contain neurotoxins, but their amount is small, and the overall contribution in snake poisoning is not great. Perhaps other toxic components are responsible for the apparent tissue damage caused by rattlesnake bites and, possibly, for

Chemistry of Venoms

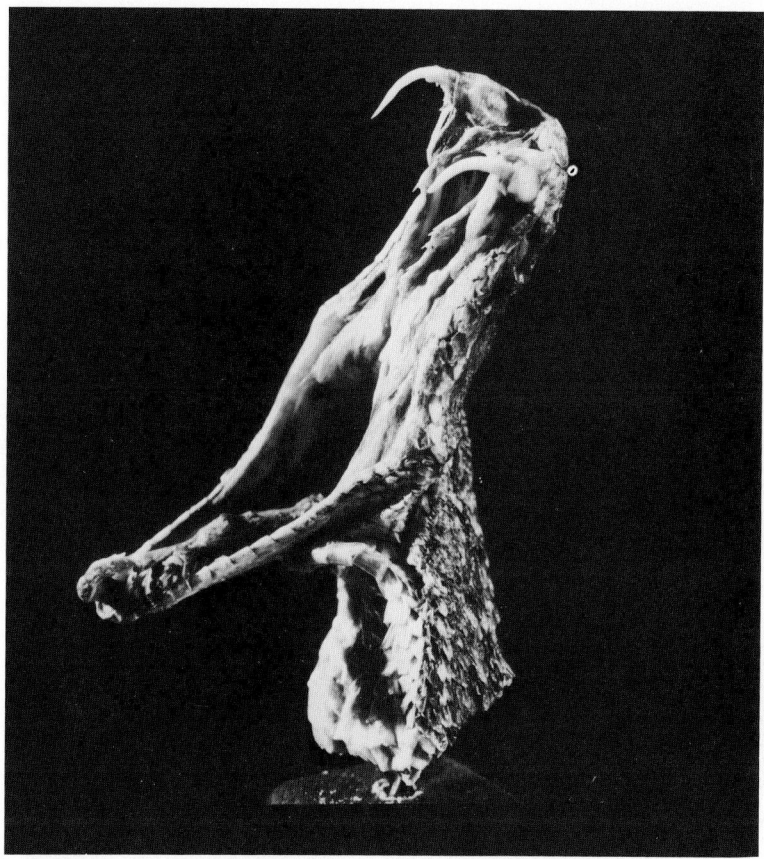

Figure 2 Normally, pit vipter (Crotalidae), including the rattlesnakes, possess a large pair of fangs through which venom is injected into the tissues of a victim.

death. Rattlesnakes possess large hinged fangs with movable jaws; thus, the mouth can open wide, stab forward, and inject a large amount of venom (Figures 2 and 3).

As compared with the venoms of the elapidae and Hydrophiidae, rattlesnake venoms are much less studied, although rattlesnakes are indigenous to the American continent. But the situation is changing, as evidenced by the many toxic and nontoxic components recently isolated in homogeneous form along with the detailed study of their chemical properties. Chemical composition of rattlesnake venoms is, however, very similar to the venoms of other Crotalidae.

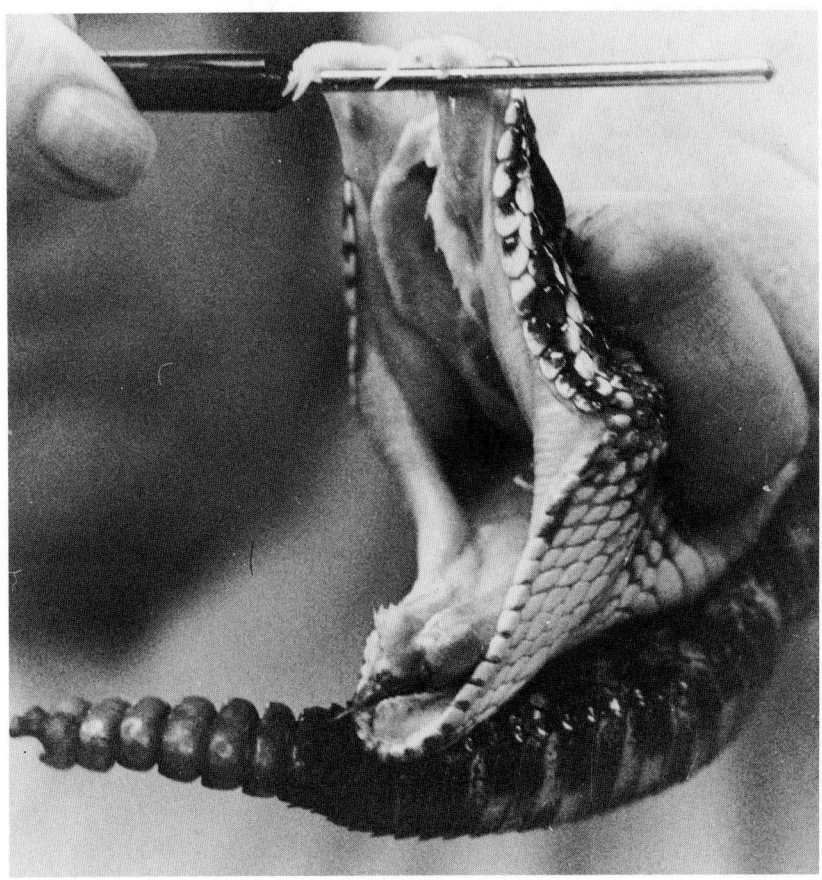

Figure 3 Movable jaws enable rattlesnakes to bite prey much larger than their own body.

The major differences between rattlesnake venoms and those of the Elapidae and Hydrophiidae are summarized in Table 1. As more study is done in the future, more precise information will be gained about the venom components of all snakes.

Because of space limitations, newer references are emphasized. Those readers who are interested in earlier works are advised to look at other books (Tu, 1977; Lee, 1979).

Table 1 Major Differences Between Crotalidae Venoms and Hydrophiidae and Elapidae Venoms

Components	Rattlesnake venoms	Hydrophiidae	Elapidae
Neurotoxins			
postsynaptic	Very small quantity. Not well studied because of few isolations	Very common	Very common
presynaptic	So far found in *C. durissus terrificus* and *C. scutulatus* venom	None	In *Bungarus* and Australian
Proteolytic enzymes	Very common	None	None or very weak
Arginine esterase	Common	None	None
Hemorrhagic toxin	Common	None	None
Myotoxins	Common	Due to phospholipase A_2	Has not been isolated yet
Bradykinin releasing enzyme	Common	None	None
Angiotensin converting enzyme inhibitor	Common	None	None
Acetylcholine esterase	None	Present, but not well studied	Common

COMPOSITION OF RATTLESNAKE VENOMS (RELATIVELY NONTOXIC COMPONENTS)

Snake venoms, especially rattlesnake venoms, contain a large number of proteins. There are 22 bands visible in the isoelectric focusing pattern for the venom of *Crotalus scutulatus* (Bieber et al., 1975).

The electrophoretic patterns of venoms from different species of rattlesnakes always show numerous bands (Bertke et al., 1966; Zwisler, 1965; Johnson, 1968; Basu et al., 1969; Moran and Geren, 1979; Possani et al., 1980; Young et

Table 2 Biologically Active Proteins Found in Rattlesnake Venoms

Hydrolytic enzymes
  Phospholipase A_2
  Phosphodiesterase
  Phosphomonoesterase
    Nonspecific phosphatase
    Specific phosphatase: $5'$-nucleotidase
  Proteolytic enzymes
    Endopeptidase
    Exopeptidase—dipeptidase, tripeptidase
    Collagenase
    Elastase
    Arginine ester hydrolase
    Bradykinin-releasing enzyme
    Enzymes involved in blood coagulation
  Hyaluronidase
  NAD nucleosidase

Nonhydrolytic enzymes
  L-amino acid oxidase

Others
  Neurotoxins—Mojave toxin, crotoxin, basic proteins
  Nerve growth factor
  Hemorrhagic toxins
  Myotoxins
  Cytotoxins, nephrotoxins—rattlesnake venoms have cytotoxic and nephrotoxic activities. Since such toxins have not been isolated yet, it is not clear whether there are such toxins, or if the actions are due to side effects of other toxins.

et al., 1980). A total of 33 electrophoretic bands are visible for the venoms of 18 species and 8 subspecies of *Crotalus* and *Sistrurus* (Foote and MacMahon, 1977). When the gel filtration pattern of venoms from Crotalidae, Viperidae, and Elapidae were compared on Sephadex G-50 and G-75, all Crotalidae venoms, including rattlesnake venoms, had elution patterns located closer to the void volume than did the venoms of other families. This indicates that rattlesnake venoms contain higher molecular weight proteins than do the others (Yang, 1963).

Usually a rattlesnake venom has multiple lethal, hemorrhagic, myotoxic, and proteolytic enzyme fractions. All together there are five lethal fractions in the venom of *Crotalus horridus horridus,* three of which are associated with hemorrhagic activities (Sullivan et al., 1979). There are five hemorrhagic toxins in *Crotalus atrox* venom (Bjarnason and Tu, 1978) and two myotoxic fractions in *Crotalus viridis viridis* venom (Cameron and Tu, 1977). Different toxins usually have different isoelectric points (Tu, 1979).

As compared with the neurotoxic Elapidae and sea snake venoms, fewer rattlesnake venom components have been isolated and characterized. The complexity of their composition is one of the reasons that rattlesnake venom is less characterized. The main protein components, enzymes, and nonenzymatic proteins are listed in Table 2. This list is by no means inclusive, as more components can be isolated and identified in the future.

Inorganic Constituents

About 90% of snake venom in dry form is protein. Major biological and toxic activities originate from proteins; therefore, protein constituents are more important than inorganic components. Since proteins are charged molecules which automatically bind with miscellaneous cations and anions, it is natural that snake venom contains a variety of inorganic compounds. Moreover, many enzymes require metal ions for their activities. In such cases, metal ions serve as cofactors without which the enzyme cannot function.

Metals

Nonspecifically bound (adsorption) metals can be removed by dialyzing venom against deionized water. However, even in prolonged dialysis (48 hr) of rattlesnake venoms, a considerable amount of metal ions still remain (Table 3). The largest amount of metal ions still present in the venom is monovalent Na(I). Sodium ions probably serve as counter ions in proteins to neutralize the negative charge of protein molecules.

Among divalent metals, calcium is the most abundant, followed by zinc. The content of magnesium is much less than that of calcium. The same result was found from the analysis of 0.6% calcium and 0.01% magnesium for the venom of *C. durissus terrificus* (Raudonat, 1963). A fair amound of divalent cations such as zinc, calcium, and magnesium are present in other rattlesnake venoms (Friederich and Tu, 1971). Similar results were obtained by Rodriquez et al. (1974), who found *Crotalus durissus cumanensis* venom contains 1.66% Na, 0.45% Ca, 0.24% K, 0.12% Mg, 0.078% Zn, 0.049% Fe, and very little Co and Mn. Analyses made by two independent groups show a surprising parallel. When the metal content was analyzed for still another venom, *Crotalus vegrandis*, comparable results were obtained (Scannone et al., 1978). Sodium is the most abundant metallic ion and has 1.4% of the dry weight. Other metal contents are 0.46% K, 0.065% Ca, 0.035% Mg, 0.04% Zn, 0.01% Fe, and less than 0.005% of Co and 0.0008% of Mn.

Table 3 Metal Contents (micrograms metal per gram venom) of Snake Venoms before and after Dialysis, Analyzed by Atomic Absorption

Venom (original)	Hr[a]	Ca	Zn	Mg	Na	K	Cu	Mn	Fe	Other metals[b]
Crotalus atrox (United States)	0	4,196	1,394	701	57,300	410	0	0	0	0
	48	3,780	1,093	344	24,600	320				
C. adamanteus (United States)	0	1,610	773	107	42,300	750	0	0	0	0
	48	1,604	452	97	8,400	750				
C. basiliscus (Mexico)	0	1,989	1,400	376	16,800	670	0	0	0	0
	48	1,990	990	310	10,200	638				
C. durissus (Central America)	0	3,003	1,203	1470	36,700	13,500	0	0	0	0
	48	2,968	700	775	12,800	3,970				
C. durissus terrificus (South America)	0	2,390	1,856	342	45,700	1,660	0	0	0	0
	48	2,280	1,380	204	1,780	1,440				
C. durissus totonacus (Mexico)	0	1,633	840	117	28,800	590	0	0	0	0
	48	1,590	680	100	1,500	550				
C. horridus horridus (United States)	0	4,930	980	973	53,000	420	0	0	0	0
	48	3,629	800	406	21,900	400				
C. horridus atricaudatus (United States)	0	150	680	129	49,900	350	0	0	0	0
	48	97	657	91	10,010	240				
C. viridis viridis (United States)	0	4,560	1,847	240	26,400	710	0	0	0	0
	48	2,730	1,050	209	1,200	600				
Sistrurus milarius barbouri (United States)	0	4,000	2,010	446	39,500	2,540	200	0	0	0
	48	2,750	1,525	297	1,550	2,159	90			

[a]Length of time that crude venoms were dialyzed against distilled water before analysis.
[b]Mo, Bi, Se, Pt, Pd, Ag, and Au.
Source: Friederich and Tu (1971).

Nonmetals

Rattlesnake venoms also contain nonemtals. In *C. durissus cumanensis* venom, there is 0.068% total phosphorus, of which 0.038% is inorganic phosphorus (Rodriguez et al., 1974). In *C. vegrandis* venom, the total phosphorus content is 0.083%, and the inorganic content is 0.046% (Scannone et al., 1978).

Organic Constituents

Rattlesnake venoms contain small amounts of free amino acids as well as small oligopeptides. Carbohydrates can be found in the venoms, but they are present as a form of glycoprotein. The carbohydrate content in *Crotalus adamanteus* and *C. atrox* venoms is 7.9 and 4.0%, respectively. Carbohydrates consist of neutral sugars, amino sugars, and sialic acid (Oshima and Iwanaga, 1969).

Biogenic amines such as serotonin (5-hydroxytryptamine) can be found in the venom of *C. atrox* (0.1 μg ml^{-1}) and *C. adamanteus* (0.35 g ml^{-1}). Serotonin is one of many biogenic amines which produce pain; it may be the pain inducer when one is bitten by rattlesnakes (Zarafonetis and Kalas, 1960).

Enzymes

Phospholipase A$_2$

Phospholipase A$_2$ is one of the most common enzymes found in snake venoms, and its presence is not restricted to rattlesnake venoms. This enzyme has been subjected to the most extensive study because of its important biological roles, especially in presynaptic toxins, in disruption of mitochondrial oxidation, and mitochondrial membranes, and in hemolysis. Presynaptic neurotoxins usually have phospholipase A$_2$ activity. Therefore, some people classify the enzyme as toxic phospholipase A$_2$ and nontoxic phospholipase A$_2$.

Chemical and enzymatic properties of rattlesnake venom phospholipase A$_2$ is indeed very similar to the one isolated from the other snake venoms. The enzyme releases a fatty acid from phosphatidylcholine (lecithin) by hydrolyzing the acyl bond at position 2. The product is lysophosphatidylcholine (lysolecithin) (Figure 4).

Nonneurotoxic phospholipase A$_2$ Although rattlesnake venom phospholipase A$_2$ is very similar to the one from Elapidae and Hydrophiidae venoms, there is one difference. Enzymatically active rattlesnake phospholipase A$_2$ is a dimer, whereas the Elapidae enzymes are monomers. For instance, sea snake venom phospholipase A$_2$ has a molecular weight of 11,000 (Tu et al., 1970b), whereas the rattlesnake venom enzyme gives 30,000 for *C. adamanteus* venom (Wells and Hanahan, 1969) and 28,850 for *C. atrox* venom (Purdon et al., 1977).

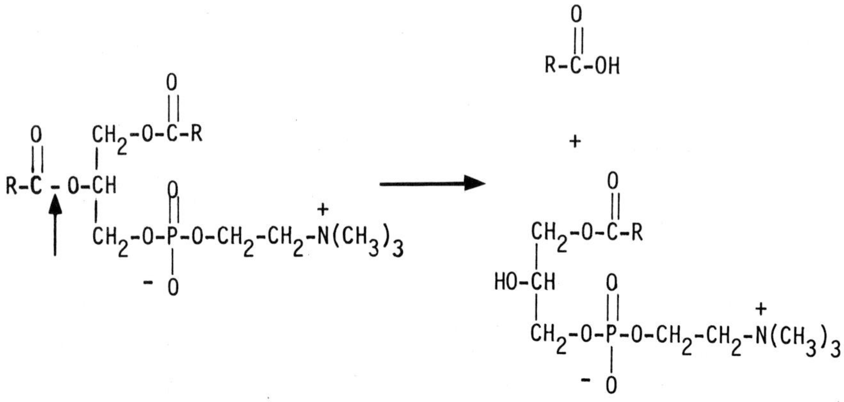

Figure 4 Site of hydrolysis of phospholipase A_2 and its products.

Rattlesnake phospholipase A_2 is a dimer at neutral pH but dissociates into two identical monomers at low pH.

The amino acid sequence of phospholipase A_2 isolated from the venom of *C. adamanteus* has been established (Heinrikson et al., 1977) and is shown in Figure 5.

Close examination of the sequence of the enzyme reveals that rattlesnake venom enzyme is not only similar to that of Elapidae and Vierpidae venoms but also has a considerable homologous region with horse and porcine pancreatic phospholipase A_2.

The mechanism of hydrolysis of phosphatidylcholine by the rattlesnake enzyme is not simple and apparently involves a two-step hydrolysis. The enzyme reaction consists of a short initial burst of hydrolysis, a long lag period of very slow reaction, followed by a dramatic increase in the reaction rate (Tinker et al., 1978; Tinker and Wei, 1979).

Phospholipase A_2 isolated from *Crotalus atrox* venom can be modified by a transamination reaction using glyoxylic acid. By this reaction protein converts to a α-keto acid and the N terminal of the α-amino acid. The modified proteins have no activity when tested with micellar substrates but retain the affinity of the enzyme for micelles of the substrate. In other words, the modified enzyme can bind to substrate but does not hydrolyze it. Two explanations are possible: One is that the active site is not properly oriented toward the interface; the other is that a specific conformation is required for binding and hydrolysis and which is stabilized by the N-terminal amino acid (Verheij et al., 1981). For the phospholipase A_2 reaction a certain orientation of the enzyme-substrate is important. D-phospholipids are actually inhibitors. It seems that the head-group

H$_2$N-Ser-Leu-Val-Gln-Phe-Glu-Thr-Leu-Ile-Met-Lys-Val-Ala-Lys-Arg-Ser-Gly-Leu-Leu-Trp-Tyr-
                    5   10   15   20

Ser-Ala-Tyr-Gly-Cys-Tyr-Cys-Gly-Trp-Gly-Gly-His-Gly-Arg-Pro-Gln-Asp-Ala-Thr-Asp-Arg-
        25   30   35   40

Cys-Cys-Phe-Val-His-Asp-Cys-Cys-Tyr-Gly-Lys-Ala-Thr-Asn-Cys-Asn-Pro-Lys-Thr-Val-Ser-
        45   50   55   60

Tyr-Thr-Tyr-Ser-Glu-Glu-Asn-Gly-Glu-Ile-Val-Cys-Gly-Gly-Asp-Asp-Pro-Cys-Gly-Thr-Gln-
        65   70   75   80

Ile-Cys-Glu-Cys-Asp-Lys-Ala-Ala-Ala-Ile-Cys-Phe-Arg-Asp-Asn-Ile-Pro-Ser-Tyr-Asp-Asn-
        85   90   95   100   105

Lys-Tyr-Trp-Leu-Phe-Pro-Pro-Lys-Asp-Cys-Arg-Gln-Glu-Pro-Glu-Pro-Cys-COOH
        100   115   120

Figure 5 Amino acid sequence of phospholipase A$_2$ isolated from *Crotalus adamanteus* venom. (From Heinrikson et al., 1977.)

region of the phospholipids in phospholipase-substrate interactions is important. Two optical isomers of the substrate may be differently oriented on the enzyme surface. Thus, phospholipase A$_2$ of *Crotalus* venom has an absolute specificity for L-phospholipids (Huang and Law, 1981).

Presynaptic neurotoxin with phospholipase A$_2$ activity It is well known that a presynaptic toxin from *Bungarus multicinctus,* in Elipidae venom, possesses weak phospholipase A$_2$ activity. Presynaptic toxins isolated from rattlesnake venoms are, so far, limited to two: Mojave toxin and crotoxin. Both toxins are composed of acidic and basic subunits. The toxin itself shows phospholipase A$_2$ activity and the enzyme activity originates from component B. Therefore, a toxin can be separated to a nonenzymatic acidic component and to a basic phospholipase A$_2$ component. The molar ratio of the two subunits in crotoxin is 1:1 (Nakazone, 1979). The toxicity of crotoxin can be neutralized by antiserum to *Crotalus* phospholipase A$_2$, which suggests that the enzyme has a role in neurotoxic action (Hanashiro et al., 1978). The acidic component actually inhibits the phospholipase A$_2$ activity of component B (Breithaupt, 1976). The neurotoxicity itself also originates from a basic component. But component A enhances the neurotoxicity of the basic subunit (Habermann and Rübsamen, 1971; Hendon and Fraenkel-Conrat, 1971; Breithaupt et al., 1975; Cate and Bieber, 1978; Jeng et al., 1978; Hendon and Tu, 1979).

Crotoxin and its basic component inhibit Na(I) efflux in acetylcholine receptor-rich *Torpedo californica* electroplaque vesicles. This may suggest that crotoxin and its basic component have postsynaptic in addition to presynaptic

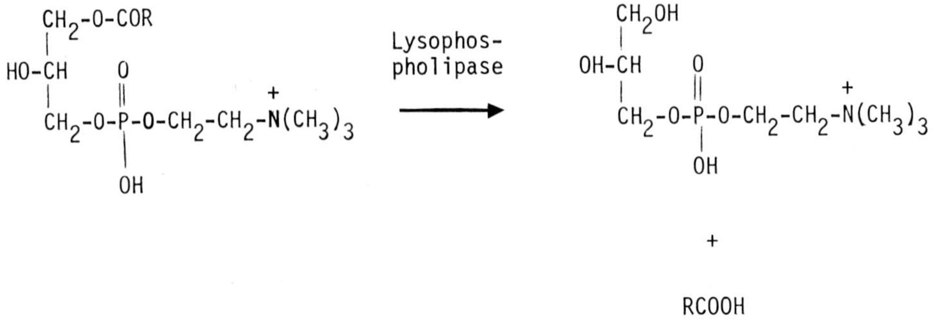

$$\begin{array}{c} CH_2\text{-O-COR} \\ | \\ HO\text{-CH} \quad\;\; O \\ | \quad\quad\;\; \| \\ CH_2\text{-O-P-O-}CH_2\text{-}CH_2\text{-}\overset{+}{N}(CH_3)_3 \\ | \\ OH \end{array} \quad\xrightarrow{\text{Lysophospholipase}}\quad \begin{array}{c} CH_2OH \\ | \\ OH\text{-CH} \quad\;\; O \\ | \quad\quad\;\; \| \\ CH_2\text{-O-P-O-}CH_2\text{-}CH_2\text{-}\overset{+}{N}(CH_3)_3 \\ | \\ OH \end{array}$$

+ RCOOH

Figure 6 Specificity of lysophospholipase activity.

effects. This postsynaptic-like effect may be due to the endogenous phospholipase A_2 activity of the B component (Hanley, 1978).

Lysophospholipase Activity

Lysophospholipase (phospholipase B) catalyzes the hydrolysis of the remaining fatty acid from lysophosphatidylcholine, which itself is the reaction product of phospholipase A_2 on phosphatidylcholine (Figure 6).

It is not clear whether lysophospholipase activity is a side reaction of phospholipase A_2 or due to an independent enzyme of lysophospholipase. Lysophospholipase activity was reported in the venoms of *C. adamanteus* and *C. atrox* but not of *C. horridus* (Fletcher et al., 1979). For the same venom, the activity of phospholipase A_2 is always higher than that of lysophospholipase. At alkaline pH, lysophospholipase activity is always higher than at lower pH.

Phosphomonoesterase

Phosphomonoesterase has specific- and nonspecific-type enzymes. The most common specific phosphomonoesterase is 5'-nucleotidase, which is specific for the 5' isomer with no reaction toward the 3' isomer or nucleoside diphosphates.

Specific phosphatase: 5'-nucleotidase *Crotalus atrox* venom hydrolyzes 5'-AMP or 5'-IMP well with no activity toward GMP, fructose-6-phosphate, mannos-6-phosphate, trehalose monophosphate, or fructose-1,6-diphosphate (Gulland and Jackson, 1938). *Crotalus adamanteus* venom has hydrolytic activity toward ribose-5-phosphate, nicotinamide-5-phosphate, and 5'-UMP (Heppel and Hilmoe, 1951). The 3'-isomers are usually quite resistant to hydrolysis. In *C. atrox*, venom 5'-nucleotidase using 5'-AMP as a substrate shows the highest enzyme activity among all venom enzymes (Jiménez-Porras, 1961). When *C. durissus cumanensis* was fractionated, 5'-nucleotidase appeared in the void column, indicating a molecular weight above 150,000 (Rodriquez and Scannone,

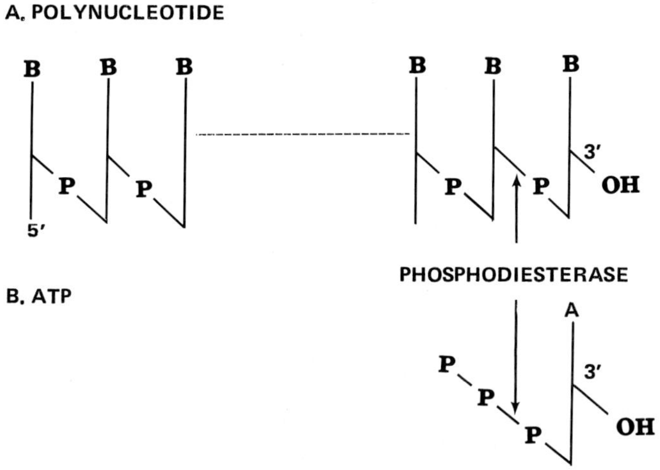

Chemistry of Venoms

A. POLYNUCLEOTIDE

B. ATP

PHOSPHODIESTERASE

Figure 7 Specificity of snake venom phosphodiesterase. A. Action on polynucleotides. B. Action on ATP.

1976). The hydrolysis of a 5'-nucleotide with snake venom 5'-nucleotidase involves a P-O bond cleavage. In order to clarify the stereochemical mechanism, the hydrolysis of adenosine 5'-thiophosphate with the *C. atrox* enzyme was investigated. It was proposed that the hydrolytic reaction proceeds with a configuration inversion at phosphorus without involving a phosphoryl-enzyme intermediate and without pseudorotation (Tsai, 1980). 5'-Nucleotidase is known to reactivate DNase I activity of actin. It was shown that concanavalin A inhibits 5'-nucleotidase isolated from *C. adamanteus* venom and also inhibits the interaction of the enzyme with actin (Mannherz and Magener, 1979).

Phosphodiesterase

This is another very common enzyme found in all species of snake venoms. Venom phosphodiesterase should be called "exonuclease" because it liberates a 5'-nucleotide from the 3' end of a polynucleotide (Figure 7). Phsophodiesterase should be used as a general name for enzymes hydrolyzing all types of diesterified phosphate. However, both terms are used extensively in the literature. When the size of the polynucleotide substrate is large, the rate of hydrolysis is fast; however, as the size decreases, hydrolysis depends more on the type of bases present in the 3'-terminal end (Dolapchiev, 1980).

The prime use of exonuclease is for the elucidation of the polynucleotide base sequence. Thus, the enzyme is extensively isolated and sold commercially. Normally a venom contains more than two different types of exonuclease. The commercially purified exonuclease has less than 1% 5'-nucleotidase activity, but,

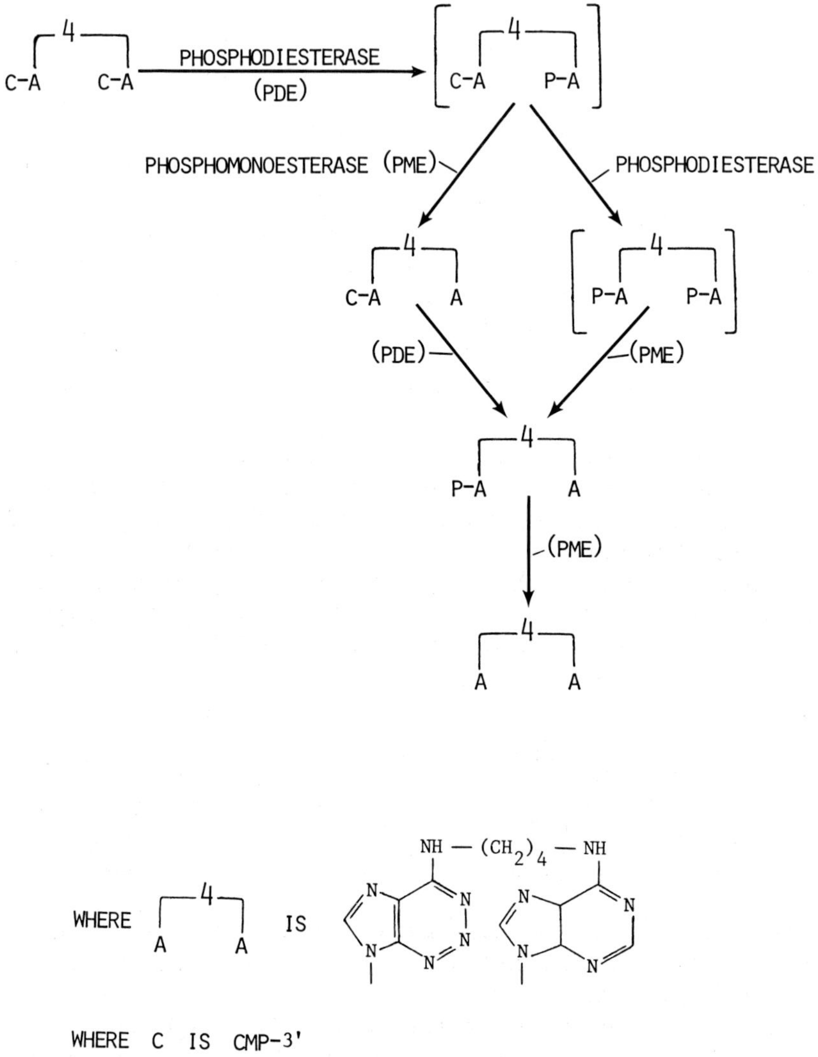

Figure 8 Enzymatic action of snake venom phosphodiesterase on bridged oligonucleotides.

for some applications, the small amount of contaminated 5′-nucleotidase activity can be a source of trouble. A number of methods have been proposed to inactivate this undesirable activity (summarized by Tu, 1977). The enzyme isolated from *C. adamanteus* venom by Worthington Biochemical Corporation was found to contain two components with isoelectric points of 6.8 and 8.7 (Dolapchiev et al., 1977).

The enzymatic action of venom exonuclease is more complicated than originally thought. It also has an intrinsic endonuclease activity. Single-stranded T7 DNA, duplex T7 DNA, and superhelical PM2 DNA are hydrolyzed by a homogeneous preparation from the venom of *C. adamanteus*. Since exonuclease hydrolyzes oligonucleotides stepwise from the 3' terminal end, the cleavage at the 5' end of duplex T7 DNA is ascribed to an endonuclease activity (Pritchard et al., 1977; Pritchard and Laskowski, 1978a). When $\phi \times 174$ DNA is used, there are at least seven specific cleavages. The hydrolysis site is not random and it seems that (A + T)-rich regions in the genome are not cleaved by the enzyme (Pritchard and Laskowski, 1978b). Snake phosphodiesterase has a considerably broad specificity. A synthetic compound with bridged oligonucleotides can be hydrolyzed (Figure 8).

The final product is A ⌐——4——⌐ A because of contaminating phosphomonoesterase (Zemlicka, 1980).

Snake venom phosphodiesterase is extremely stereoselective; therefore it hydrolyzes only one isomer of a diastereomeric phosphorothioate diester about 1000 times faster than the other isomer (Burgers and Eckstein, 1978).

When ATPαS is hydrolyzed with *C. adamanteus* phosphodiesterase, 49.4% of the anticipated pyrophosphate is achieved. The remaining ATPαS is mainly composed of isomer S. This indicates that the hydrolysis of ATPαS is highly specific for the R diastereomer (Bryant and Benkovic, 1979).

```
                    S                           S
                    ‖                           ‖
                  ,P—O—Adenosine              ,P—O—Adenosine
 Ⓟ-O-Ⓟ-O´  |                           O´  |
                   O                          O-Ⓟ-O-Ⓟ

        R isomer                              S isomer
```

A very similar result on the high stereospecificity of hydrolysis was obtained on the phosphorothioate compound by *C. durissus terrificus* venom phosphodiesterase (Burgers et al., 1979a,b).

Because the importance of venom exonuclease lies in the application to nucleic acids, the enzyme is isolated and frequently used. But molecular characterization of the enzyme itself has been neglected for a long time. Phosphodiesterase isolated from *C. adamanteus* venom has seven disulfide bonds with no free -SH groups. It is both a metalloenzyme, containing Mg(II), Zn(II), and Ca(II), and a glycoprotein (Vassileva and Dolapchiev, 1979; Dolapchiev et al., 1980)

Frequently the presence of the enzyme ATPase is reported. Sometimes the presence of ATPase is listed as the component of snake venoms in a review of the literature. This is somewhat misleading and the so-called

ATPase action is due to phosphodiesterase. Normal ATPase hydrolyzes ATP to form ADP + P_i, whereas snake venom hydrolyzes ATP to PP and AMP. This is because

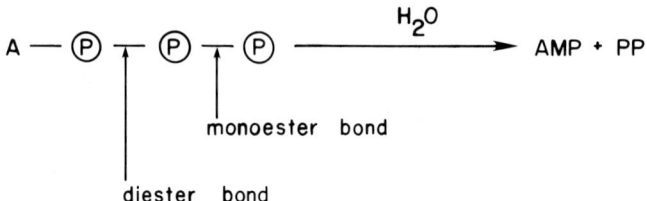

True ATPase should be specific for ATP, but the *C. adamanteus* venom enzyme also hydrolyzes UTP. Thus, it is clear that the hydrolysis of ATP is due to venom phosphodiesterase (Figure 7).

When crude venom is used, the actual hydrolysis product of ATP is P_i and adenosine. The reason is that

$$\text{ATP} \xrightarrow{\text{phosphodiesterase}} \text{AMP} + \text{PP}$$

The rattlesnake venom also contains phosphomonoesterase which hydrolyzes AMP to inorganic phosphate as shown here

$$\text{AMP} \xrightarrow{\text{phosphomonoesterase}} \text{adenosine} + P_i$$

The overall reaction is the formation of adenosine, P_i, and pyrophosphate.

Proteolytic Enzymes

Proteolytic enzymes are present in some snake venoms, and their presence is apparently family specific. Venoms of Crotalidae, Viperidae, and Colubridae families show proteolytic activity, whereas those of Elapidae and Hydrophiidae do not, or show very low activities (Tu et al., 1966, 1967; Kress and Paroski, 1978).

It is not clear why some snake venoms possess such enzymes. It has been speculated that proteolytic enzymes assist the digestion of prey before a snake can swallow it; thus the enzymes serve in the role of exodigestion. The enzymes also inhibit bacterial activity and thereby reduce the risk of prey putrefactive action before it can be digested (Thomas and Pough, 1979). The exact role of these enzymes is hard to define, but in general rattlesnake venoms are rich in proteolytic enzymes.

Chemistry of Venoms

Table 4 Purified Proteolytic Enzymes from Rattlesnake Venoms

Species	Name	Nt. Wt.	Metal requirement	References
C. atrox	α-protease	23,000	—	Zwilling and Pfleiderer (1967)
	Hemorrhagic toxin a	68,000 to about 83,000	Zn	Bjarnason and Tu (1978)
	Hemorrhagic toxin b	19,000 to about 29,300	Zn	Bjarnason and Tu (1978)
	Hemorrhagic toxin c	20,500 to about 27,900	Zn	Bjarnason and Tu (1978)
	Hemorrhagic toxin d	21,000 to about 27,900	Zn	Bjarnason and Tu (1978)
	Hemorrhagic toxin e	22,000 to about 25,700	Zn	Bjarnason and Tu (1978)
C. adamanteus	Proteinase I	24,600	Zn or Ca	Kurecki et al. (1978)
	Proteinase II	23,700	Zn or Ca	Kurecki et al. (1978)

Endopeptidase Usually a rattlesnake venom contains more than one proteolytic enzyme; for instance, three different proteolytic enzymes were separated from *C. atrox* venom (Pfleiderer and Sumyk, 1961). There were five proteolytic enzymes with hemorrhagic activity isolated from the same venom by Bjarnason and Tu (1978). Two proteinases were isolated and characterized from *C. adamanteus* venom (Kurecki et al., 1978). *Crotalus scutulatus* venom shows at least seven proteinase fractions; some of them are glycoprotein and some proteases are nonglycoprotein.

Many of these enzymes contain or require metal for their activities. All hemorrhagic proteases isolated from *C. atrox* venom have been shown to be zinc enzymes (Bjarnason and Tu, 1978). Both proteinase I and II from *C. adamanteus* venom require zinc, calcium, or both for proteolytic activity (Kurecki et al., 1978). In hemorrhagic toxin *e*, it was shown that zinc is essential for both structural integrity and biological activities (protease and hemorrhagic activities). This will be further discussed in the section on hemorrhagic toxins.

Table 5 Amino Acid Composition of Rattlesnake Venom Proteolytic Enzymes

Amino acid	Crotalus atrox hemorrhagic toxin					C. adamanteus Proteinase	
	a	b	c	d	e	I	II
Lysine	27	10	7	7	9	9	9
Histidine	18	8	8	8	9	6	6
Arginine	33	12	10	11	8	15	16
Aspartic acid	85	26	33	33	30	30	27
Threonine	36	7	11	10	10	8	8
Serine	35	14	16	16	14	18	16
Glutamic acid	55	16	24	25	26	21	21
Proline	27	6	6	6	8	7	7
Glycine	43	13	9	9	14	14	13
Alanine	36	8	10	9	7	9	8
Valine	35	11	11	11	12	13	13
Methionine	11	6	5	5	8	6	6
Isoleucine	36	12	17	18	21	15	15
Leucine	49	19	25	25	14	21	21
Tyrosine	21	7	7	7	11	8	8
Phenylalanine	17	7	7	7	6	7	7
Half-cystine	66	4	2	2	8	4	4
Tryptophan	6	4	5	5	4	3	2
Total residue	635	200	213	214	209	214	207

Sources: Bjarnason and Tu (1978) and Kurecki et al. (1978).

Many snake venoms contain very special proteases such as kininogenase or the enzymes involved in blood coagulation. Kininogenase is the enzyme which liberates bradykinin from its precursor bradykininogen and is also called "bradykinin releasing enzyme," which is discussed in the section on autopharmacologic action. Even the released bradykinin can be further hydrolyzed by *C. adamanteus* and *C. atrox* venoms. The first liberation of bradykinin is due to a specific enzyme of bradykininogenase, whereas further hydrolysis is due to other proteases present in the venoms (Suzuki et al., 1966).

All proteolytic enzymes isolated from rattlesnake venoms are shown in Table 4. Amino acid compositions of these venom proteases are shown in Table 5. The specificity of proteolysis on oxidized insulin B is shown in Figure 9 together with other venom proteases. Apparently, all Crotalidae venom proteases show similar specificity which is quite different from the well known enzymes of

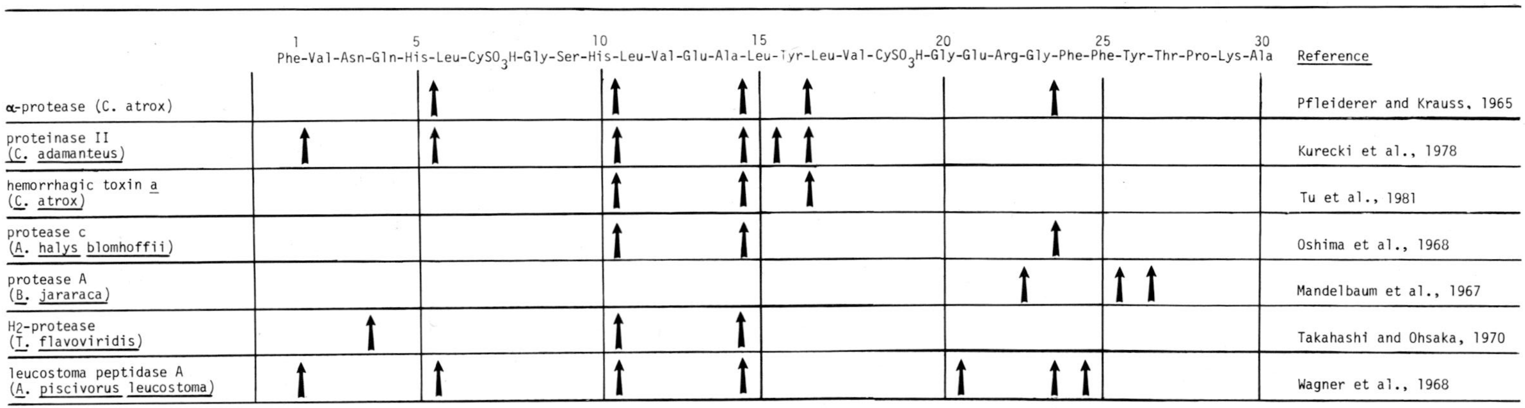

Phe-Val-Asn-Gln-His-Leu-CySO$_3$H-Gly-Ser-His-Leu-Val-Glu-Ala-Leu-Tyr-Leu-Val-CySO$_3$H-Gly-Arg-Gly-Phe-Phe-Tyr-Thr-Pro-Lys-Ala

Enzyme	Cleavage sites (residue numbers)	Reference
α-protease (*C. atrox*)	4, 14	Pfleiderer and Krauss, 1965
proteinase II (*C. adamanteus*)	6, 10, 14, 15, 16	Kurecki et al., 1978
hemorrhagic toxin a (*C. atrox*)	14, 15	Tu et al., 1981
protease c (*A. halys blomhoffii*)	10, 14	Oshima et al., 1968
protease A (*B. jararaca*)	5, 23, 25	Mandelbaum et al., 1967
H2-protease (*T. flavoviridis*)	10, 15, 20	Takahashi and Ohsaka, 1970
leucostoma peptidase A (*A. piscivorus leucostoma*)	4, 10, 22, 23, 25	Wagner et al., 1968

Figure 9 Specificity of purified snake venom proteases on the oxidized B chain of insulin.

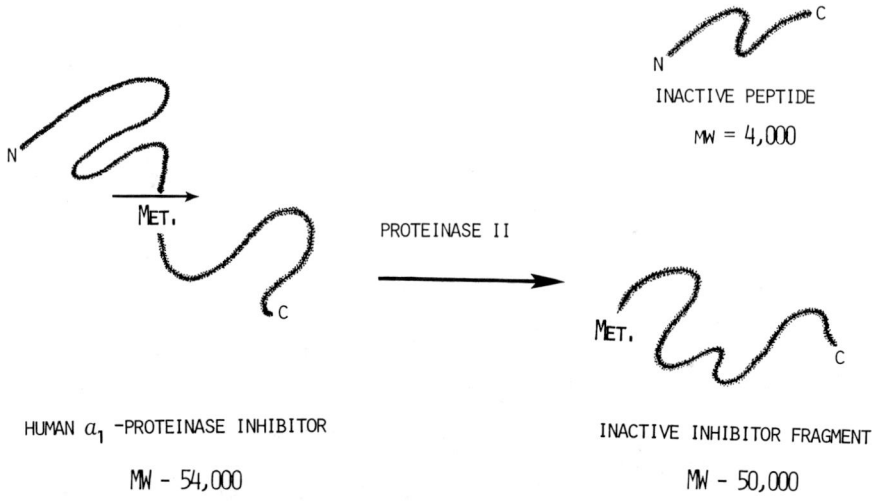

Figure 10 Cleavage of human α_1-proteinase inhibitor by proteinase II isolated from *Crotalus adamanteus*.

trypsin, chymotrypsin, and pepsin. The venom proteases are unique and different from proteases of other sources.

The proteolysis specificity of proteinase II on human α_1-proteinase inhibitor is particularly interesting (Kress et al., 1979). A single bond in the α_1-proteinase inhibitor is cleaved, and the cleavage produces the loss of inhibitory action on trypsin or chymotrypsin (Figure 10). Human α_2-macroglobulin is a plasma proteinase inhibitor which forms complexes with many proteolytic enzymes. The complexes are stable and retain the ability to digest small peptied substrates but are inactive toward large proteins. The proteinase II also forms a complex with α_2-macroglobulin but does not digest α_1 proteinase inhibitor; however, active proteinase II is released from the complex in the presence of serum or a high molecular weight *C. adamanteus* proteinase fraction (Kress and Kurecki, 1980). Antithrombin III (AT III) is a plasma proteinase inhibitor which forms complexes with many serine proteinases. Incubation of Crotalidae, Viperidae, and Colubridae venoms results in the inactivation of AT III, whereas Elapidae venoms have no effect. The inactivation of AT III is due to limited proteolysis which is accelerated in the presence of heparin (Kress and Catanese, 1980). These examples illustrate the complex action of snake venom proteases on human plasma proteins.

Exopeptidase A synthetic Substrate, L-leucyl-β-naphthylamide, is frequently used as an assay for leucine amino peptidase activity. Venom from all four families of snakes show hydrolytic activity toward this substrate (Michl and Molzer, 1965; Tu and Toom, 1967). Rattlesnake venoms which show this activity are *Crotalus basiliscus, Crotalus durissus totonacus, Crotalus durissus atricaudatus,* and *Sistrurus miliarius barbouri.*

When tripeptides are hydrolyzed with rattlesnake venoms, the peptide bond involving the amino terminal is hydrolyzed first. This suggests that tripeptides are hydrolyzed initially by some kind of amino peptidase present in the venoms, with further cleavage of peptide bonds by dipeptidases. Some dipeptides are also hydrolyzed by rattlesnake venoms (Tu et al., 1967). So far carboxypeptidase has not been reported in any venoms.

Special Proteases Specific enzymes such as collagenase and elastase, have not been isolated in pure form yet. However, enzymatic activities have been detected using crude venoms from both Crotalidae and Viperidae.

Collagenase. Collagen is a fibrous protein commonly present in connective tissue; it is large enough to be seen under the electron microscope.

Since *C. atrox* venom hydrolyzes mesenteric collagen fibers but not other proteins, it is believed that venom collagenolytic enzyme is a true collagenase (Simpson et al., 1971). Collagenase activity is strongest in the venoms of Crotalidae and least potent in those of Viperidae. Most Elapidae venoms contain minimal or no activity. It is evident from the enzyme distribution that there is a correlation between families of poisonous snakes and collagenase activity (Simpson, 1975).

Elastase. Elastin is a yellow scleroprotein present in elastic tissues. Rattlesnake venoms and other Crotalidae and Viperidae venoms hydrolyze the synthetic elastase substrate t-BOC-L-alanine-p-nitrophenol (Simpson and Taylor, 1973). Elastase activity is not found in the venoms of Elapidae; thus, its presence has a biotaxonomical significance at the family level of poisonous snakes (Bernick and Simpson, 1976).

Arginine Ester Hydrolase (Arginine Esterase)

Synthetic substrates of argine esters are convenient substrates for trypsin activity as the hydrolysis can be followed readily by spectrophotometric methods. However, great caution must be maintained in snake venom assays. Snake venoms hydrolyze arginine esters; however, the enzyme is not trypsin and usually possesses clotting, bradykinin-releasing, and capillary-diameter-increasing activity.

Both arginine esterase and kallikrein (bradykinin-releasing enzyme) activities of *C. adamanteus, C. durissus terrificus,* and *C. horridus horridus* venom can be inhibited by proteinase inhibitors (Geiger and Kortmann, 1977). Kallikrein is a special enzyme releasing bradykinin from bradykininogen; thus, the esterase activity in this case must be the side reacion of kallikrein. Snake venom arginine esterases are probably very special proteolytic enzymes that do not hydrolyze or casein, substrates commonly used for nonspecific proteases.

In snake venoms, the arginine esterase activity is most pronounced in Crotalidae and Viperidae. For the venoms of Elapidae and Hydrophiidae, the activity is either absent or very low. Thus the ability to hydrolyze arginine ester has taxonomic significance at the family level. The enzyme is commonly found in rattlesnake venoms (Tu et al., 1967; Oshima et al., 1969; Scannone et al., 1978).

The enzyme has been isolated from *Crotalus adamanteus* venom and separated from caseinolytic and thrombin-like activities (Kress et al., 1978). Thus, the enzyme is very similar to the one isolated from *Agkistrodon contortrix laticinctus* (broadbanded copperhead) venom (Toom et al., 1970).

Hyaluronidase

Hyaluronidase is commonly found in the venoms of Elapidae, Viperidae, and Crotalidae. Hyaluronic acid is an important mucopolysaccharide found in connective tissues and serves to promote intercellular adhesion. The enzyme was reported in the venoms of the following rattlesnakes:

C. adamanteus (Duran-Reynals, 1939)
C. atrox (Duran-Reynals, 1939; Madinaveitia, 1941)
C. durissus terrificus (Tarabini-Castellani, 1938; Duran-Reynals, 1939; Favilli, 1940; Madinaveitia, 1941; Slotta and Ballester, 1954)

The enzyme is often called "spreading factor" because it is believed that the hydrolysis of hyaluronic acid facilitates toxic penetration into the interior of tissues (Duran-Reynals, 1936, 1939). However, with careful reading of the original papers, it is found that the so-called spreading factor role of hyaluronidase was a proposed theory with no experimental basis. Therefore, it is surprising that this role has been assumed true without a second thought by most investigators. It is imperative that one establish this fact with solid experimental proof.

L-Amino Acid Oxidase

All amino acid oxidase found in snake venoms is of the L-type and is different from the D-amino acid oxidase found in bacteria and fungi. The presence of this enzyme is common in snake venoms and is responsible for giving a yellow color owing to the presence of FAD as the prosthetic group. Most enzymes present in snake venoms are of the hydrolytic type; however, L-amino acid oxidase is a nonhydrolytic enzyme that oxidatively deaminates amino acids.

The enzyme purified from the venom of *Crotalus adamanteus* (Wellner, 1966; Wellner and Meister, 1960) was found to be relatively unstable (Curti et al.,

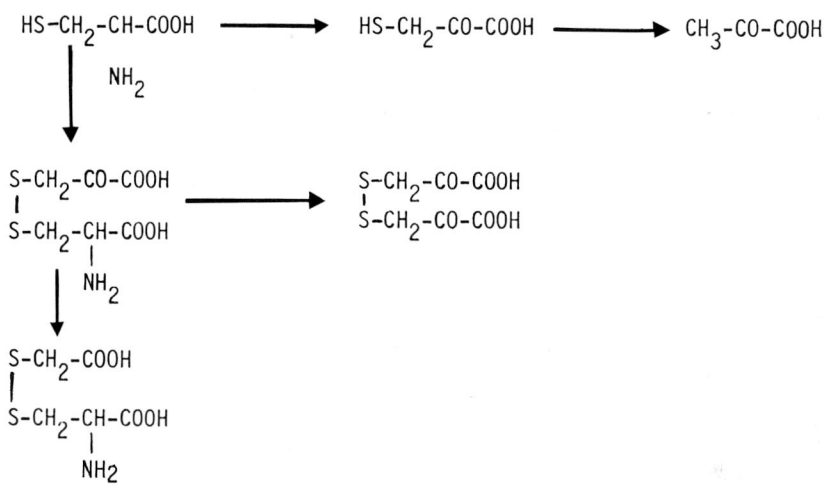

Figure 11 Enzymatic action of ophio-L-amino acid oxidase on L-cysteine.

1968; Paik and Kim, 1967, 1975). The molecular weight is about 128,000-153,000 (Meister and Wellner, 1966; Wellner, 1971) and is a glycoprotein containing about 2-5% carbohydrate (deKok and Rawitch, 1969). The L-amino acid oxidase isolated from *C. adamanteus* venom consists of isozymes with many different isolectric points (Hayes and Wellner, 1969). The enzyme has an optimum activity at pH 7-9 (Paik and Kim, 1975), requiring Mq(II) but inhibited by Ca(II), PO_4^{3-}, and *p*-chloromercuribenzoate (Paik and Kim, 1967).

Basically the produce is α-ketoacid through the oxidative deaminating reaction.

$$\text{R-CH-COOH} \longrightarrow \left[\begin{array}{c} \text{R-C-COOH} \\ \| \\ \text{NH} \end{array} \right] \longrightarrow \begin{array}{c} \text{R-C-COOH} \\ \| \\ \text{O} \end{array}$$
$$\underset{NH_2}{|}$$

The imino acid is found as an intermediate compound (Hafner and Wellner, 1971; Page and Van Etten, 1971a,b).

The enzyme has a wide substrate specificity and it oxidatively deaminates common amino acids, as well as many not so common ones. L-Cysteine undergoes complicated reactions to form a variety of products using *C. durissus terrificus* venom enzyme (Chen et al., 1971; Ubuka and Yao, 1973) (Figure 11).

Some compounds such as S-adenosyl-L-methionine, cysteic acid, homocysteic acid, lanthionine, and cystathionine have no reaction, but L-methionine, L-hemocystine, L-homocysteine, S-adenosyl-L-homocysteine, S-ribosyl-L-homocysteine, L-adjenkolic acid, L-cysteine, and L-cystine can be oxidated with L-amino acid oxidase isolated from *Crotalus durissus terrificus* venom (Chen et al.,

1971). Large substituents in phenylalanine result in lower velocities (Zeller et al., 1974). One unusual substrate, thialysine, a sulfur analog of lysine, produces an unusual product, 5-6-dihydro-Δ^3, 1,4-thiazine-3-carboxylic acid (TZCA) through the following mechanism (Cini et al., 1978):

$$H_2N\text{-}CH_2\text{-}Ch_2\text{-}S\text{-}CH_2\text{-}\underset{\underset{NH_2}{|}}{CH}\text{-}COOH \longrightarrow [H_2N\text{-}CH_2\text{-}CH_2\text{-}S\text{-}CH_2\text{-}\underset{\underset{O}{\|}}{C}\text{-}COOH]$$

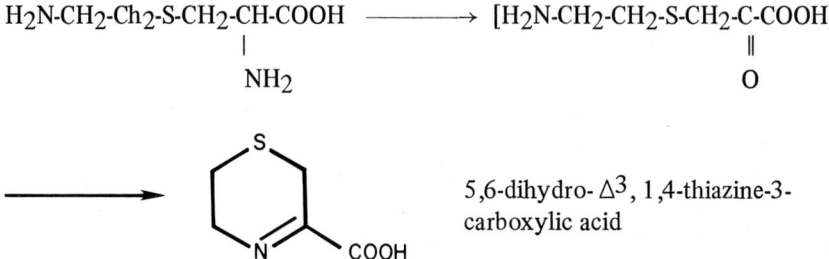

5,6-dihydro- Δ^3, 1,4-thiazine-3-carboxylic acid

When the S atom is replaced with Se, the oxidative deamination can still take place with the treatment of *C. adamanteus* L-amino acid oxidase without cleavage of C-Se bond (Cini and DeMarco, 1979).

L-Amino acid oxidase itself is slightly toxic as shown by the LD_{50} value of 9.13 mg/kg in mice, approximately half the lethal value of the original venom. However, because of the low content of the enzyme in the whole venom, the contribution is less than 1% of the total lethality (Russell et al., 1963).

Other Enzymes

Other enzymes commonly present in snake venoms, including rattlesnake venom, are NAD nucleosidase (NAD glycohydrolase, NADase).

Absence of Enzymes

One enzyme commonly found in other venoms but not found in rattlesnake and other crotalid venoms is acetylcholinesterase. This enzyme is found mainly in the venoms of Hydrophiidae and of Elapidae such as cobra, kraits, and mambas (Tu, 1977).

TOXIC ACTIONS OF RATTLESNAKE VENOMS

Although rattlesnakes are indigenous to the American continents, their venom is poorly studied. Unlike Elapidae and sea snakes, the majority of rattlesnakes in North America are nonneurotoxic, although they contain small amounts of neurotoxins which can be isolated by fractionation techniques. In actual envenomation, there is little neurotoxic effect, but the overall effect is complicated by local, as well as systemic, symptoms. Notable exceptions are the venoms of *C. durissus terrificus* (tropical rattlesnake) and *C. scutulatus* (Mojave rattlesnake)

Chemistry of Venoms

which contain potent presynaptic neurotoxins. Both Mojave toxin and crotoxin are composed of acidic and basic subunits. Because of their close similarity in chemical structure and actions, these two toxins will not be discussed here; they are discussed together in a separate chapter, and readers are advised to see Chapter 4.

Local Tissue Damage

Local tissue damage of rattlesnake poisoning is quite common, with swelling, blister formation, hemorrhaging and, myonecrosis being the most common occurrences. What component is responsible for such actions? There is not a simple answer because there is not enough experimental data to make conclusins as yet. So far some hemorrhagic toxins without myonecrotic activity have been isolated. Myotoxin *a* from *C. viridis viridis* venom causes only myonecrosis without hemorrhagic activity. On the other hand, some pure toxins can produce both myotoxic and hemorrhagic effects. When crude rattlesnake venoms are examined, most show hemorrhagic, myonecrotic, and proteolytic activities (Homma and Tu, 1971; Gutierrez and Chaves, 1980). To pin down the exact cause of these toxic actions is a real challenge to venom scientists, and already considerable progress has been made. Some of this progress will be reviewed here briefly.

Myonecrosis

This histopathology of rattlesnake poisoning was recognized and documented as early as 1905 by Flexner and Noguchi. On envenomation, rattlesnakes cause severe local tissue damage such as swelling, muscle degeneration, and bleeding (Figure 12). Eventually, toxins responsible for these toxic effects were isolated; they initially cause swelling of the sarcoplasmic reticulum (Figure 13). Actually, myotoxin *a* has a high affinity for the sacroplasmic reticulum (Figure 14).

Myotoxin *a* is a component isolated from the venom of *C. viridis viridis* (prairie rattlesnake) and is a basic protein with an isoelectric point of 9.6. The molecular weight was estimated as 4100, and it causes degeneration of muscle (Cameron and Tu, 1977). The initial reaction of this toxin is to dilatate the sarcoplasmic reticulum, but eventually the myofibrils are disorganized (Ownby et al., 1976). The final effect is very similar to that of whole venom (Stringer et al., 1972). The amino acid sequence and position of disulfide bonds were determined and are shown in Figure 15 (Fox et al., 1979). Computer analyses indicate that the sequence of myotoxin *a* is quite different from Type I neurotoxin of Elapidae and Hydrophiidae venom but has considerable similarity to crotamine. Apparently, both crotamine and myotoxin *a* are new types of toxins whose chemical structures are distinctly different from well-known neurotoxins.

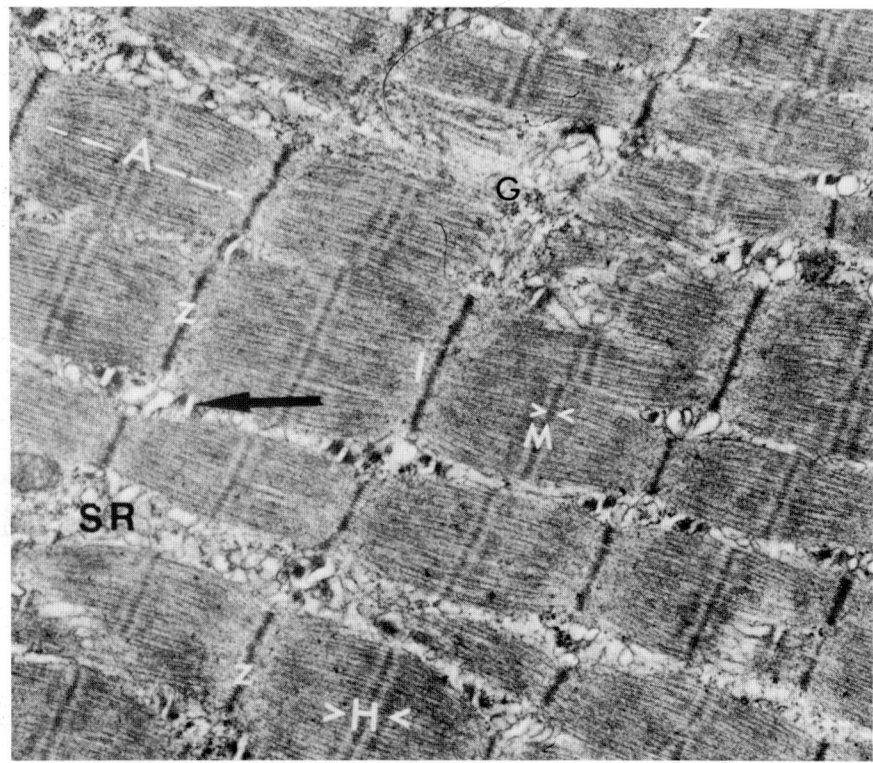

Figure 12A Electron micrograph of normal muscle. Notice the regular repeating pattern of a sarcomere. A, A-band; I, I-band; M, M-gand; Z, Z-line; H, H-band. The portion from Z-line to Z-line is a sarcomere.

Crotamine is a basic substance isolated from the venom of *C. durissus terrificus* (Gonçalves and Deutsch, 1956). However, not all snakes of this species contain crotamine. The same species from the northern and central parts of Brazil do not possess this component, and the venom from *C. durissus durissus* lacks crotamine (Schenberg, 1959).

The primary structure of crotamine was identified by Laure (1975) and was found to consist of 42 residues. Myotoxin *a* differs from crotamine at three positions: Ile(19) → Leu, Leu(25) → Phe, and Lys(33) → Arg. Actually, these replacements are chemically conservative substitutions. Direct comparison of these two toxins indicates that they have almost identical electrophoretic mobility, isoelectric points, and circular dichroism patterns which are reflections of identical protein conformation (Cameron and Tu, 1978). Myotoxin *a* has a

Figure 12B Complete destruction of muscle by injection of *C. viridis viridis* venom into mice. Notice that no sarcomere can be seen. A more detailed process of myonecrosis can be seen in Chapter 3 of this book.

peptide backbone conformation of β-sheet and β-turn structure (Bailey et al., 1979). A toxin isolated from *C. viridis helleri* also has an amino acid sequence very similar to that of myotoxin *a* (Maeda et al., 1978). Thus, it is firmly established that myotoxin *a*, crotamine, and peptide c are a group of similar toxins distinct from sea snake and Elapidae neurotoxins.

Like myotoxin *a*, crotamine also has myotoxic activity (Cameron and Tu, 1978). Brazil et al. (1979) observed the interesting effect of crotamine to produce myotonias, which is the failure of muscle to relax normally owing to the high frequency discharge of action potentials. The cause of myotonias is not well established yet, but it may be due to an alteration in sodium conductance (Filho et al., 1978; Chang and Tseng, 1978).

Because of its small size, it is logical that crotamine is a very stable protein. Heating crotamine at 80°C for 30 min in 8 M urea does not affect its biological activity (Hampe et al., 1978).

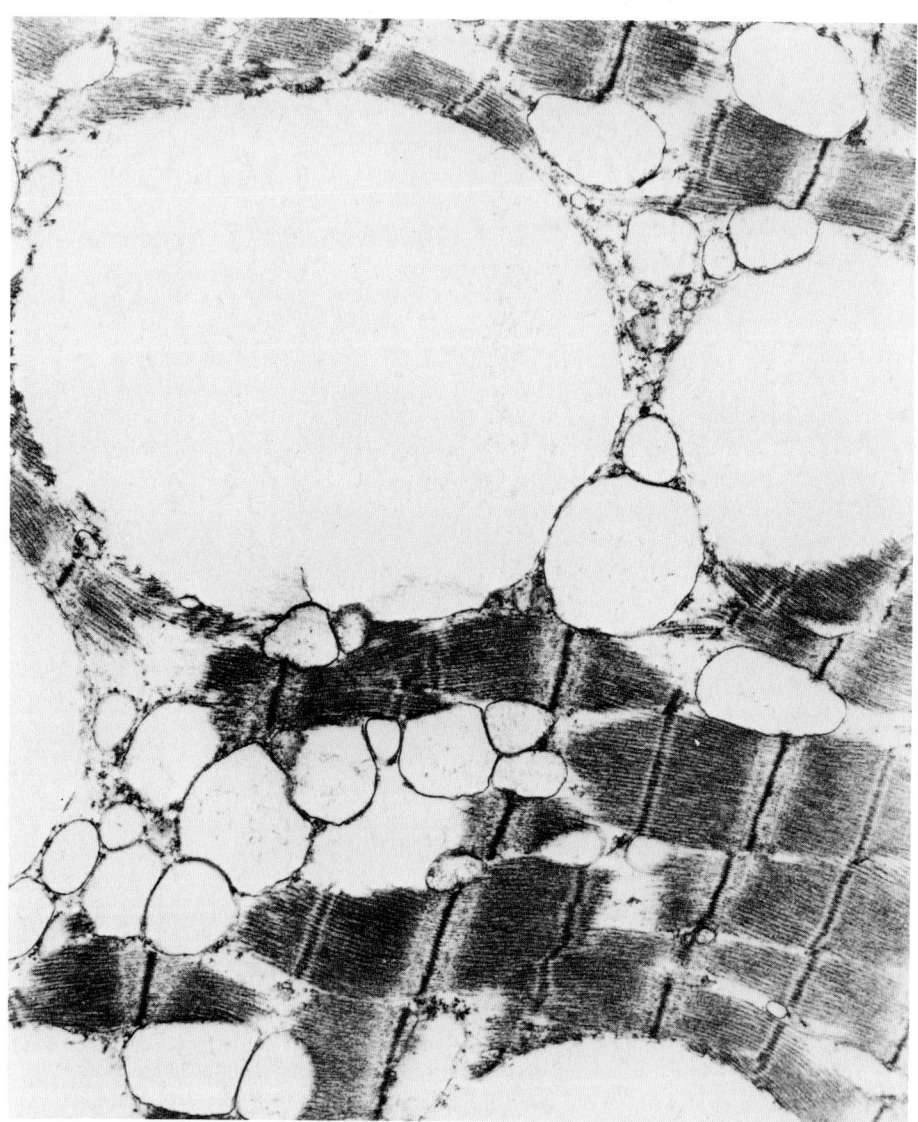

Figure 13 Sarcoplasmic reticulum is greatly dilated by the intramuscular injection of myotoxin *a* into mice. (Courtesy of C. L. Ownby, Oklahoma State University, Stillwater, Oklahoma.)

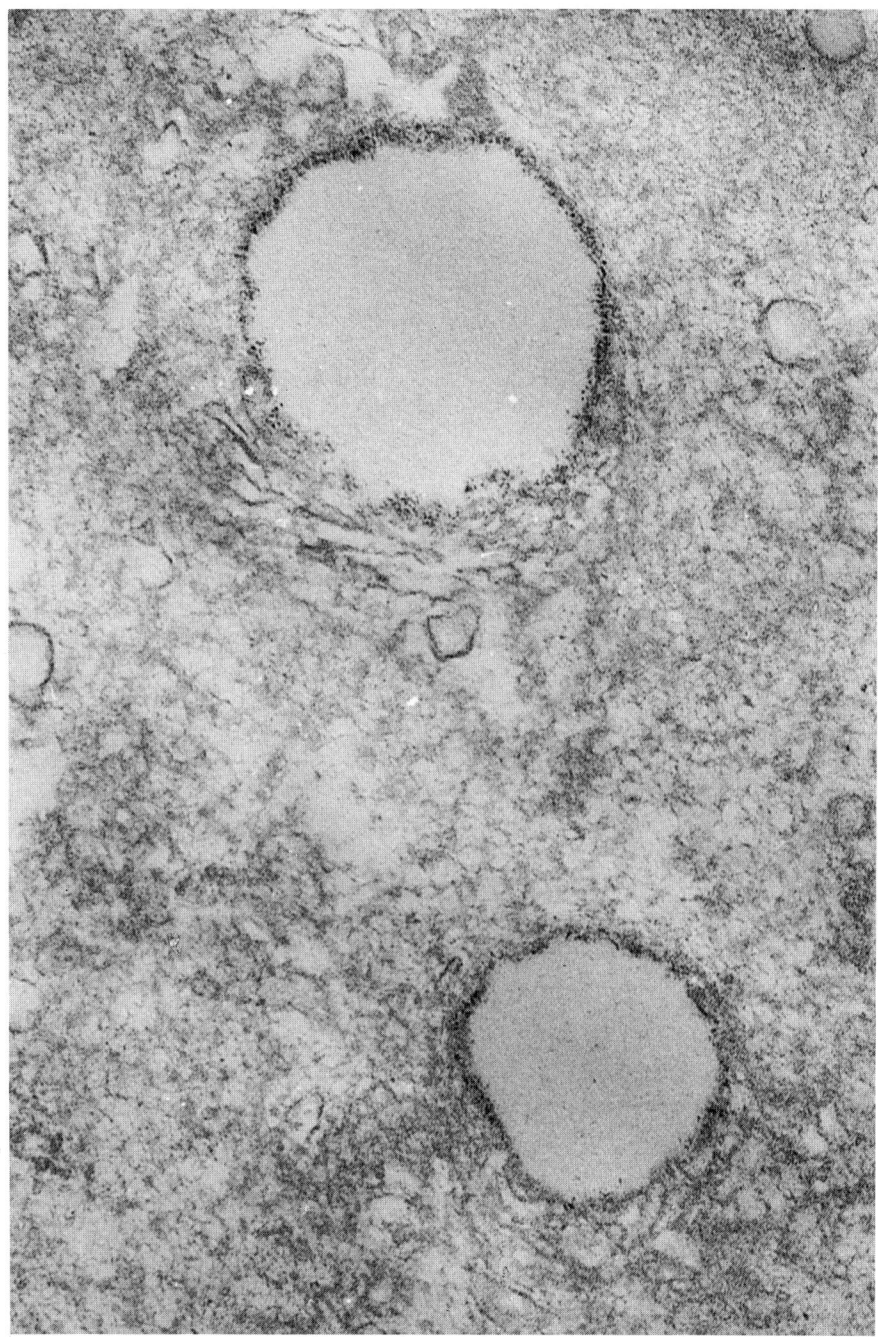

Figure 14 Effect of myotoxin *a* on human frozen muscle tissue can be seen from the specific attachment of peroxidase conjugated myotoxin *a* to the surface of the sarcoplasmic reticulum. (Unpublished data.)

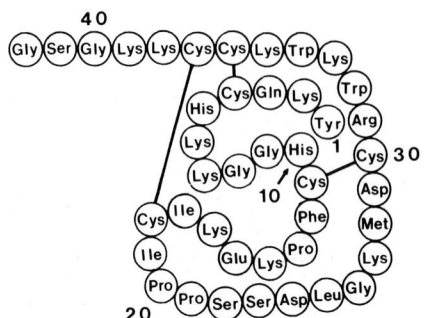

Figure 15 Amino acid sequence of myotoxin *a*. (From Fox et al., 1979.)

It is well known that commercial antivenin does not neutralize the myotoxic effect in rattlesnake poisoning. This is probably due to the low titer of antibody against myotoxic components. This is evidenced by the fact that antibody against pure myotoxin *a* can be produced (Ownby et al., 1979).

Serum creatine kinase levels are frequently used to monitor the cellular damage to heart tissue following a myocardial infarction. This method utilizes the release of creatine kinase into the circulatory system as a result of damage to the muscle cells of the cardiac or skeletal muscles. Using experimentally envenomated mice, the increase in serum creatine kinase levels was observed for a necrotic toxin isolated from tarantula venom (Lee et al., 1974) and for viriditoxin (Fabiano and Tu, 1981). The use of serum creatine kinase and histologic assays in conjunction with one another is the most effective means of studying the myotoxicity of snake venoms.

Hemorrhagic Action

Because of the extensive use of antivenin in the United States, many people survive the bite but suffer severe hemorrhage and necrosis, with resulting dysfunction or loss of a finger and/or hand.

The pathogenesis of hemorrhage induced by *C. atrox* venom was investigated at the electron microscopic level, and it was found that hemorrhage was induced by rhexis; that is, the rupture of capillary endothelial cells rather than the escaping of red cells through a tight junction (Ownby et al., 1974). Eventually, five hemorrhagic toxins were isolated (Bajarnason and Tu, 1978). Ultrastructural study using hemorrhagic toxins *a*, *b*, and *e* indicated that hemorrhagic toxins induced hemorrhage by the same mechanism as the crude venom. While toxins *a* and *e* are hemorrhagic only, toxin *b* produced myonecrotic, in addition to hemorrhagic, action (Ownby et al., 1978). Rattlesnake venom hemorrhagic activity can be neutralized by various chelating agents, suggesting that the hemorrhage-inducing factor may be a metalloprotein (Tu et al., 1970a; Ownby et al., 1975). They are indeed metalloproteins and are zinc-containing proteolytic

Chemistry of Venoms

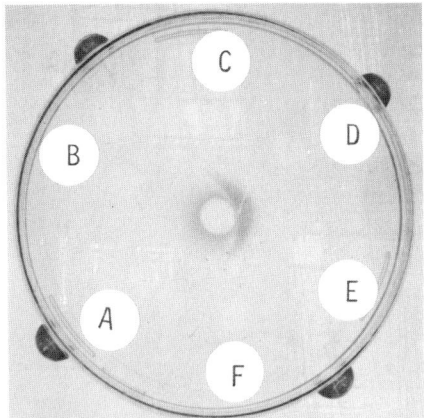

Figure 16 Immunodiffusion of viriditoxin (*C. viridis viridis*) against antisera for hemorrhagic toxin *c* (*C. atrox*). Center well, antisera for hemorrhagic toxin *c*; A and B, viriditoxin; C and F, saline control; D and E, hemorrhagic toxin *c*. (Reprinted with permission from R. J. Fabiano and A. T. Tu, Purification and biochemical study of viriditoxin, tissue damaging toxin, from prairie rattlesnake venom, *Biochemistry* 20:21. Copyright 1981. American Chemical Society.)

enzymes. When zinc was removed from hemorrhagic toxin *e* with 1,10-phenanthroline, both the proteolytic and hemorrhagic activities were equally inhibited. When the apohemorrhagic toxin thus produced was incubated with zinc, the hemorrhagic and proteolytic activities were regenerated to the same extent (Bjarnason and Tu, 1978).

The pathologic aspect of hemorrhage is discussed in detail by Dr. Ownby in ths book (see Chapter 3).

Toxins showing both Myotoxic and Hemorrhagic Activities

In the two previous sections I discussed toxins which show either myotoxic or hemorrhagic action. However, there are some toxins which have both myotoxic and hemorrhagic activities. For instance, hemorrhagic toxin *b* isolated from *C. atrox* venom has both activities (Ownby et al., 1978). It is not uncommon for a toxin to exhibit more than one biological activity; for example, some cytotoxins have been shown to also have cardiotoxic activity. Additionally, both myotoxin *a* and crotamine demonstrate myotoxic as well as hemolytic activity (Cameron and Tu, 1978). The biological versatility of these toxins is probably a function of the toxin's membrane specificity.

Viriditoxin from *C. viridis viridis* venom also has both myotoxic and hemorrhagic activities (Fabiano and Tu, 1981). Viriditoxin has an LD_{50} value of approximately 5.0 µg/g in mice, indicating the protein is also moderately lethal in addition to its tissue-damaging effect.

Both viriditoxin (*C. viridis viridis*) and hemorrhagic toxin *c* (*C. atrox*) are acidic proteins capable of inducing hemorrhage, but they are not identical immunologically and do not cross-react with each other (Fabiano and Tu, 1981) (Figure 16).

Table 6 Comparison of Tissue-Damaging Toxins Isolated from Rattlesnake Venom

	Crotalus viridis viridis		*Crotalus atrox* Hemorrhagic toxins				*Crotalus durissus durissus* Crotamine
	viriditoxin	myotoxin *a*	*a*	*b*	*c*	*e*	
Hemorrhage	+	−	+	+	+	+	−
Myonecrosis	+	+	−	+[a]	−	−	+[b]
Proteolytic	+	−	+	+	+	+	−
Total residues	1018	42[c]	636	200	213	219	24
Isoelectric pH	4.8	9.6	Acidic	Basic	Acidic	5.6	Basic
Reference	Fabiano and Tu (1981)	Cameron and Tu (1977)	Bjarnason and Tu (1978)				Laure (1975)

[a]Ownby et al. (1978).
[b]Cameron and Tu (1978).
[c]Cameron and Tu originally reported 39 residues; however, when myotoxin *a* was sequenced 42 residues were found (Fox et al., 1979).
Note: A + indicates that the biological activity has been detected for that toxin; a − indicates that the activity has not been observed.

Summary of Tissue-Damaging Toxins

A number of tissue-damaging toxins have been isolated. Some toxins are only myotoxic without hemorrhagic activity and some are hemorrhagic but not myotoxic. Some toxins have both myotoxic and hemorrhagic activities. It is not uncommon for a snake toxin to exhibit more than one type of biological activity. The tissue-damaging toxins which have been isolated from the venoms of rattlesnakes are summarized in Table 6.

Lethal Toxins

On envrnomation, most North American rattlesnakes do not cause neurotoxic symptoms, although neurotoxins can be isolated by fractionating venoms. A very small basic protein with a molecular weight of only 3300 was isolated from the venom of *C. adamanteus*. The LD_{50} of this basic protein is 5.2 μg/g, indicating that it is not as toxic as Hydrophiidae and Elapidae neurotoxins (Bonilla et al., 1971). Similar basic proteins can also be obtained from the venom of *Crotalus horridus atricaudatus* (canebrake rattlesnake). These basic proteins elevate plasma glutamic-oxalacetic transaminase, hydroxybutyric dehydrogenase, and aldolase activities. Since the first two enzymes will be elevated in the event of cardiac muscle injury and the latter because of skeletal muscle damage, it is believed that the basic proteins cause damage to the heart and skeletal muscle (Bonilla et al., 1971-1972).

Protein E was isolated from *C. horridus horridus* venom which has a molecular weight of 15,000 and an LD_{50} of 6-7 μg/g. Protein E has no effect on the neuromuscular junction, indicating a nonneurotoxic nature of the toxin. It causes paralysis of limbs and respiratory distress prior to death of the tested animals. It has no phospholipase A_2 or hemorrhagic activity. The site of the action of this toxin has not been elucidated (Sullivan and Geren, 1979; Sullivan et al., 1979).

Norepinephrine is absent in the venom of *C. adamanteus* (Anton and Gennero, 1965), and acetylcholine is also absent in rattlesnake venoms of *C. horridus* and *Sistrurus miliarius barbouri* (Welsh, 1966).

Cardiovascular Effect

Rattlesnake venom also has a profound effect on the cardiovascular system. A protein, myocardial depressor protein (MDP), was isolated from the venom of *C. atrox*. The intravenous injection of this peptide to dogs and cats produces an immediate and profound decrease in the cardiac output, the left ventricular systolic and mean pressures, the velocity of shortening of the contractile element, and the systemic arterial pressure, and it causes an elevation in the left ventricular end-diastolic and pulmonary wedge pressure (Bonilla, 1976; Bonilla and Rammel, 1976). Rattlesnake envenomation induces an immediate drop in blood pressure. This temporary hypotensive action is fairly common in other Crotalidae and Viperidae envenomations. This action probably originates from bradykinin released by the autopharmacologic action of snake venoms.

Hemolysis

Hemolysis is the lysis of erythrocytes, and the result is that hemoglobin originally confined to erythrocytes diffuses into the medium. The term hemolysis should not be confused with hemorrhage, which is the breakage of capillary tubes with the blood penetrating to surrounding tissues.

There are two types of hemolytic factors present in snake venoms. One is a direct hemolytic factor (DHF) [it is also called direct lytic factor (DLF)], which itself can hemolyze the red cell. Indirect hemolytic factor lyses the red cell very slowly, but its lytic activity can be enhanced by the addition of phosphatidylcholine or some basic compounds including DHF. However, one should be aware that there is no distinct difference to distinguish them. Indirect hemolytic factor is identified as phospholipase A_2. However, in some special cases, a basic phospholipase A_2 itself is hemolytic (Tu, 1977). Nevertheless, these two terms are convenient divisions when we discuss the phospholipase A_2 effect of hemolysis.

One should also be aware that the degree of hemolysis is not only dependent upon the hemolytic factors but also depends on erythrocytes of different species. The hemolytic factor found in rattlesnake venoms is mainly an indirect one and found to be phospholipase A_2. However, upon close examination of hemolytic activity and phospholipase A_2 activity of rattlesnake venom, the two activities do not parralel. For instance, *C. durissus durissus* venom has a very high indirect hemolytic activity with low phospholipase A_2 activity. On the other hand, *C. enyo enyo* venom has low indirect hemolytic activity with high phospholipase A_2 activity. This may suggest that other factors which are unknown to us are involved in the hemolytic process (Sosa et al., 1979).

Crotoxin, a presynaptic toxin isolated from the venom of *C. durissus terrificus*, has weak phospholipase A_2 activity and strong indirect hemolytic activity. Crotoxin is composed of an acidic component (A) and a basic component (B). The basic component possesses both indirect hemolytic activity and enzymatic activity. The function of the acid component is believed to diminish the nonspecific binding of the basic component (Jeng et al., 1978; Hendon and Tu, 1979).

Some toxins have diverse actions. For instance, myotoxin *a* also possesses direct hemolytic activity even though it is not an enzyme (Cameron and Tu, 1977).

Chemistry of Venoms

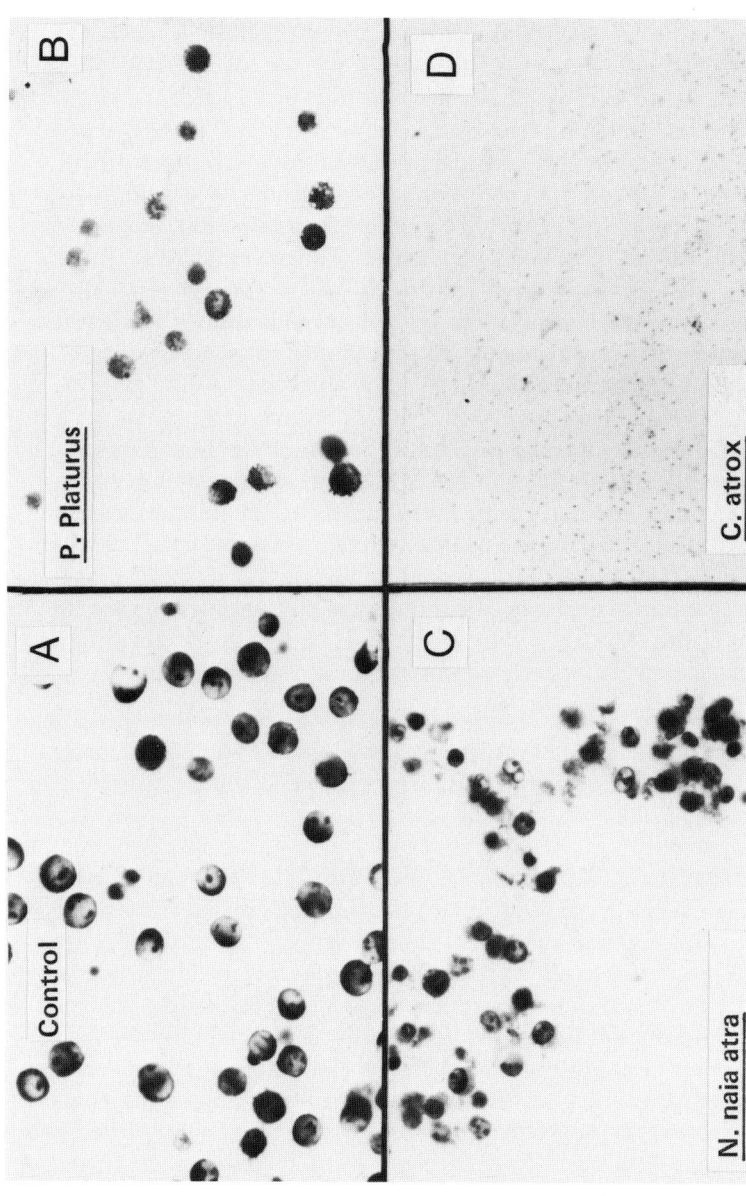

Figure 17 Cytotoxicity of snake venoms on Yoshida sarcoma cells. A. Yoshida sarcoma cell control. B. Addition of 1 mg/ml sea snake, *Pelamis platurus*, venom showing no effect. C. Addition of 1 mg/ml cobra, *Naja naja atra*, venom showing plasma membrane lysis and subsequent clumping. D. Addition of 1 mg/ml rattlesnake, *C. atrox*, venom showing complete lysis of plasma and nuclear membranes. Chromatin granules can still be seen. (From Tu and Giltner, 1974.)

Other Toxins

Acute renal hypotension and shock are frequent complications in rattlesnake envenomation. The bite of *C. durissus terrificus* causes respiratory failure in addition to the above symptoms (López et al., 1972; Silva et al., 1979). Renal failure was attributed to nephrotoxic and hemolytic venom action, although a specific nephrotoxin has not been isolated (Hadler and Brazil, 1966; Rosenfeld et al., 1960). Gyroxin is a nonlethal toxin isolated from *C. durissus terrificus* venom and comprises only 0.8 to 1.0% of the crude venom (Seki et al., 1980). The toxin is responsible for evoking the syndrome of a labyrinthic lesion. It has a molecular weight of 33,000 without arginine esterase activity.

Convulxin is a toxin also isolated from *C. durissus terrificus* venom and causes convulsions in poisoned animals (Prado-Franceschi and Brazil, 1969). Convulxin also induces aggregation of guinea pig platelets. The chemical nature of convulxin has not yet been identified.

Rattlesnake venoms have cytolytic activity. For instance, the venom of *C. viridis viridis* lyses spleen cells and Yoshida Sarcoma cells (Tu and Passey, 1969; Tu and Giltner, 1974). It is not clear whether the lytic action is due to a specific cytotoxin or to nonspecific agents such as venom proteases. In the future, this question can be solved by assay of the fractionated components for cytotoxic action. Unlike rattlesnake venoms, cytotoxins isolated from Elapidae venoms are structurally similar to cardiotoxins and postsynaptic neurotoxins (Tu, 1977) (Figure 17).

Mechanism of Overall Toxic Action

At the present stage of our knowledge it is premature to draw definite conclusions about the mechanism of toxic action for rattlesnake venoms. Despite this shortage, we can picture the following possible mechanism based on the knowledge we have so far.

In the case of the neurotoxic venoms of *C. durissus terrificus* and *C. scutulatus,* the cause of death can be attributed to presynaptic toxins, crotoxin, and Mojave toxins (see Chapter 4). Other rattlesnake venoms do contain neurotoxic components, but their content is too small to account for lethality in cases of actual poisoning. Besides tropical rattlesnake and Mojave rattlesnake venoms, the site of action of neurotoxins from other rattlesnake venoms has not been elucidated. Nevertheless, these neurotoxins probably play relatively minor roles in causing death. One characteristic of rattlesnake envenomation is the induction of considerable damage to the tissues surrounding the site of the bite. The most common of these local tissue effects is swelling, myonecrosis, and hemorrhage. Since antivenin treatments may not prevent myonecrosis and hemorrhage unless treatment is administered immediately following envenomation, these effects remain a source of serious clinical concern. Relatively minor local tissue damage is, of itself, no danger to one's life; however, severe tissue damage may become fatal to a victim. For instance, in minor tissue damage, myonecrosis and hemorrhage are localized in the limited area. In severe poisoning, extensive hemorrhage can be observed in the lung, heart, mesenteries, small intestine, kidneys, and many other internal organs. Hemorrhage is due to breakage of the endothelial cells of capillary tubes which causes the loss of circulating fluid, erythrocytes, neutrophils, eosinophils, basophils, platelets, and many other compounds associated with blood. Prolonged hemorrhage certainly will cause an imbalance in the hemodynamic state of a victim and may even cause the loss of circulating blood volume. Hypovolemia may, indeed, be the cause of death on some occasions. Unlike the poisoning of cobra, kraits, and sea snakes which causes rapid neurotoxic symptoms after the bite, rattlesnake bite may be painful but death may not come until 2 or 3 days after the bite. Since hypovolemia is a relatively slow process, death does not come immediately.

Initial transient hypotension is common in all snake envenomations. Such a drop in blood pressure usually does not last more than 20 min. In rattlesnake venoms, this drop in blood pressure is probably due to bradykinin released from the victim's plasma protein by the action of kallikrein or bradykinin-releasing enzyme present in the venoms. Therefore, the hypotensive action is a secondary autopharmacologic action rather than a primary venom action. Cobra and other Elapidae venoms do not contain such an enzyme; therefore, they do not release bradykinin, yet they show similar immediate drops in blood pressure. In the case of Elapidae venoms, this action is probably due to endogeneous histamine present in the venoms.

Neither bradykinin nor histamine are lethal factors as bradykinin- and histamine-releasing enzymes can be separated from the lethal fraction. Moreover, antihistamine drugs do not reduce the lethal action of rattlesnake venoms.

Rattlesnake venoms are rich in a variety of enzymes. In Elapidae and Hydrophiidae venoms, enzymes probably play a less important role than potent neurotoxins, cardiotoxins, and cytotoxins. In rattlesnake venoms, enzymes play a more important role in toxic actions. All five hemorrhagic toxins isolated from *C. atrox* venoms have proteolytic activity (Bjarnason and Tu, 1978). Recent infestigation by Kress and his coworkers indicates that venom protease hydrolyzes human plasma proteinase inhibitor which binds to all proteases (Kress et al., 1979; Kress and Catanese, 1980). This may suggest that venom protease activates all proteases in the victim. Proteases are known to cause a variety of tissue damage. Thus, venom protease may play a primary role as well as a secondary action in tissue damage. Rattlesnake venom also contains procoagulant and anticoagulant fractions. Individual components are not lethal, as each fraction can be separated from the lethal fraction. Rattlesnake venoms are also known to produce thrombosis. Thrombosis formation can become lethal depending on the place it is produced, but one should keep in mind that even though the venom is potentially thrombotic it does not always produce thrombosis every time the venom is injected.

As mentioned before, the toxins responsible for hemorrhage and myonecrosis may be responsible for lethal action. There are experimental data to support this assumption. For instance, myotoxin *a* and viriditoxin are not as toxic as crude venom, but the LD_{50} values of 3.13 and 5.0 µg/g in mice are still fairly toxic. Hemorrhagic toxins are much weaker in lethality than myotoxins, as can be seen from Table 6. However, the data in the table are based on the occurrence of death in 24 hr. As mentioned before, the average time from rattlesnake bite to death is 2 to 3 days. Therefore, lethality of the hemorrhagic toxin (observed in 24 hr) may be weak, and prolonged bleeding may eventually cause hypovolemia and subsequent death by shock after 2 or 3 days. Thus, both hemorrhagic toxins and myotoxins certainly contribute to the lethal action significantly.

Hemclysis is the rupture of the erythrocyte membrane so that the hemoglobulin diffuses into the surrounding media. Rattlesnake venoms usually cause hemolysis, because of the presence of an indirect hemolytic factor, phospholipase A_2. Minor hemolysis does not cause death of a victim, but it is one type of toxic action. Hemolytic fractions, indeed, can be separated from lethal fractions. Nevertheless, hemolytic action contributes to the overall toxic action in rattlesnake envenomation.

NONLETHAL ACTIONS OF RATTLESNAKE VENOMS

As mentioned before, rattlesnake venom is a mixture of different proteins with diverse biological actions. Unlike the venoms of Elapidae and Hydrophiidae, most rattlesnake venoms do not contain potent neurotoxins and cardiotoxins. Myotoxins and hemorrhagic toxins are lethal, but their toxicities are weaker than the original venoms (Table 7). This is highly suggestive of synergistic effects of the different components of rattlesnake venoms. Therefore, if one pure component is isolated from the venom, the toxicity of that component may be very weak. Nonetheless, these relatively weak toxic substances may contribute to the overall toxic action in the case of rattlesnake envenomation. In this section I will discuss the nonlethal action of rattlesnake venoms. However, one should keep in mind the division of toxic and nontoxic is sometimes arbitrary. I have selected components which are relatively nontoxic; however, strictly speaking, all components discussed in this section can be toxic if a large quantity is injected.

Table 7 Toxicity of Rattlesnake Venoms and their Components in Mice

Venom or component	LD_{50} ($\mu g/g$)	Route of injection
Crotalus atrox		
Venom	3.6	i.v.
	3.7	i.p.
	4.2	i.v.
Hemorrhagic toxin *b*	12.8	i.v.
Hemorrhagic toxin *c*	Nonlethal at 35 $\mu g/g$	i.v.
Hemorrhagic toxin *d*	Nonlethal at 35 $\mu g/g$	i.v.
Hemorrhagic toxin *e*	27.6	i.v.
Crotalus viridis viridis		
Venom	1.29	i.v.
Myotoxin *a*	Lethal at 3.13	i.v.
Viriditoxin	5.0	i.v.

Nerve Growth Factor

Snake venoms contain nerve growth factor (NGF) which promotes the growth of the sympathetic chain ganglia in vivo and a halo-like outgrowth of nerve fibers from embryonic sensory ganglia cultured in vitro (Figure 18). NGF isolated from cobra venom has an amino acid sequence homology similar to that of NGF from mouse submaxillary glands.

The presence of NGF in snake venoms is quite common and so far has been found in the venoms of Elapidae, Viperidae, and Crotalidae. NGF was reported in the following rattlesnake venoms:

Crotalus adamanteus (Cohen, 1959; Angeletti, 1968b)
C. atrox (Angeletti, 1968a)
C. durissus terrificus (Angeletti, 1968a)
C. horridus (Cohen, 1959)

The molecular weight of NGF from *C. adamanteus* venom is 30,000 (Angeletti, 1968b). Sulkowski et al. (1975) isolated a NGF-like nontoxic basic protein weight of 24,300 and pI of 9.8 from the same venom.

Microcomplement fixation shows that there is a cross-reactivity for *C. atrox* NGF and mouse salivary gland NGF (Zanini et al., 1968). *Crotalus adamanteus*

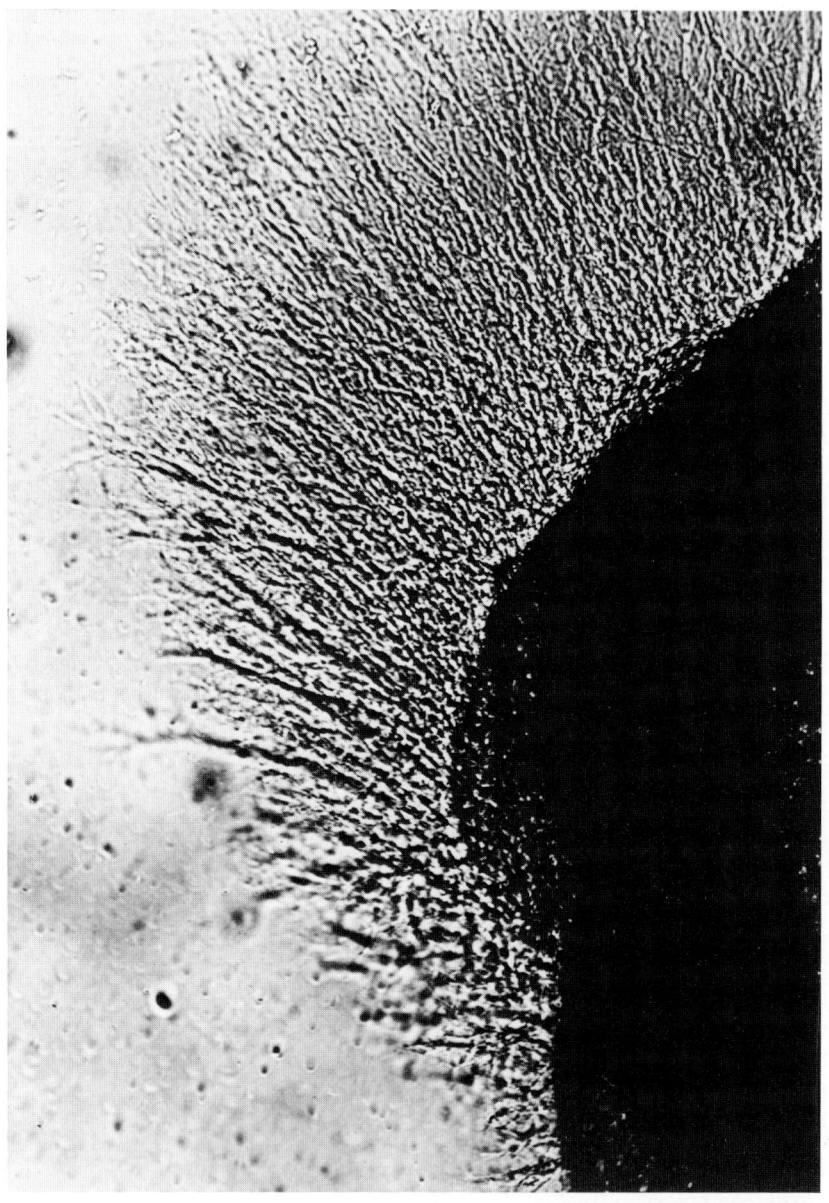

Figure 18 Effect of *Crotalus adamanteus* NGF on sensory ganglia of 9-day-old chick embryos. (Courtesy of J. R. Perez-Polo, University of Texas Medical Branch, Galveston, Texas.)

NGF activity can be suppressed by antibody to NGF made from mouse salivary gland NGF, indicating there is some molecular similarity between rattlesnake NGF and mouse NGF. *Crotalus adamanteus* β NGF is a basic protein of 14,500 daltons with an isoelectric point of 9.5. The $\gamma\beta$ NGF is a high molecular weight NGF complex which can also be isolated from the same venom. The γ protein has arginine-esterase activity, whereas the basic β unit has none. Snake venom β NGF is very similar to mouse β NGF, indicating that there is a strong evolutionary preservation of the NGF molecule in vertebrates (Pérez-Polo et al., 1978). The function of the γ component is to cleave a pro-NGF chain of larger size (Levi-Montalcini and Calissano, 1979).

Proteolytic enzyme activity associated with the 116,000 molecular weight form of mouse NGF is an intrinsic property of the NGF protein (Young, 1979). It is logical to assume that snake venom NGF also possesses such a proteolytic activity.

Blood Coagulation

Unlike neurotoxic venoms of Elapidae and sea snakes, rattlesnake venom has a profound effect on blood coagulation. Snake venoms are quite frequently referred to as "coagulant" or "anticoagulant" depending on their primary effect. Such a division is a convenient way to describe venoms, but the actual situation is much more complicated. Some venoms contain both factors but each factor can be separated when they are fractionated. Crotalase is a thrombin-like enzyme isolated from *C. adamanteus* venom. It converts fibrinogen to fibrin; therefore, one may say crotalase is a coagulant. However, the clot formed is different from ordinary fibrin and forms microclots which are rapidly lysed by the fibrinolytic action of tissues. The result is a prolonged defibrinogenated state and a prevention of coagulation. Thus, crotalase is a coagulant in vitro but an anticoagulant in vivo.

A further complication is that coagulability of blood may also depend on the age of the snake. Reid and Theakston (1978) studied this aspect for the venoms of *C. atrox*. They found that, up to 8 months, the venoms clotted fibrinogen solutions directly. At 9 to 10 months, plasma was clotted but not fibrinogen. Subsequently, the venoms no longer clotted plasma. This may explain some of the conflicting reports both in the clinical aspects of snakebite in man and in experimental studies. Some reports show that *C. atrox* venom is noncoagulant (Eagle, 1937; Didisheim and Lewis, 1956), while others show it as a coagulant (Oshima et al., 1969; Copley et al., 1973) (Figure 19).

Thrombin-like Enzymes

Thrombin-like enzymes in rattlesnake venoms are common. As early as 1964, Nahas et al. (1964) observed that *C. durissus terrificus* venom showed thrombin-

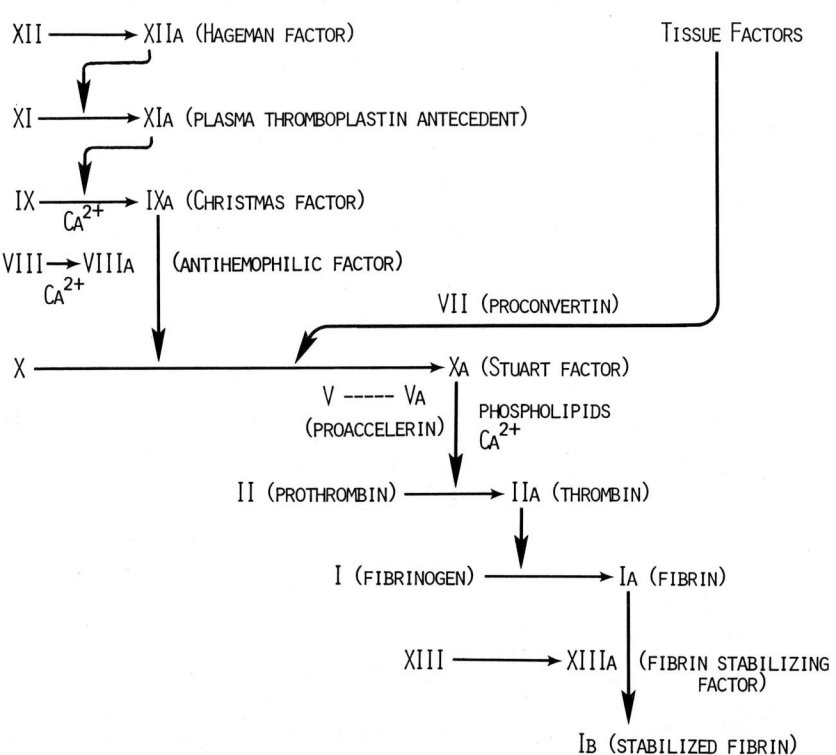

Figure 19 Scheme of blood coagulation.

like activity but is much weaker than that of *Agkistrodon rhodostoma* (Malayan pit viper) venom. Prothrombin activator activity is absent in the venom of *C. durissus terrificus*, although it is strong in the Viperidae venom of *Echis colorata* (Nahas et al., 1964).

Clinical observation of afibrinogenemia following the bite of an Eastern diamondback rattlesnake (*C. adamanteus*) was reported by Weiss et al. (1969). They even observed the single fibrinogen degradation product associated with the condition of defibrinogenation.

Isolation of Arvin (Ancrod), a thrombin-like enzyme, from the Malayan pit viper (*Agkistrodon rhodostoma*) venom greatly stimulated similar enzyme isolation from other venoms. In rattlesnake venom the isolation was made from the venoms of *C. adamanteus* (crotalase) and *C. horridus horridus* (Markland and Damus, 1971; Bonilla, 1975).

A fibrinogen molecule contains two subunits; each subunit consists of three polypeptide chains, namely $\alpha(A)$, $\beta(B)$ and γ. The molecular weights of these chains are 64,000, 57,000, and 48,000 respectively. Two subunits are connected by a number of disulfide bonds. For simplicity, a fibrinogen molecule is expressed diagramatically in Figure 20.

When thrombin acts upon fibrinogen, two fibrinopeptides, A and B, are released. The remaining portion of the molecule (fibrin monomer) undergoes extensive polymerization to produce a fibrin clot. The polymerization of fibrin apparently involves cross-links between lysine and the glutamide side chain.

$$(\gamma)CONH_2 \quad\quad (\epsilon)NH_2$$
$$| \quad\quad\quad\quad\quad |$$
$$CH_2 \quad\quad\quad\quad CH_2$$
$$| \quad\quad + \quad\quad | \quad\quad\quad Ca(II)$$
$$CH_2 \quad\quad\quad\quad CH_2 \quad\quad \xrightarrow{\text{fibrin stabilizing factor}} \quad Glu(\gamma)CONH(\epsilon)\,Lys=$$
$$| \quad\quad\quad\quad\quad |$$
$$CH \quad\quad\quad\quad CH_2$$
$$\quad\quad\quad\quad\quad\quad |$$
$$\quad\quad\quad\quad\quad\quad CH_2$$
$$\quad\quad\quad\quad\quad\quad |$$
$$(Gln) \quad\quad\quad\quad CH$$
$$\quad\quad\quad\quad\quad (Lys)$$

In contrast to thrombin, crotalase cleaves only the A peptides from human fibrinogen by splitting the Arg-Gly bond. Furthermore, crotalase slowly degrades the β (B) chains producing smaller fragments. This property distinguishes crotalase from Arvin (Markland and Pirkle, 1977a,b) (Figure 21).

There are, in addition, other differences between crotalase and thrombin. Unlike thrombin, crotalase has no effect on platelet aggregation, platelet release, factor VIII activation, and factor V activation. Crotalase hydrolyzes prothrombin but does not produce thrombin (Pirkle et al., 1976).

Crotalase is a glycoprotein containing sialic acid. The isoelectric point of the apoprotein is 4.6. Coagulant activity is not affected by neuraminidase treatment (Bajwa and Markland, 1979).

A partial amino acid sequence of crotalase was identified. From the data of two fragments, 36% of the positions were occupied by amino acids identical to those in thrombin, and an additional 21% of the positions were occupied by amino acids chemically similar to those in thrombin (Baumgartner et al., 1980).

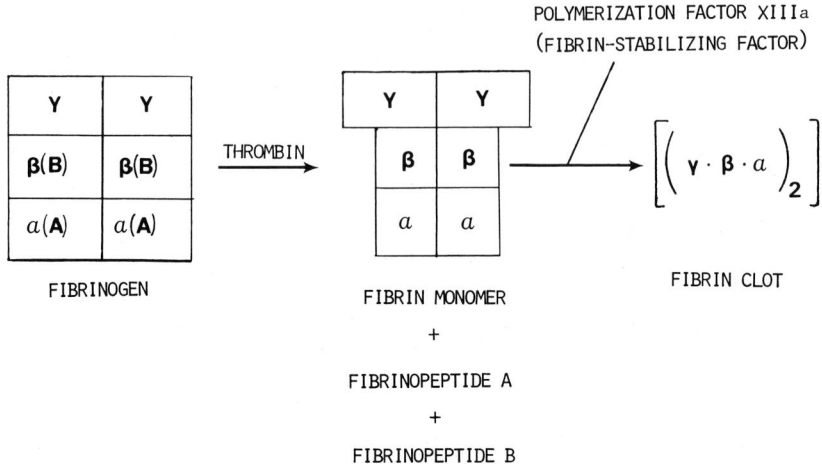

Figure 20 Schematic diagram of thrombin action on fibrinogen.

Crotalase is not only similar to thrombin in amino acid sequence but also similar to factor IXa, factor Xa, and kallikrein. The N-terminal sequence of these four enzymes is compared (Pirkle et al., 1979, 1981).

Crotalase	Val	Ile	Gly	Gly	Asp	Glu
Factor IXa	Val	Val	Gly	Gly	Glu	Asp
Factor Xa	Ile	Val	Gly	Gly	Arg	Asp
Thrombin	Ile	Val	Glu	Gly	Ser	Asp

The defibrinogenation state in rabbits owing to injections of crotalase lasts only a few hours but can be prolonged to several days by daily infusion (Bajwa and Markland, 1978).

Thrombin-like enzyme is commonly present in other rattlesnake venoms such as *Crotalus durissus durissus, C. horridus, C. terrificus basilicus, C. durissus terrificus,* and *Crotalus viridis helleri* (Janszky, 1950; Denson, 1969; Denson et al., 1972; Stocker, 1978).

Fibrinogenolytic and Fibrinolytic Action

Fibrinogenolytic action refers to the hydrolysis of fibrinogen, and fibrinolytic action refers to the hydrolysis of fibrin. It is not clear whether these actions are due to specific or nonspecific proteases present in snake venoms. As discussed

Chemistry of Venoms

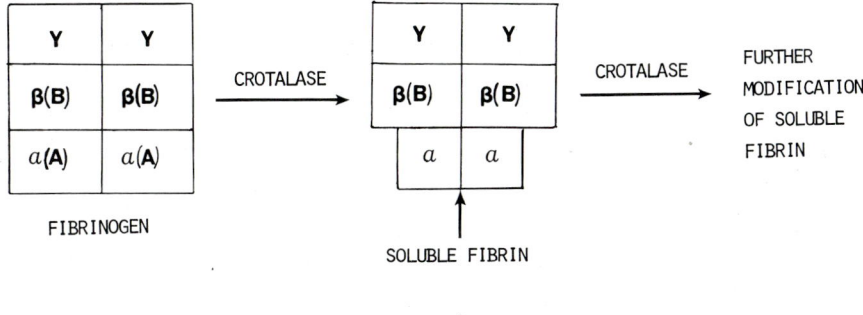

Figure 21 Schematic diagram of crotalase action on fibrinogen.

in the section on proteolytic enzymes, rattlesnake venoms are rich in proteolytic enzymes. If one can isolate a specific fibrinolytic enzyme, it may be useful in dissolving blood clots of thrombosis and myocardial infarction.

Fibrinolytic venoms reported for rattlesnakes are *C. basiliscus, C. atrox, C. horridus, C. adamanteus, C. viridis hellerii, C. durissus terrificus* (Didisheim and Lewis, 1956; Bajwa et al., 1980; Ruiz et al., 1980).

Factor X Activating Enzyme

Factor X (Stuart factor, autoprothrombin C) is involved in the blood coagulation mechanism. Any enzyme that activates factor X into its active form should promote blood coagulation. Extensive study has been made of factor X from Russell's viper (*Vipera russellii*) venom. Factor X activating factor has been reported in *C. viridis helleri* venom but not in the venoms of *C. durissus durissus, C. adamanteus, C. atrox, C. scutulatus scutulatus,* and *Sistrurus miliarus barbouri* (Denson et al., 1972). This activity is also absent in the venom of *C. durissus terrificus,* although it is commonly found in the venoms of the closely related pit viper *Bothrops* (Nahas et al., 1964).

Platelet Aggregation

Rattlesnake venom causes aggregation of platelets. Strictly speaking, platelet aggregation is outside the blood coagulation system. However, it is better to discuss it here than in any other place. Original venom of *C. atrox* causes the aggregation of platelets in vivo (Ownby et al., 1974). Rattlesnake envenomation induces severe hemorrhage. Platelets produce blood coagulation enzymes. Platelet aggregation is probably a part of the victim's defense mechanism trying

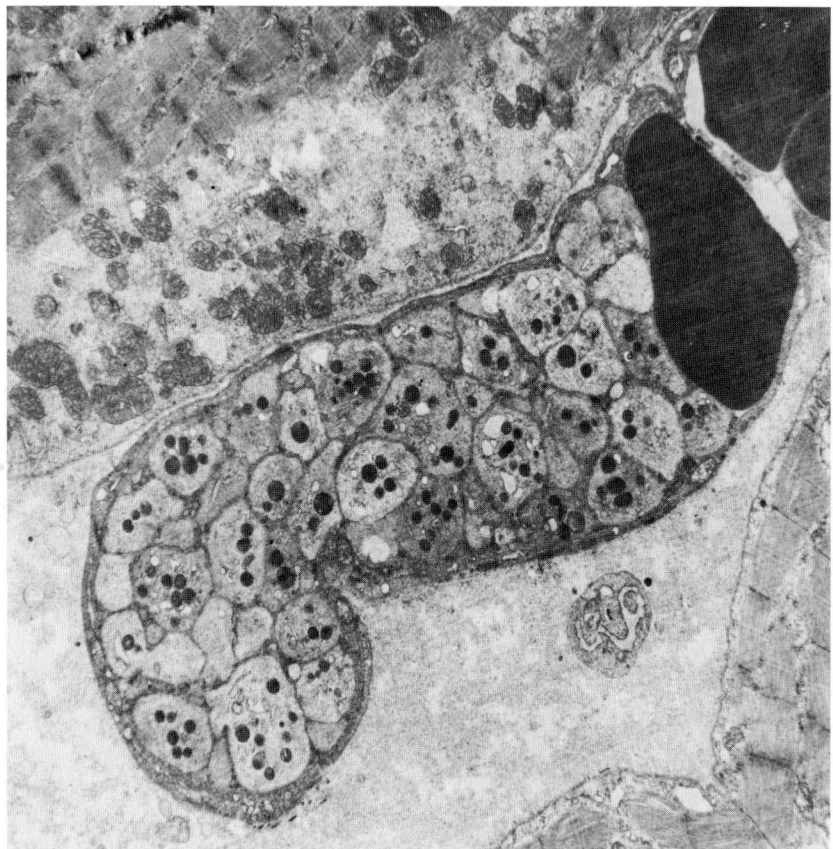

Figure 22 Platelet aggregation in the lumen of a capillary induced by *Crotalus atrox* venom. (Courtesy of C. L. Ownby, Oklahoma State University, Stillwater, Oklahoma.)

to prevent further leakage of blood from the vessels by forming a hemostatic plug. The result is often the occlusion of the capillary lumen (Figure 22). This hypothesis is further substantiated by the fact that pure hemorrhagic toxins have exactly the same effect on platelets as crude venom (Ownby et al., 1978).

Such an aggregation causes a decrease in platelet count as indicated by a case of *Crotalus ruber ruber* poisoning (Lyons, 1971). In severe rattlesnake poisoning, platelet counts can be as low as 4000 per cubic millimeter. The low platelet count is believed to be due to the destruction of platelets by the direct action of snake venom (Glass, 1976).

Chemistry of Venoms

Crotalocytin is a platelet-aggregating protein isolated from timber rattlesnake venom. It is a serine protease possessing proteolytic activity. EDTA blocks the ability to activate platelets (Schmaier and Colman, 1980). Thrombocytopenia after timber rattlesnake bite appears to be due to a protein that directly activates platelets (Schmaier et al., 1980). Convulxin also induces platelet aggregation (Vargaftig et al., 1980).

Antivenom Activity of Blood

Snakes are not immune to venoms, as they can be killed by the injection of their own venom. Frequently snakes in a snake farm are killed by being bitten by other snakes of the same species. However, a snake has more resistance to its own venom because of the presence of a plasma protein which neutralizes its own venom (Nichol et al., 1933; Keegan and Andrews, 1942; Clark and Voris, 1969).

The antivenom factors of *C. atrox* and *C. adamanteus* also have a neutralization capacity for other rattlesnake venoms (*Crotalus viridis concolor, C. durissus terrificus,* and *C. scutulatus scutulatus* (Straight et al., 1976). The protein responsible for the neutralization of venom has not been isolated from rattlesnake venom but has a molecular weight range of from 50,000 to 100,000. The fraction with neutralization activity also exhibits an inhibitory action on venom proteases (Philpot et al., 1978). Thus, the neutralization of venom by the serum protein is at least partly due to inhibition of venom protease action.

The protective capacity of the serum against pit viper venom is not restricted to that of venomous snakes. For insance, the serum of a nonpoisonous snake, the king snake, can neutralize pit viper venom (Philpot and Smith, 1950).

Even warm-blooded animal serum possesses a resistant activity toward venoms. The Virginia opossum (*Didelphis virginiana*) showed no local tissue damage when the shaven portion of the animal was bitten by *C. adamanteus, C. atrox,* and *C. horridus horridus* (Kilmon, 1976; Werner and Vick, 1977). Different species of mouse, rat, and racoon serum also have antihemorrhagic activity against *C. atrox* venom (Pérez et al., 1978, 1979).

It is easier to imagine snake serum possessing antivenom activity towards its own venom than to explain antivenom activity in the serum of warm-blooded animals. The exact reason will never be known, but these animals are prey for rattlesnakes. It may be that the natural selection rule applies in this case; thus, the animals with the highest resistance activity survive. Over many years such prey eventually possess antivenom activity through evolution.

Anticomplement Action

It is well known that cobra venom factor (CVF) isolated from cobra venoms activate the complement system without antigen-antibody interaction. Actually

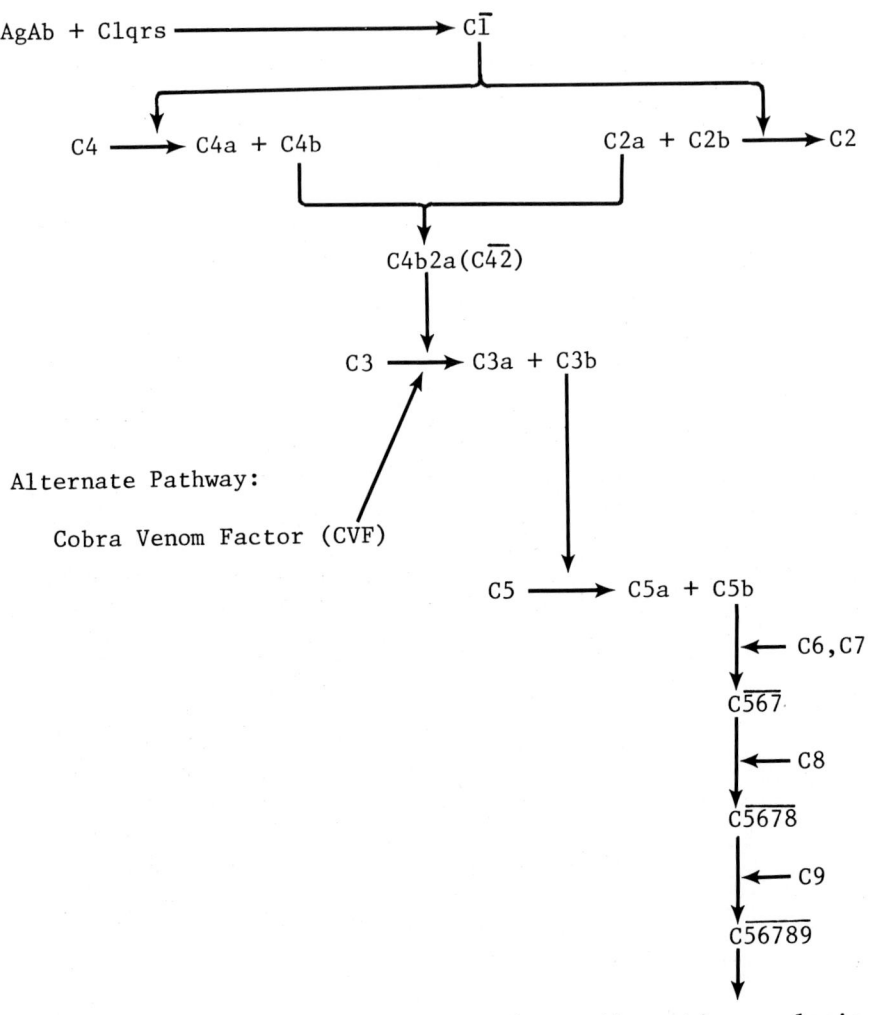

Figure 23 Diagram showing the complement system.

CVF bipasses the normal complement pathway by activating C3 (Tu, 1977; Alper, 1979) as shown in Figure 23.

Venoms of other snakes are also known to have an effect on the complement system, although their mechanisms are not well defined (Birdsey et al., 1971).

Crotalus atrox (Western diamondback rattlesnake) venom has an anticomplementary effect; thus, the incubation of serum with venom results in a loss of hemolytic activity (Minta et al., 1977). The complement factors C1t, C2, C4, C3, C5, and factor B are cleaved by *C. atrox* venom, indicating that the mechanism of anticomplementary activity of different venoms differ. There are 4 anticomplementary factors (Q_1-Q_4) in *C. atrox* venom, and they differ in molecular size and specificity. The molecular weight of Q_1, Q_2, Q_3, and Q_4 are 19,200, 14,500, 12,600, and 55,000, respectively. Although Q_2 and Q_3 possess esterase activity, the others do not when t-butyloxycarbonyl-alanine-*p*-nitrophenol is used as substrate. All 4 factors are proteolytic enzymes cleaving complement substrates (Man and Minta, 1977).

Serum sickness in the case of snakebite after receiving antivenin treatment is probably related to complement activation. A Danish youth who was bitten by his pet *C. atrox* showed signs of intramuscular complement activation in the form of low C3, C4, and C5 values. In this case, the mechanism is complicated because, presumably, two mechanisms take place simultaneously. One is the classic pathway activation of the complement system by immune complexes, and the other is the activation via the alternative pathway owing to the direct action of venoms. Nielsen et al. (1978) believe that immediate lowering of C3, C4, and C5 is due to a transient activation of the classic pathway by rapidly formed snake venom-antivenin complexes. However, the appearance of active factor B on day 3 or day 4 after the snakebite is due to the induced activation of the complement system via the alternative pathway initiated by envenomation.

Crotalus atrox anticomplementary factor is very similar to the cobra venom factor. The subunit compositions of the two factors are similar and are composed of α and β chains held together by covalent and noncovalent bonds. So the *C. atrox* factor cross-reacts with antiserum to CVF. On an equal weight basis, the *C. atrox* factor has one-third the anticomplementary activity of CVF (Minta and Man, 1980).

Autopharmacologic Action

Rattlesnake venom possesses a variety of autopharmacologic actions. On envenomation, substances which are not present in venom are released from the tissues of envenomated animals. Bradykinin, histamine, serotonin, and ATP are released. Among these the first compound is the most extensively studied.

Bradykinin

Bradykinin is released by the proteolytic action of venom kallikrein on bradykininogen, a precursor in the globulin fraction of plasma, intestine, uterus, and smooth muscle (Rocha e Silva et al., 1949). Bradykinin is a nonapeptide and has a sequence of

Arg-Pro-Pro-Gly-Phe-Ser-Pro-Phe-Arg

In some cases, kallidin (lysylbradykinin can also be released. Kallikrein is also called bradykininogenase or bradykinin-releasing enzyme. Trypsin also liberates bradykinin.

Therefore, the bradykininogen level in rattlesnake bite victims is low (Margolis et al., 1965; Russell, 1965). Bradykinin-releasing activity of a number of snake venoms was investigated by Deutsch and Diniz (1955), Mebs (1968, 1970), and Oshima et al. (1969), and it was found that such activity is common in *Crotalus* and *Sistrurus* venoms (Habermann, 1961).

The kinin-releasing activity can be inhibited with DFP (diisopropyl-fluorophosphate), indicating the enzyme is a serine-protease. The enzyme activity is also inhibited by the arginine ester hydrolase.

Bradykinin is a potent hypotensive agent which causes an immediate drop in blood pressure. When an animal is injected with a snake venom, there is an immediate drop in blood pressure, which is caused both by bradykinin and histamine. Elapidae snake venoms do not contain bradykinin-releasing enzyme and hence do not produce bradykinin on envenomation. This is due to the histamine present in Elapidae venoms.

Bradykinin-induced hypotensive action is transient, and an animal recovers to normal blood pressure in a short time. Bradykinin is also a pain-producing substance. When one is bitten by rattlesnakes or other Crotalidae and Viperidae snakes, one feels more pain than when bitten by Elapidae or sea snakes. Both Crotalidae and Viperidae venoms produce bradykinin on envenomation, whereas the snakes of the other two families do not. Thus, it is logical to conclude bradykinin contributes to the pain of rattlesnake bites.

Release of bradykinin is not a cause of death. This can be shown from the fact that the lethal fraction can be separated from bradykinin-releasing component.

Potentiation of Bradykinin Action

Certain peptides having a pyroglutamic acid as the amino terminal enhance the bradykinin activity. This is an angiotensin-converting enzyme inhibitor. A potent hypertensive compound, angiotensin II, is produced from its precursor, angiotensin I, by the action of the angiotensin-converting enzyme. If the formation of angiotensin II is inhibited, it naturally inhibits hypertensive action or increases the hypotensive action. Thus, it appears that angiotensin-converting

Chemistry of Venoms

A. Absence of the enzyme inhibitor:

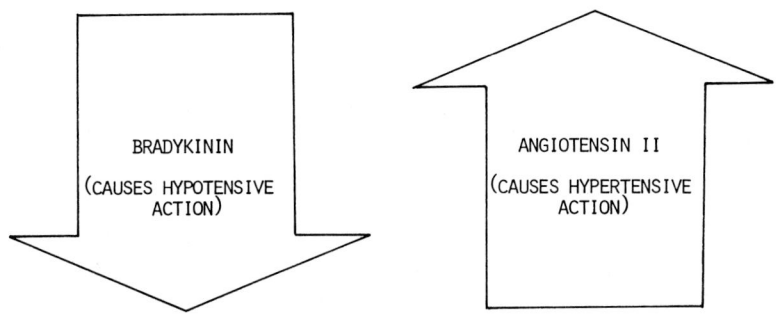

B. In the presence of enzyme inhibitor, formation of angiotensin II is inhibited:

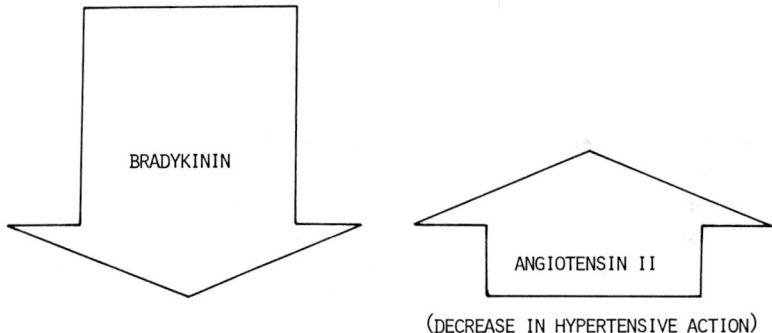

Figure 24 Actions of bradykinin (hypotensive agent) and angiotensin (hypertensive agent) in relation to angiotensin-converting enzyme inhibitor. In the presence of angiotensin-converting enzyme inhibitor, the formation of angiotensin II is inhibited; thus the net result is an increase in the hypotensive action.

enzyme inhibitor enhances the bradykinin action which is a hypotensive agent. The mutual relationship of bradykinin, angiotensins, and the inhibitor is shown in Figure 24.

Angiotensin-converting enzyme inhibitor is also called "Bradykinin Potentiating Factor (BPF)," but the former is more correct. The enzyme inhibitor can be found in the venoms of many Crotalidae, including rattlesnakes. It was reported in the venom of *C. viridis viridis, C. durissus terrificus, C. atrox, C.*

adamanteus, and *C. viridis helleri,* but the activity was not as strong as that of *Bothrops jararaca* (Ferreira, 1966).

Histamine Release

Histamine is a hypotensive agent and also a powerful vasodilating compound. On envenomation, one receives the histamine effect from two sources. One is endogenous histamine present in venom, while the other is exogenous histamine released from the victim's tissue by the action of histamine-releasing factor present in the venom. Histamine release is common among different snake venoms (Feldberg and Kellaway, 1937). Histamine can be released in many organs such as the lungs, liver, mesenteries, skin, platelet, ileum, and diaphragm. Thus, histamine levels increase in the blood of a victim. Histamine liberation occurs in the whole animal or from isolated tissues in vitro. Histamine release was observed in the following rattlesnake venoms:

C. adamanteus (Rocha e Silva and Essex, 1942; Högberg and Üvnas, 1957).
C. atrox (Dragstedt et al., 1938; Feldberg and Kellaway, 1937; Högberg and Üvnas, 1957)
C. horridus (Chute and Waters, 1941)
C. durissus terrificus (Rocha e Silva, 1949; Markwardt et al., 1966; Gonçalves, 1956; Rothschild, 1966; Högberg and Üvnas, 1957)

Histamine liberation is probably induced by different factors. One cause is apparently due to lysophosphatidylcholine liberated from phosphatidylcholine by the action of venom phospholipase A_2. For instance, histamine can be released from mast cells by the action of lysophosphatidylcholine.

The mode of histamine release has not been elucidated yet, but it is not entirely due to phospholipase A_2 as the two enzymes can be separated from each other by fractionation (Rothschild, 1966, 1967). Crotamine also releases histamine (Gonçalves, 1956). This is a rather isolated case and other enzymes must be responsible for histamine release as most venoms do not contain crotamine. From all these observations, it is reasonable to conclude that histamine can be liberated from different tissues by multiple mechanisms.

Although histamine lowers the blood pressure, it is not resonsible for lethal activity as antihistamine drugs do not reduce the lethal effects.

Many snake venoms including those of rattlesnakes increase vascular permeability which can be tested readily by the use of a dye. Normally dye will not leak from blood vessels owing to its impermeability. When an animal is injected with dye intravenously, the blue spot of the dye around the site of a venom injection in the skin appears because of increasing vascular permeability by venom action. The vascular permeability increase is probably due to histamine rather than bradykinin release. There is an excellent review on histamine release owing to different snake venoms; readers are advised to read Rothschild and Rothschild (1979) for more detailed information. For contribution to the overall toxic effect of rattlesnake venom, released histamine probably plays a relatively minor role.

Other Pharmacologically Active Substances

Serotonin can be released by rattlesnake venoms. The only well-studied work was done by Markwardt et al. (1966). Serotonin-releasing factor is a protein and can be separated from phospholipase A_2 and blood coagulation activity.

Serotonin has only a weak vascular permeability effect; therefore, a contribution to the overall toxic effect in rattlesnake envenomation is very minor. However, like bradykinin, serotonin also causes pain upon injection. Serotonin liberated by the action of rattlesnake venoms may enhance the feeling of pain.

Many other substances are probably released on rattlesnake envenomation. It was reported that cobra venom releases prostaglandin (Vogt et al., 1969). It is not clear yet whether rattlesnake venom will produce such a compound because no one has investigated this area as yet. Similarly, epinephrine release from the adrenal glands with cobra venom was reported (Feldberg, 1940). Again, it is not clear whether this is also the case for rattlesnake venoms because of the lack of investigation.

Aggregation of platelets was observed in mice after the injection of *C. atrox* venom and hemorrhagic toxin (Ownby et al., 1974; 1978). It is believed that the aggregation of platelets is induced by liberated ADP.

ACKNOWLEDGMENT

This research was supported by NIH grant 5R01 GM15591.

REFERENCES

Alper, C. A. (1979). Snakes in the complement system. In *Snake Venoms*, C. Y. Lee (Ed.). Springer-Verlag, Berlin, pp. 863-880.
Angeletti, R. H. (1968a). Studies on the nerve growth factor (NGF) from snake venom: Gel filtration patterns of crude venoms. *J. Chromatogr. 36*:535-537.
Angeletti, R. H. (1968b). Nerve growth factor (NGF) from snake venom: Molecular heterogeneity. *J. Chromatogr. 37*:62-69.
Anton, A. H., and Gennaro, J. F. (1965). Norepinephrine and serotonin in the tissues and venoms of two pit vipers. *Nature 208*:1174-1175.

Bailey, G. S., Lee, J., and Tu, A. T. (1979). Conformational analysis of myotoxin *a* (muscle degenerating toxin) or prairie rattlesnake venom. Predictions from amino acid sequence, circular dichroism, and Raman spectroscopy. *J. Biol. Chem. 254*:8922–8926.

Bajwa, S. S., and Markland, F. S. (1978). Defibrinogenation studies with crotalase: Possible clinical applications. *Proc. West. Pharmacol. Soc. 21*:461

Bajwa, S. S., and Markland, F. S. (1979). A new method for purification of the thrombin-like enzyme from the venom of the eastern diamondback rattlesnake. *Thromb. Res. 16*:11–23.

Bajwa, S. S., Markland, F. S., and Russell, F. E. (1980). Fibrinolytic enzyme(s) in western diamondback rattlesnake (*Crotalus atrox*) venom. *Toxicon 18*: 285–290.

Basu, A. S., Parker, R., and O'Connor, R. (1969). Disc electrophoresis of snake venoms. I. A qualitative comparison of the protein patterns from the species of Crotalidae and Elapidae. *Can. J. Biochem. 47*:807–810.

Baumgartner, R., Fletcher, T., Theodor, I., Bajwa, S. S., Markland, F. S., and Pirkle, H. (1980). Amino acid sequences in crotalase, a thrombin-like enzyme from the venom of *Crotalus adamanteus. Fed. Proc. 39*:1027.

Bernick, J. J., and Simpson, J. W. (1976). Distribution of elastaselike enzyme activity among snake venoms. *Comp. Biochem. Physiol. 54B*:51–54.

Bertke, E. M., Watt, D. D., and Tu, T. (1966). Electrophoretic patterns of venoms from species of Crotalidae and Elapidae snakes. *Toxicon 4*:73–76.

Bieber, A. L., Tu, T., and Tu, A. T. (1975). Studies of an acidic cardiotoxin isolated from the venom of Mojave rattlesnake (*Crotalus scutulatus*). Biochim. Biophys. Acta 400:178–188.

Birdsey, V., Londorfer, J., and Gewurz, H. (1971). Interaction of toxic venoms with the complement system. *Immunology 21*:299–310.

Bjarnason, J. B., and Tu, A. T. (1978). Hemorrhagic toxins from Western diamondback rattlesnake (*Crotalus atrox*) venom: Isolation and characterization of five toxins and the role of zinc in hemorrhagic toxin e. *Biochemistry 17*:3395–3404.

Bonilla, C. A. (1975). Defibrinating enzyme from timber rattlesnake (*Crotalus horridus horridus*) venom: A potential agent for therapeutic defibrination. I. Purification and properties. *Thromb. Res. 6*:151–169.

Bonilla, C. A. (1976). Hypotensin—a hypotensive peptide isolated from *Crotalus atrox* venom: Purification, amino acid composition and terminal amino acid residues. *FEBS Lett. 68*:297–302.

Bonilla, C. A., and Rammel, O. J. (1976). Comparative biochemistry and pharmacology of salivary gland secretions. III. Chromatographic isolation of a myocardial depressor protein (MDP) from the venom of *Crotalus atrox. J. Chromatogr. 124*:303–314.

Bonilla, C. A., Fiero, M. K., and Frank, L. P. (1971). Isolation of a basic protein neurotoxin from *Crotalus adamanteus* venom. In *Toxins of Animal and Plant Origin*, A. DeVries and E. Kochva (Eds.). Gordon and Breach, New York, pp. 343–360.

Bonilla, C. A., Fiero, M. K., and Novak, J. (1971-1972). Serum enzyme activities following administration of purified basic proteins from rattlesnake venoms. *Chem. Biol. Interact. 4*:1-10.

Brazil, O. V., Prado-Franceschi, J., and Laure, C. J. (1979). Repetitive muscle responses induced by crotamine. *Toxicon 17*:61-67.

Breithaupt, H. (1976). Enzymatic characteristics of *Crotalus* phospholipase A_2 and the crotoxin complex. *Toxicon 14*:221-233.

Breithaupt, H., Omori-Satoh, T., and Lang, J. (1975). Isolation and characterization of three phospholipase A from the crotoxin complex. *Biochim. Biophys. Acta 403*:355-369.

Bryant, F. R., and Benkovic, S. J. (1979). Stereochemical course of the reaction catalyzed by $5'$-nucleotide phosphodiesterase from snake venom. *Biochemistry 18*:2825-2828.

Burgers, P. M. J., and Eckstein, F. (1978). Absolute configuration of the diastereomers of adenosine 5'-*O*- (1-thiotriphosphate): consequences for the stereochemistry of polymerization by DNA-dependent RNA polymerase from *Escherichia coli. Proc. Natl. Acad. Sci. USA 75:*4798-4800.

Burgers, P. M. J., Eckstein, F., and Hunneman, D. H. (1979a). Stereochemistry of hydrolysis by snake venom phosphodiesterease. *J. Biol. Chem. 254*:7476-7478.

Burgers, P. M. J., Sathyanarayana, K., Saenger, W., and Eckstein, F. (1979b). Crystal and molecular structure of adenosine $5'O$-phosphophorothioate-*O*-*P*-nitrophenyl ester (*Sp* diastereomer). Substrate stereospecificity of snake venom phosphodiesterase. *Eur. J. Biochem. 100*:585-591.

Cameron, D. L., and Tu, A. T. (1977). Characterization of myotoxin *a* from the venom of prairie rattlesnake (*Crotalus viridis viridis*). *Biochemistry 16*:2546-2553.

Cameron, D. L., and Tu, A. T. (1978). Chemical and functional homology of myotoxin *a* from prairie rattlesnake venom and crotamine from South American rattlesnake venom. *Biochim. Biophys. Acta 532*:147-154.

Cate, R. L., and Bieber, A. L. (1978). Purification and characterization of Mojave (*Crotalus scutulatus scutulatus*) toxin and its subunits. *Arch. Biochem. Biophys. 189*:397-408.

Chang, C. C., and Lee, J. D. (1977). Crotoxin, the neurotoxin of South American rattlesnake venom, is a presynaptic toxin acting like β-bungarotoxin. *Naunyn Scheidbergs Arch. Pharmacol. 296*:159-168.

Chang, C. C., and Tseng, K. H. (1978). Effect of crotamine, a toxin of South American rattlesnake venom, on the sodium channel of murine skeletal muscle. *Br. J. Pharmacol. 63*:551-559.

Chen, S. S., Walgate, J. H., and Duerre, J. A. (1971). Oxidative deamination of sulfur amino acids by bacterial and snake venom L-amino acid oxidase. *Arch. Biochem. Biophys. 146*:54-63.

Chute, A. L., and Waters, E. T. (1941). Effect of rattlesnake venom (crotalin) on plasma histamine of the rabbit. *Am. J. Physiol. 132*:552-554.

Cini, C., and DeMarco, C. (1979). Oxidative deamination of ε-N-acetylthialysine and ε-N-acetylselenolysine by snake venom L-amino acid oxidase. *Ital. J. Biochem. 28*:221-231.

Cini, C., Foppoli, C., and DeMarco, C. (1978). Oxidative deamination of thialysine by snake venom L-amino acid oxidase. *Ital. J. Biochem. 25:*305–320.

Clark, W. C., and Voris, H. K. (1969). Venom naturalization by rattlesnake serum albumin. *Science 164*:1402–1404.

Cohen, S. (1959). Purification and metabolic effects of a nerve growth-promoting protein from snake venom. *J. Biol. Chem. 234*:1129–1137.

Copley, A. L., Banerjee, S., and Levi, A. (1973). Studies of snake venoms on blood coagulation. I. The thromboserpentin (thrombin-like) enzyme in the venoms. *Thromb. Res. 2*:487–508.

Curti, B., Massey, V., and Zmudka, M. (1968). Inactivation of snake venom L-amino acid oxidase by freezing. *J. Biol. Chem. 243*:2306–2314.

DeKok, A., and Rawitch, A. B. (1969). Studies on L-amino acid oxidase. II. Dissociation and characterization of its subunits. *Biochemistry 8*:1405–1411.

Denson, K. W. E. (1969). Coagulant and anticoagulant action of snake venoms. *Toxicon 7*:5–11.

Denson, K. W. E., Russell, F. E., and Almagro, D., and Bishop, R. C. (1972). Characterization of the coagulant activity of some snake venoms. *Toxicon 10*:557–562.

Deutsch, H. F., and Diniz, C. R. (1955). Some proteolytic activities of snake venoms. *J. Biol. Chem. 216*:17–26.

Didisheim, P., and Lewis, J. H. (1956). Fibrinolytic and coagulant activities of certain snake venoms and proteases. *Proc. Soc. Exp. Biol. Med. 93*:10–13.

Dolapchiev, L. B. (1980). Venom exonuclease–kinetics and specificity towards different nucleotide substrates. *Int. J. Biol. Macromol. 2*:7–12.

Dolapchiev, L., Ostrowski, W., Wasylewska, E., Wasylewski, Z., and Weber, M. (1977). Heterogeneity of exonuclease from the venom of *Crotalus adamanteus*. *Bull. Acad. Pol. Sci. Ser. Sci. Biol. 25*:359–362.

Dolapchiev, L. B., Vassileva, R. A., and Koumanov, K. S. (1980). Venom exonuclease. II. Amino acid composition and carbohydrate, metal ion and lipid content in the *Crotalus adamanteus* venom exonuclease. *Biochim. Biophys. Acta. 622*:331–336.

Dragstedt, C. A., Mead, F. B., and Eyer, S. W. (1938). Role of histamine in circulating effects of rattlesnake venom. *Proc. Soc. Exp. Biol. 37*:709–710.

Duran-Reynals, F. (1936). The invasion of the body by animal poisons. *Science 83*:286–287.

Duran-Reynals, F. (1939). A spreading factor in certain snake venoms and its relation to their mode of action. *J. Exp. Med. 69*:69–81.

Eagle, H. (1937). The coagulation of blood by snake venoms and its physiologic significance. *J. Exp. Med. 65*:613–639.

Fabiano, R. J., and Tu, A. T. (1981). Purification and biochemical study of viriditoxin, tissue damaging toxin, from prairie rattlesnake venom. *Biochemistry 20*:21.

Favilli, G. (1940). Mucolytic effect of several diffusing agents and of a diasotized compound. *Nature 145*:866.

Feldberg, W. (1940). The action of bee venom, cobra venom and lysoelecithin on the adrenal medulla. *J. Physiol. 99*:104–118.

Feldberg, W., and Kellaway, C. H. (1937). Liberation of histamine from the perfused lung by snake venoms. *J. Physiol. 90*:257-279.

Ferreira, S. H. (1966). Bradykinin-potentiating factor. In *Hypotensive Peptides,* E. G. Erdos, N. Back, and F. Sicateri (Eds.). Florence, pp. 356-367.

Filho, A. P., Brazil, O. V., Fontana, M. D., and Laure, C. J. (1978). The action of crotamine on skeletal muscle: An electrophysiological study. *Toxicon* (Suppl. 1):375-382.

Fletcher, J. E., Elliott, W. B., Ishay, J., and Rosenberg, P. (1979). Phospholipase A and B activities of reptile and hymenoptera venoms. *Toxicon 17*:591.

Flexner, S., and Noguchi, H. (1905). On the plurality of cytolysins in snake venoms. *J. Pathol. Bact. 10*:111-124.

Foote, R., and MacMahon, J. A. (1977). Electrophoretic studies of rattlesnake (*Crotalus* and *Sistrurus*) venom: Taxonomic implications. *Comp. Biochem. Physiol. [B] 57*:235-241.

Fox, J. W., Elzinga, M., and Tu, A. T. (1979). Amino acid sequence and disulfide bond assignment of myotoxin *a* isolated from the venom of prairie rattlesnake (*Crotalus viridis viridis*). *Biochemistry 18*:678-684.

Friederich, C., and Tu, A. T. (1971). Role of metals in snake venoms for hemorrhagic, esterase, and proteolytic activities. *Biochem. Pharmacol. 20*:1549-1556.

Geiger, R., and Kortmann, H. (1977). Esterolytic and proteolytic activities of snake venoms and their inhibition by proteinase inhibitors. *Toxicon 16*:257-259.

Glass, T. G. (1976). *Management of Poisonous Snakebite.* San Antonio, Texas.

Gonçalves, M. (1956). Purification and properties of Crotamine. In *Venoms,* E. E. Buckley and N. Porges (Eds.). Am. Assoc. Adv. Sci., Washington, D.C., pp. 261-274.

Gonçalves, J. M., and Deutsch, H. F. (1956). Ultracentrifugal and zohe electrophoresis studies of some Crotalidae venoms. *Arch. Biochem. Biophys. 60*: 402-411.

Gopalakrishnakone, P., Hawgood, B. J., Holbrooke, S. E., Marsh, N. A., Santana de Sa, S., and Tu, A. T. (1980). Sites of action of Mojave toxin isolated from the venom of the Mojave rattlesnake. *Br. J. Pharmacol. 69*:421-431.

Gulland, J. M., and Jackson, E. M. (1938). 5'Nucleotidase. *Biochem. J. 32*:597-601.

Gutierrez, J. M., and Chaves, F. (1980). Effectos proteolitico, hemorragico y mionecrotico de los venenos de serpientes Costarricenses de los generos *Bothrops, Crotalus* y *Lachesis. Toxicon 18*:315-321.

Habermann, E. (1961). Zuordnung pharmakologischer und enzymatischer Wirkungen von Kallikrein and Schlangengiften mittels Diisoprophyl-Fluorophosphat und Electrophorese. *Naunyn Schmiedebergs Arch. Exp. Pharmakol. 240*:552-572.

Habermann, E., and Rübsamen, K. (1971). Biochemical and pharmacological analysis of the so-called crotoxin. In *Toxins of Animal and Plant Origin,* vol. 1, A. deVries and E. Kochva (Eds.). Gordon and Breach, New York, pp. 333-342.

Hadler, W. A., and Brazil, V. O. (1966). Pharmacology of crystalline crotoxin. IV. Nephrotoxicity. *Mem. Inst. Butantán 33*:1001-1008.

Hafner, E. W., and Wellner, D. (1971). Demonstration of amino acids as products of the reactions catalyzed by D- and L-amino acid oxidase. *Proc. Natl. Acad. Sci. 68*:987-991.

Hampe, O. G., Vozari-Hampe, M. M., and Gonçalves, J. M. (1978). Crotamine conformation: Effect of pH and temperature. *Toxicon 16*:453-460.

Hanashiro, M. A., Da Silva, M. H., and Bier, O. G. (1978). Neutralization of crotoxin and crude venom by rabbit antiserum to *Crotalus* phospholipase A. *Immunochemistry 15*:745-750.

Hanley, M. R. (1978). Crotoxin effects on *Torpedo californica* cholinergic excitable vesicles and the role of its phospholipase A activity. *Biochim. Biophys. Res. Comm. 82*:392.

Hayes, M. B., and Wellner, D. (1969). Microheterogeneity of L-amino acid oxidase: Separation of multiple components by polyacrylamide gel electrofocusing. *J. Biol. Chem. 244*:6636-6644.

Heinrikson, R. L., Krueger, E. T., and Keim, P. S. (1977). Amino acid sequence of phospholipase A_2-α from the venom of *Crotalus adamanteus*. *J. Biol. Chem. 252*:4913-4921.

Hendon, R. A., and Fraenkel-Conrat, H. (1971). Biological roles of the two components of crotoxin. *Proc. Natl. Acad. Sci. 68*:1560-1563.

Hendon, R. A., and Tu, A. T. (1979). The role of crotoxin subunits in tropical rattlesnake neurotoxic action. *Biochim. Biophys. Acta 578*:243-252.

Heppel, L. A., and Hilmoe, R. J. (1951). Purification and properties of 5'-nucleotidase. *J. Biol. Chem. 188*:655-676.

Hogberg, B., and Uvnas, B. (1957). The mechanism of the disruption of mast cells produced by compound 48/80. *Acta. Physiol. Scand. 41*:345.

Homma, M., and Tu, A. T. (1971). Morphology of local tissue damage in experimental snake envenomation. *Br. J. Exp. Pathol. 52*:538-542.

Huang, K.-S., and Law, J. H. (1981). Photoaffinity labeling of *Crotalus atrox* phospholipase A_2 by a substrate analogue. *Biochemistry 20*:181-187.

Janszky, B. (1950). The relation between the proteolytic and blood clotting activity of snake venoms. *Arch. Biochem. Biophys. 28*:139-140.

Jeng, T. W., Hendon, R. A., and Fraenkel-Conrat, H. (1978). Search for relationship among the hemolytic, phospholipolytic, and neurotoxic activities of snake venoms. *Proc. Natl. Acad. Sci. 75*:600-604.

Jiménez-Porras, J. M. (1961). Biochemical studies on venom of the rattlesnake, *Crotalus atrox atrox*. *J. Exp. Zool. 148*:251-258.

Johnson, B. D. (1968). Selected Crotalidae venom properties as a source of taxonomic criteria. *Toxicon 6*:5-10.

Keegan, H. L., and Andrews, T. F. (1942). Effects of crotalid venom on North American snakes. *Copeia 4*:251-254.

Kilmon, J. A. (1976). High tolerance to snake venom by the Virginia opossum, *Didelphis virginiana*. *Toxicon 14*:337-340.

Kress, L. F., and Catanese, J. (1980). Enzymatic inactivation of human antithrombin III: Limited proteolysis of the inhibitor by snake venom proteinases in the presence of heparin. *Biochim. Biophys. Acta 615*:178-186.

Kress, L. F., and Paroski, E. A. (1978). Enzymatic inactivation of human serum proteinase inhibitors by snake venom proteinases. *Biochem. Biophys. Res. Comm. 83*:649-656.

Kress, L. F., and Kurecki, T. (1980). Studies on the complex between human α_2-macroglobulin and *Crotalus adamanteus* proteinase II, release of active proteinase from the complex. *Biochim. Biophys. Acta 613*:469-475.

Kress, L. F., DaRoza, D., and Laskowski, M. (1978). Purification of an arginine ester hydrolase from *Crotalus adamanteus* venom. *Toxicon* (Suppl. 1): 141-146.

Kress, L. F., Kurecki, T., Chan, S. K., and Laskowski, M. (1979) Characterization of the inactive fragment resulting from limited proteolysis of human α_1-proteinase inhibitor by *Crotalus adamanteus* proteinase II. *J. Biol. Chem. 254*:5317-5320.

Kurecki, T., Laskowski, M., and Kress, L. F. (1978). Purification and some properties of two proteinases from *Crotalus adamanteus* venom that inactivate human α_1-proteinase inhibitor. *J. Biol. Chem. 253*:8340-8345.

Laure, C. J. (1975). Die Primarstruktur des Crotamins. *Hoppe Seylers Z. Physiol. Chem. 356*:213-215.

Lee, C. Y. (1979). *Snake Venoms.* Springer-Verlag, Berlin, 1130 pp.

Lee, C. K., Chan, T. K., Ward, B. C., Howell, D. E., and Odell, G. V. (1974). The purification and characterization of a necrotoxin from tarantula. *Arch. Biochem. Biophys. 164*:341-350.

Levi-Montalcini, R., and Calissano, P. (1979). The nerve-growth factor. *Sci. Am. 240*:68-78.

Lòpez, M., Foscarini, L. G., Àlvarez, J. M., Diniz Filho, I., Marra, U. D., and Procópio, N. M. M. (1972). Tratamento intensivo das complicações do acidente ofídico. *Rev. Ass. Med. Minas Gerais. 23*:107.

Lyons, W. J. (1971). Profound thrombocytopenia associated with *Crotalus ruber ruber* envenomation: A clinical case. *Toxicon 9*:237-240.

Madinaveitia, J. (1941). Diffusion factors. VII. Concentration of mucinase from testicular extracts and from *Crotalus atrox* venom. *Biochem. J. 35*:447-452.

Maeda, N., Tamiya, N., Pattabhiraman, T. R., and Russell, F. E. (1978). Some chemical properties of the venom of the rattlesnake, *Crotalus viridis helleri*. *Toxicon 16*,431-441.

Man, D. P., and Minta, J. O. (1977). Purification, characterization and analysis of the mechanism of action of four anti-complementary factors in *Crotalus atrox* venom. *Immunochemistry 14*:521-527.

Mandelbaum, F. R., Carrillo, M., and Henriques, S. B. (1967). Proteolytic activity of *Bothrops* protease A on the B chain of oxidized insulin. *Biochim. Biophys. Acta 132*:508-510.

Mannherz, H. G., and Magener, M. (1979). Concanavalin A inhibits the interaction of snake venom 5'-nucleotidase and actin. *FEBS Lett. 103*:77-80.

Margolis, J., Bruce, S., Starzecki, B., Horner, G. J., and Halmagyi, D. F. J. (1965). Release of bradykinin-like substance in sheep by venom of *Crotalus atrox*. *Aust. J. Exp. Biol. Med. Sci. 43*:237-244.

Markland, F. S., and Damus, P. S. (1971). Purification and properties of a thrombinlike enzyme from the venom of *Crotalus adamanteus* (Eastern diamondback rattlesnake). *J. Biol. Chem. 246*:6460–6473.

Markland, F. S., and Pirkle, H. (1977a). Thrombin-like enzyme from the venom of *Crotalus adamanteus* (Eastern diamondback rattlesnake). *Thromb. Res. 10*:487–494.

Markland, F. S., and Pirkle, H. (1977b). Biological activities and biochemical properties of thrombin-like enzymes from snake venoms. In *Chemistry and Biology of Thrombin*, R. L. Lundblad, J. W. Fenton, and K. G. Mann (Eds.). Ann Arbor Science, Ann Arbor, pp. 71–89.

Markwardt, F., Barthel, W., Glusa, E., and Hoffman, A. (1966). Über die Freisetzung biogener Amine aus Blutplattchen durch tierische Fifte. *Naunyn Schmiedebergs Arch. Exp. Pathol. Pharmak. 252*:297–304.

Mebs, D. (1968). Vergleichende Enzymuntersuchungen an Schlangengiften unter besonderer Berücksichtigung ihrer caseinspaltenden Proteasen. *Hoppe Seylers Z. Physiol. Chem. 349*:1115–1125.

Mebs, D. (1970). A comparative study of enzyme activities in snake venoms. *Int. J. Biochem. 1*:335–342.

Meister, A., and Wellner, D. (1966). L-Amino acid oxidase. In *Flavins and Flavoproteins*, E. C. Slater (Ed.). Elsevier North-Holland, Amsterdam, pp. 226–241.

Michl, H., and Molzer, H. (1965). The occurrence of L-leucyl-β-naphthylamide (LNA) splitting enzymes in some amphibia and reptile venoms. *Toxicon 2*: 281–282.

Minta, J. O., Man, M. P., Wasi, S., and Painter, R. H. (1977). Interaction of *Crotalus atrox* venom with serum complement: Kinetic analysis. *Immunochemistry 14*:513–519.

Minta, J. O., and Man, D. (1980). Immunological, structural and functional relationships between an anti-complementary protein from *Crotalus atrox* venom, cobra venom factor and human C3. *Immunology 39*:503–509.

Moran, J. B., and Geren, C. R. (1979). A comparison of biological and chemical properties of three North American (Crotalidae) snake venoms. *Toxicon 17*: 237–244.

Nahas, L., Denson, K. W. E., and MacFarlane, R. G. (1964). A study of the coagulant action of eight snake venoms. *Thromb. Diath. Haemorrh. 12*:355–367.

Nakazone, A. K. (1979). Immunochemical aspects of crotoxin and its subunits. Report 1979, IEA-DT-112, 79 pp. INIS Atomindex 1979, 10(20), Abstr. No. 482182.

Nichol, A. A., Doublas, V., and Peck, L. (1933). On the immunity of rattlesnakes to their venom. *Copeia 4*:211–213.

Nielsen, H., Sørensen, H., Faber, V., and Svehag, S. E. (1978). Circulating immune complexes, complement activation kinetics and serum sickness following treatment with heterologous anti-snake venom globulin. *Scand. J. Immunol. 7*:25–33.

Oshima, G., and Iwanaga, S. (1969). Occurrence of glycoproteins in various snake venoms. *Toxicon* 7:235-238.

Oshima, G., Matsuo, Y., Iwanaga, S., and Suzuki, T. (1968). Studies on snake venoms. XIX. Purification and some physicochemical properties of proteinase A and C from the venom of *Agkistrodon halys blomhoffii. J. Biochem.* 64:227-238.

Oshima, G., Sato-Ohmori, T., and Suzuki, T. (1969). Proteinase, arginine ester hydrolase and a kinin releasing enzyme in snake venoms. *Toxicon* 7:229-233.

Ownby, C. L., Kainer, R. A., and Tu, A. T. (1974). Pathogenesis of hemorrhage induced by rattlesnake venom: An electron microscopic study. *Am. J. Pathol.* 76:401-414.

Ownby, C. L., Tu, A. T., and Kainer, R. A. (1975). Effect of diethylenetriaminepentaacetic acid and procaine on hemorrhage induced by rattlesnake venom. *J. Clin. Pharmacol.* 15:419-26.

Ownby, C. L., Cameron, D., and Tu, A. T. (1976). Isolation of myotoxin component from rattlesnake (*Crotalus viridis viridis*) venom. *Am. J. Pathol.* 85: 149-166.

Ownby, C. L., Bjarnason, J., and Tu, A. T. (1978). Hemorrhagic toxins from rattlesnake (*Crotalus atrox*) venom. Pathogenesis of hemorrhage induced by three purified toxins. *Am. J. Pathol.* 93:201-210.

Ownby, C. L., Woods, W. M., and Odell, G. V. (1979). Antiserum to myotoxin from prairie rattlesnake (*Crotalus viridis viridis*) venom. *Toxicon* 17:373-380.

Page, D. S., and Van Etten, R. L. (1971a). L-Amino acid oxidase. II. Deuterium isotope effects and the action mechanism for the reduction of L-amino acid oxidase by L-leucine. *Biochim. Biophys. Acta* 227:16-31.

Page, D. S., and Van Etten, R. L. (1971b). L-Amino acid oxidase. III. Substrate substituent effect upon the reaction of L-amino acid oxidase with phenylalanine. *Bioorg. Chem.* 1:361-373.

Paik, W. K., and Kim, S. (1967). Studies on the stability of L-amino acid oxidase of snake venom. *Biochim. Biophys. Acta* 139:49-55.

Paik, W. K., and Kim, S. (1975). A factor which counteracts the stabilizing activity of acetate ion on ophioxidase at alkaline pH. *Experientia* 31:150-151.

Pérez, J. C., Haws, W. C., Garcia, V. E., and Jennings, B. M., III. (1978). Resistance of warm-blooded animals to snake venoms. *Toxicon* 16:375-383.

Pérez, J. C., Pichyangkul, S., and Garcia, V. E. (1979). The resistance of three species of warm-blooded animals to Western diamondback rattlesnake (*Crotalus atrox*) venom. *Toxicon* 17:601-608.

Pérez-Polo, J., Bomar, H., Beck, C., and Hall, K. (1978). Nerve growth factor from *Crotalus adamanteus* snake venom. *J. Biol. Chem.* 253:6140-6148.

Pfleiderer, G., and Krauss, A. (1965). Wiikungsspezifitat von Schlangengift-Proteasen (*Crotalus atrox*). *Biochem. Z.* 342:85-94.

Pfleiderer, G., and Sumyk, G. (1961). Investigation of snake venom enzymes. I. Separation of rattlesnake venom proteinase by cellulose ion-exchange chromatography. *Biochim. Biophys. Acta* 51:482-493.

Philpot, V. B., and Smith, R. G. (1950). Neutralization of pit viper venom by kingsnake serum. *Proc. Soc. Exp. Biol. Med.* 74:521-523.

Philpot, V. B., Ezekiel, E., Laseter, Y., Yaeger, R. G., and Stjernholm, R. L. (1978). Neutralization of crotalid venoms by fractions from snake venoms. *Toxicon* 16:603-609.

Pirkle, H., Markland, F. S., and Theodor, I. (1976). Thrombin-like enzymes of snake venoms: Action of prothrombin. *Thrombosis Res.* 8:619-627.

Pirkle, H. C., Markland, F. S., Jr., Theodor, I., and Baumgartner, R. (1979). Thrombin-like enzyme amino acid sequence homology. *Thrombos. Haemostas.* 42:441.

Pirkle, H., Markland, F. S., Theodor, I., Baumgartner, R., Bajwa, S. S., and Kirakossian, H. (1981). The primary structure of Crotalase, a thrombin-like venom enzyme, exhibits closer homology to kallikrein than to other serine proteases. *Biochem. Biophys. Res. Commun.* 99:715-721.

Possani, L. D., Sosa, B. P., Alagon, A. C., and Burchfield, P. M. (1980). The venom from the snakes *Agkistrodon bilineatus taylori* and *Crotalus durissus totonacus*: Lethality, biochemical and immunological properties. *Toxicon* 18:356-360.

Prado-Franceschi, J., and Brazil, O. V. (1969). Convulxina, uma nova neurotoxina da peconha de *Crotalus durissus terrificus*. *Congr. Latamer. Cienc. Fisiolo.*, Abstract IX, Brazil.

Pritchard, A. E., and Laskowski, M., Sr. (1978a). Discrete fragments produced by limited digestion of superhelical PM2 DNA with venom phosphodiesterase. *J. Biol. Chem.* 253:6606-6613.

Pritchard, A. E., and Laskowski, M., Sr. (1978b). Specific cleavages inflicted by venom phosphodiesterase on superhelical ϕ X 174 DNA. *J. Biol. Chem.* 253:7989-7992.

Pritchard, A. E., Kowalski, D., and Laskowski, M. (1977). An endonuclease activity of venom phosphodiesterase specific for single-stranded and superehlical DNA. *J. Biol. Chem.* 252:8652-8659.

Purdon, A. D., Tinker, D. O., and Spiro, L. (1977). The interaction of *Crotalus atrox* phospholipase A_2 with calcium ion and 1-anilino naphthalene-8-sulfonate, *Can. J. Biochem.* 55:205-214.

Raudonat, H. W. (1963). Zur Biochemie und Pharmakologie der Schlangengifte mit einem Beitrag uber ihre chemischen Eigenschaften. In *Die Giftschlangen der Erde*. N. G. Universitats-und Verlags-Buchhandlung, Marburg/Lahn, pp. 11-30.

Reid, H. A., and Theakston, R. D. G. (1978). Changes in coagulation effects by venoms of *Crotalus atrox* as snake ages. *Am. J. Trop. Med. Hyg.* 27: 1053-1057.

Rocha e Silva, M. (1949). Autofarmacologia e Venenos animals. *Arq. Inst. Biol.* (Sao Paulo) 19:1-22.

Rocha e Silva, M., and Essex, H. E. (1942). The effects of animal poisons (rattlesnake venom and trypsin) on blood histamine of guinea pigs and rabbits. *Am. J. Physiol.* 135:372-377.

Rocha e Silva, M., Beraldo, W. T., and Rosenfeld, G. (1949). Bradykinin, a hypotensive and smooth muscle stimulating principle released from plasma globulin by snake venoms and by trypsin. *Am. J. Physiol.* 156:261-272.

Rodriguez, O. G., and Scannone, H. R. (1976). Fractionation of *Crotalus durissus cumanensis* venom by gel filtration. *Toxicon* 14:400-403.

Rodriguez, O. G., Scannone, H. R., and Parra, N. D. (1974). Enzymatic activities and other characteristics of *Crotalus durissus cumanensis* venom. *Toxicon* 12:297-302.

Rosenfeld, G., Kelen, E. M. A., and Nudel, F. (1960). Hemolytic activity of animal venoms. I. Classification in different types and activities. *Mem. Inst. Butantan* 30:103-116.

Rothschild, A. M. (1966). Mechanisms of histamine release by animal venoms. *Mem. Inst. Butantan* 33:467-476.

Rothschild, A. M. (1967). Chromatographic separation of phospholipase A from a histamine releasing component of Brazilian rattlesnake venom (*Crotalus durissus terrificus*). *Experientia* 23:741-742.

Rothschild, A. M., and Rothschild, Z. (1979). Liberation of pharmacologically active substances by snake venoms. In *Snake Venoms*, C. Y. Lee (Ed.). Springer-Verlag, Berlin, pp. 591-628.

Ruiz, C. E., Schaeffer, R. C., Jr., Weil, M. H., and Carlson, R. W. (1980). Hemostatic changes following rattlesnake (*Crotalus viridis helleri*) venom in the dog. *J. Pharmacol. Exp. Ther.* 21:414-417.

Russell, F. E. (1965). Bradykininogen levels following *Crotalus* envenomation. *Toxicon* 2:277-279.

Russell, F. E., Buess, F. W., Woo, M. Y., and Eventov, R. (1963). Zootoxicological properties of venom L-amino acid oxidase. *Toxicon* 1:229-234.

Scannone, H. R., Rodriguez, O. G., and Lancini, A. R. (1978). Enzymatic activities and other characteristics of *Crotalus vegrandis* snake venom. *Toxicon* (Suppl. 1):223-229.

Schenberg, S. (1959). Geographical pattern of crotamine distribution in the same rattlesnake subspecies. *Science* 129:1361-1363.

Schmaier, A. H., and Colman, R. W. (1980). Crotalocytin: Characterization of the timber rattlesnake platelet activating protein. *Blood* 56:1020-1028.

Schmaier, A. H., Claypool, W., and Colman, R. W. (1980). Crotalocytin: Recognition and purification of a timber rattlesnake platelet aggregating protein. *Blood* 56:1013-1019.

Seki, C., Vidal, J. C., and Barrio, A. (1980). Purification of gyroxin from a South American rattlesnake (*Crotalus durissus terrificus*) venom. *Toxicon* 18:235-247.

Silva, O. A., Lopez, M., and Godoy, P. (1979). Intensive care unit treatment of acute renal failure following snake bite. *Am. J. Trop. Med. Hyg.* 28:401-407.

Simpson, J. W. (1975). Distribution of collagenolytic enzyme activity among snake venoms. *Comp. Biochem. Physiol [B]* 51:425-428.

Simpson, J. W., and Taylor, A. C. (1973). Elastrolytic activity from the venom of the rattlesnake *Crotalus atrox*. *Proc. Soc. Exp. Biol. Med.* 144:380-383.

Simpson, J. W., Taylor, A. C., and Levy, B. M. (1971). Collagenolytic activity in some snake venoms. *Comp. Biochem. Physiol.* 39B:963-967

Slotta, K., and Ballester, A. (1954). Determinacao colorimetrica de hialuronidase das venenos ofidicos. *Mem. Inst. Butantan* 26:311.

Sosa, B. P., Alagon, A. C., Possani, L. D., and Julia, J. Z. (1979). Comparison of phospholipase activity with direct and indirect lytic effects of animal venoms upon human red cells. *Comp. Biochem. Physiol. [B]* 648:231-234.

Stocker, K. (1978). Defibrinogenation with thrombin-like snake venom enzymes. In *Handbook of Experimental Pharmacology* 46:451-484.

Straight, R., Glenn, J. L., and Snyder, C. C. (1976). Antivenom activity of rattlesnake blood plasma. *Nature* 261:259-260.

Stringer, J. M., Kainer, R. A., and Tu, A. T. (1972). Myonecrosis induced by rattlesnake venom: An electron microscopic study. *Am. J. Pathol.* 67:127-134.

Sulkowski, E., Kress, L. F., and Laskowski, M. (1975). Crystalline basic protein from venom of *Crotalus adamanteus*. *Toxicon* 13:149-157.

Sullivan, J., and Geren, C. R. (1979). Isolation, stabilization, and characterization of a toxin from timber rattlesnake venom. *Prep. Biochem.* 9:321-333.

Sullivan, J. A., Farr, E., and Geren, C. R. (1979). Fractionation and partial characterization of toxic components of timber rattlesnake venom. *Toxicon* 17:269-277.

Suzuki, T., Iwanaga, W., Nagasawa, S., and Sato, T. (1966). Purification and properties of bradykinin and of the bradykinin-releasing and destroying enzymes in snake venom. In *Hypotensive Peptides*, E. G. Erös, N. Back, F. Sicuteri, and A. F. Wilde (Eds.). Springer-Verlag, New York, pp. 149-160.

Takahashi, T., and Ohsaka, A. (1970). Purification and characterization of a proteinase in the venom of *Trimeresurus flavoviridis*. *Biochim. Biophys. Acta* 198:293-307.

Tarabini-Castellani, G. (1938). Presence of diffusion factors in some animal venoms. *Arch. Ital. Med. Sper.* 2:969-978.

Thomas, R. G., and Pough, F. H. (1979). The effect of rattlesnake venom on digestion of prey. *Toxicon* 17:221.

Tinker, D. O., and Wei, J. (1979). Heterogeneous catalysis by phospholipase A_2: Formation of a kinetic description of surface effects. *Can. J. Biochem.* 57:97-106.

Tinker, D. O., Purdon, A. D., Wei, J., and Mason, E. (1978). Kinetics of hydrolysis of dispersions of saturated phosphatidylcholines by *Crotalus atrox* phospholipase A_2. *Can. J. Biochem.* 56:552-558.

Toom, P. M., Solie, T. N., and Tu, A. T. (1970). Characterization of a nonproteolytic arginine ester-hydrolyzing enzyme from snake venom. *J. Biol. Chem.* 245:2549-2555.

Tsai, M.-D. (1980). Stereochemistry of the hydrolysis of adenosine 5'-thiophosphate catalyzed by venom 5'-nucleotidase. *Biochemistry* 19:5310-5316.

Tu, A. T. (1977). *Venoms: Chemistry and Molecular Biology*. John Wiley, New York, 560 pp.

Tu, A. T. (1979). Application of electrofocusing to snake venom research. In *Electrofocus/78*, H. Haglund, J. G. Westerfeld, and J. T. Ball (Eds.). Elsevier North-Holland, New York, pp. 169-178.

Tu, A. T., and Toom, P. M. (1967). The presence of L-leucyl-β-naphthylamide hydrolyzing enzyme in snake venoms. *Experientia 23*:439-443.

Tu, A. T., and Giltner, J. B. (1974). Cytotoxic effects of snake venoms of KB and Yoshida sarcoma cells. *Res. Comm. Chem. Pathol. Pharmacol. 9:*783-786.

Tu, A. T., and Passey, R. B. (1969). Effects of snake venoms on mammalian cells in tissue culture. *Toxicon 6:*277-280.

Tu, A. T., Chua, A., and James, G. P. (1966). Proteolytic enzyme activities in a variety of snake venoms. *Tox. Appl. Pharmacol. 8*:218-223.

Tu, A. T., Homma, M., Hong, B., and Terrill, J. B. (1970a). Neutralization of rattlesnake venom toxicities by various compounds. *J. Clin. Pharmacol. 10*: 323-329.

Tu, A. T., Toom, P. M., and Murdock, D. S. (1967). Chemical difference in the venoms of genetically different snakes. In *Animal Toxins*, F. E. Russell and P. R. Saunders (Eds.). Pergamon Press, Oxford, pp. 351-362.

Tu, A. T., Passey, R. B., and Toom, P. M. (1970b). Isolation and characterization of phospholipase A from sea snake, *Laticauda semifasciata* venom. *Arch. Biochem. Biophys. 140*:96-106.

Tu, A. T., Stermitz, J., Ishizaki, H., and Nonaka, S. (1980). Comparative study of pit viper venoms of genera *Trimeresurus* from Asia and *Bothrups* from America: An immunological and isotachophoretic study. *Comp. Biochem. Physiol. [B] 66*:249-254.

Tu, A. T., Nikai, T., and Baker, J. O. (1981). Proteolytic specificity of hemorrhagic toxin *a* isolated from western diamondback rattlesnake (*Crotalus atrox*) venom. *Biochemistry*, in press.

Ubuka, T., and Yao, K. (1973). Oxidative deamination of L-cysteine by L-amino acid oxidase from snake venom: Formation of S-(s-oxo-2-carboxyethylthio) cysteine and S-(carboxymethylthio) cysteine. *Biochem. Biophys. Res. Commun. 55*:1305-1310.

Vargaftig, B. B., Prado-Franceschi, J., Chignard, M., Lefort, J., and Marlas, G. (1980). Activation of guinea-pig platelets induced by convulxin, a substance extracted from the venom of *Crotalus durissus cascavella*. *J. Pharmacol. 68*: 451-464.

Vassileva, R., and Dolapchiev, L. (1979). Significance of different metal ions for pyrophosphatase activity of venom exonuclease. *Dokl. Bolg. Akad. Nauk. 32*: 1109-1112.

Verheij, H. M., Egmond, M. R., and de Haas, G. H. (1981). Chemical modification of the α-amino group in snake venom phospholipase A_2. A comparison of the interaction of pancreatic and venom phospholipases with lipid-water interfaces. *Biochemistry 20*:94-99.

Vogt, W., Meyer, U., Kunze, H., Luft, E., and Babilli, S. (1969). Entstehung von SRS-C in der durchstromten Meerschweinchenlungs durch Phospholipase A. Identifizierung mit Prostaglandin. *Naunyn Schmiedebergs Arch. Pharmak. Exp. Pathol. 262*:124-134.

Wagner, F. W., Spiekerman, A. M., and Prescott, J. M. (1968). *Leucostoma* peptidase A: Isolation and physical properties. *J. Biol. Chem. 348*:519.

Weiss, H. J., Allan, S., Davidson, E., and Kochva, S. (1969). Afibrinogenemia in man following the bite of a rattlesnake (*Crotalus adamanteus*). *Am. J. Med.* 47:625-634.

Wellner, D. (1966). Evidence for conformational changes in L-amino acid oxidase associated with reversible inactivation. *Biochemistry* 5:1585-1591.

Wellner, D. (1971). L-amino acid oxidase (snake venom). *Methods Enzymol.* 17: (Part B):597-600.

Wellner, D., and Meister, A. (1960). Crystalline L-amino acid oxidase of *Crotalus adamanteus*. *J. Biol. Chem.* 235:2013-2018.

Wells, M. A., and Hanahan, D. J. (1969). Studies on phospholipase A. I. Isolation and characterization of two enzymes from *Crotalus adamanteus* venom. *Biochemistry* 8:414-424.

Welsh, J. H. (1966). Acetylcholine in snake venoms. In *Animal Toxins*, F. E. Russell and B. P. R. Saunders (Eds.). Pergamon Press, Elmsford, N.Y., pp. 363-368.

Werner, R. M., and Vick, J. A. (1977). Resistance of the opossum (*Didelphis virginiana*) to envenomation by snakes of the family Crotalidae. *Toxicon* 15: 29-33.

Yang, C. C. (1963). Fractionation of snake venom on Sephadex. *J. Formosan Med. Assoc.* 62:611.

Young, M. (1979). Yes, NGF is an enzyme. *Nature* 281:15.

Young, R. A., Miller, D. M., and Ochsner, D. C. (1980). The Grand Canyon rattlesnake (*Crotalus viridis abyssus*): Comparison of venom profiles with other *viridis* subspecies. *Comp. Biochem. Physiol. [B]* 66:601-603.

Zanini, A., Angeletti, P. U., and Levi-Montalcini, R. (1968). Immunochemical properties of the nerve growth factor. *Proc. Natl. Acad. Sci.* 61:835-842.

Zarafonetis, C. J., and Kalas, J. P. (1960). Serotonin, catecholamine, and amine oxidase activity in the venoms of certain reptiles. *Am. J. Med. Sci.* 240: 764-768.

Zeller, E. A., Clauss, L. M., and Ohlsson, J. T. (1974). Interaction of ophidian L-amino acid oxidase with its substrates and inhibitors: Role of molecular geometry and electron distribution. *Helv. Chim. Acta* 57:261-262.

Zemlicka, J. (1980). Synthesis of dicytidylyl-3(3′-5′)-1,2-di(adenosin-N^6-yl) ethane and dicytidylyl-(3′-5′)-1,4-di(adenosin-N^6-yl)butane: Covalently joined terminals of two transfer ribonucleic acids and their behavior toward snake venom phosphodiesterase. *Biochemistry* 19:163-168.

Zwilling, V. R., and Pfleiderer, G. (1967). Eigenschaften der α-Protease aus dem Gift von *Crotalus atrox*. *Hoppe Seylers Z. Physiol. Chem.* 348:519-524.

Zwisler, V. O. (1965). Investigation of snake venoms by electrophoresis and polyacrylamide gel. *Z. Imm. Allerg.* 129:444-451.

part **II**
Clinical Aspects

6

Treatment of Rattlesnake Bites

ROBERT E. ARNOLD
University of Louisville School of Medicine, Louisville, Kentucky

Introduction   315

First Aid   319

Capturing the Snake   321

Factors Affecting Medical Treatment   323
Tourniquets • Incisions and Suction • Cryotherapy • Alcohol • Antivenin • Transportation

Definitive Medical Treatment   326
Reassurance of the Patient • Estimation of the Severity of the Bite • Planning a Rational Treatment

Associated Factors of Medical Treatment   330
Pain • Cortisone • Kidney • Tetanus Prophylaxis • Antibiotics • Duration of Treatment

Summary   336

References   336

INTRODUCTION

Psychology may be as important as pharmacology in the treatment of rattlesnake bites. Snakes are fearsome creatures. They have a sinister appearance, are stealthy, strike quickly, and have always been associated with sudden disaster. Our ancestors believed snakes were supernatural; their steady gaze and unblinking

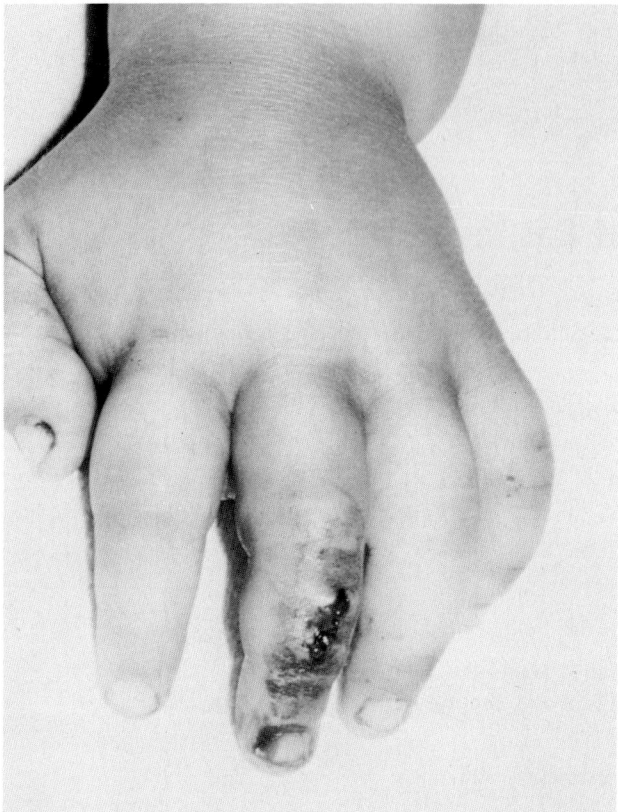

Figure 1 An excellent demonstration of fang marks. The twin puncture wounds oozing blood are easily identified and should not be confused with the scratch marks characteristic of a nonpoisonous snake. The edema of the hand is also apparent and is pathognomonic of envenomation.

eyes were thought to hypnotize and their bite was thought to be fatal. The inevitable myths followed, and the destructive capabilities of snakes were exaggerated until most people believed that immediate treatment was necessary to prevent death. Many doctors did not have much experience treating snakebites, and when their patients lived, their survival was incorrectly attributed to the therapeutic measures employed instead of the natural history of the disease.

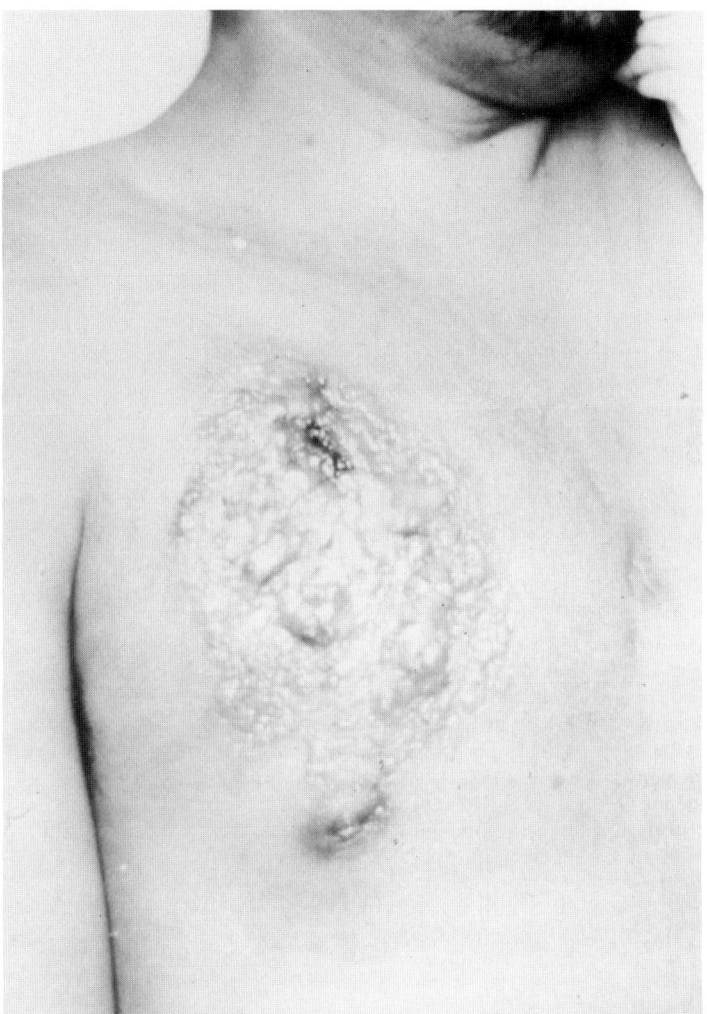

Figure 2 This is an example of the tissue destruction caused by a rattlesnake bite. Note that the most severe damage is in the immediate vicinity of the bite, and the venom is diluted as it spreads through the tissues, leaving all of the tissue more than a few inches from the bite unharmed. Edema in this area will not compromise the circulation as it does in the extremity. (Courtesy of Center for Disease Control, Atlanta, Georgia.)

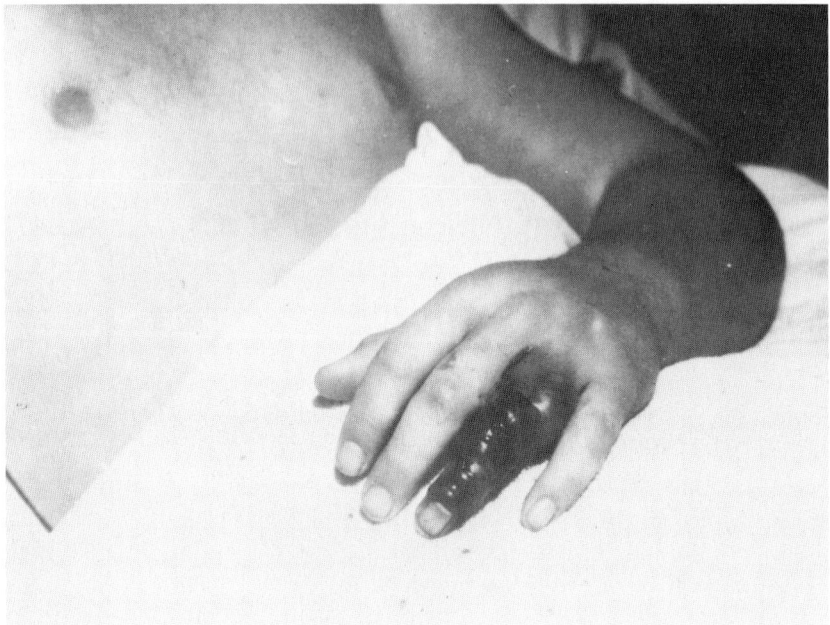

Figure 3 An example of anatomy contributing to the morbidity of a rattlesnake bite. The finger is a closed space and the edema was severe enough to reduce the circulation and also keep the venom concentrated in a small volume of tissue. The damage to this finger was so extensive that an amputation was necessary. Notice the tourniquet effect as demonstrated by the sharp line between viable and nonviable tissue at the metacarpal-phalangeal joint. The hand is edematous and discolored but not permanently damaged because the circulation of the hand is better and the venom is diluted. (Courtesy of T. Wilson, Louisville, Kentucky.)

This has led to many recommendations in the medical literature which have had no demonstrable benefit but which do have significant hazards (Arnold, 1979; Hopkins, 1976). Survival of the patient was considered a therapeutic success; the concomitant loss of a few fingers or toes was considered acceptable. That is no longer true. As our knowledge of pharmacology and surgery increased, it became apparent that many of these procedures were just as dangerous as the rattlesnakes. Several examples of envenomation, and poor treatment of the envenomation, are shown in Figures 1-8.

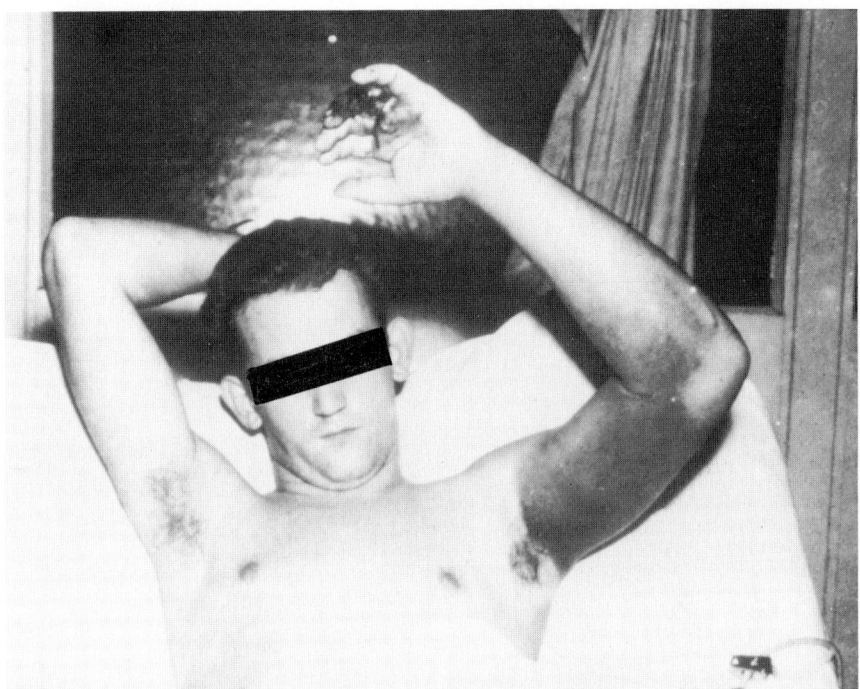

Figure 4 The same patient as in Figure 3. The arm and forearm are swollen and discolored from the blood, tissue fluid, and venom in the extremity. Again the tissue damage is not enough to cause permanent damage because the circulation is adequate and the venom is diluted. (Courtesy of T. Wilson, Louisville, Kentucky.)

FIRST AID

The object of the emergency treatment of rattlesnake bites is to get the patient to a medical facility as soon as possible and in the best condition possible. There are two fundamental principles involved. First, do no harm; then transport the patient safely and rapidly to a hospital for definitive treatment.

First, do no harm. This is the cardinal rule of all emergency treatment. Snakebites are unique because of our inherent dread of them and most peoples' lack of knowledge of the natural history of snakebites. Many people incorrectly assume that immediate action is necessary to avoid death or severe disability. This has led to many questionable recommendations; and the subsequent recovery of the patient has been incorrectly attributed to these procedures. Over

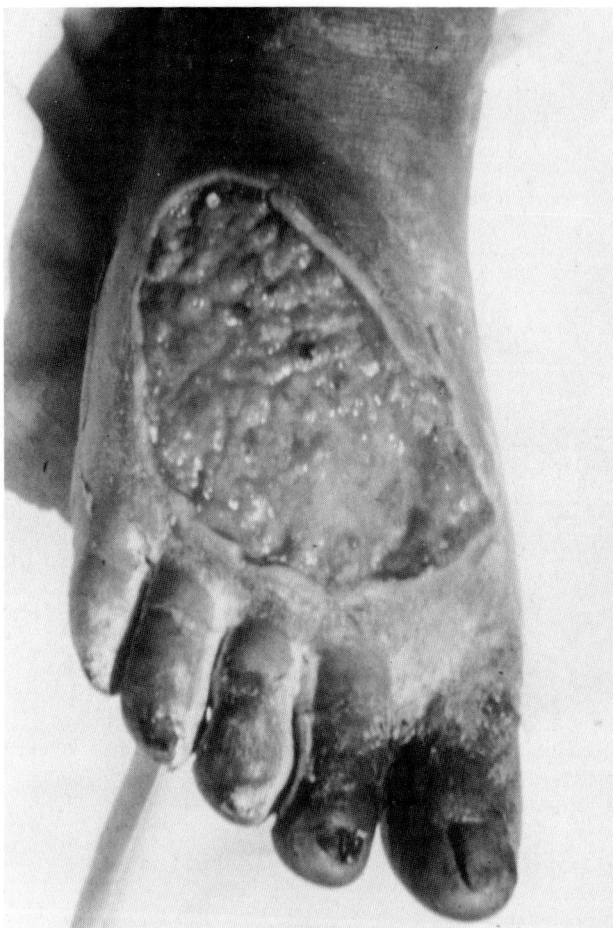

Figure 5 Tissue slough caused by a rattlesnake bite. The fascial bands of the ankle predispose to vascular insufficiency when edema is present. (Courtesy of Center for Disease Control, Atlanta, Georgia.)

200 specific recommendations have resulted, but most of them are so hazardous or ridiculous that they have been abandoned. There is a hard core of inappropriate recommendations which remain because of literary inertia instead of proven medical benefit, and these are found in many recent publications.

Treatment of Bites

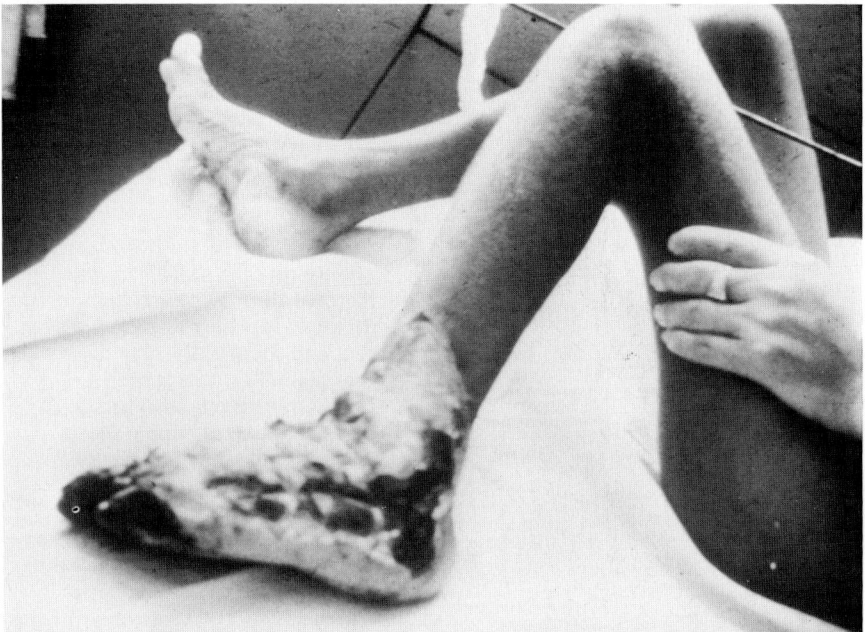

Figure 6 Severe tissue slough resulting from a rattlesnake bite. There is considerable tissue damage of the foot and ankle, but the leg is undamaged. (Courtesy of Center for Disease Control, Atlanta, Georgia.)

CAPTURING THE SNAKE

Many first aid manuals recommend capturing the snake and bringing it with the patient because identification of the offending reptile will enable physicians to use the proper antivenin. This may be useful in areas where the snake population is so diverse that identification of the snake is necessary for selection of the correct antivenin. Rattlesnakes, however, are found only in the Western hemisphere, and the only other venomous reptiles occupying the same area are other pit vipers and coral snakes. Identification is unnecessary for the following reasons.

Coral snakes are found in some of the same areas as rattlesnakes. Coral snakes are elapids, with small fixed fangs and neurotoxic venoms. They will hold and chew when they bite, and their venom causes paralysis instead of tissue destruction. These features are so different from rattlesnakes that confusion of coral snake bites with rattlesnake and other pit viper bites has not been a problem.

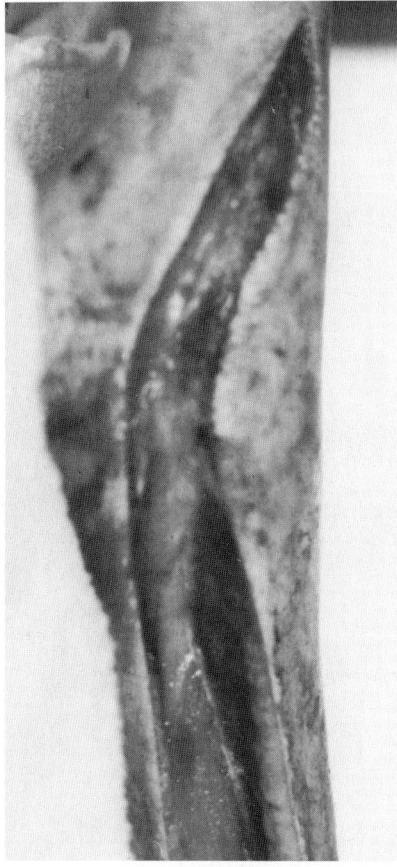

Figure 7 A fasciotomy through a snakebite on the ankle. This was performed to improve circulation and permit rapid egress of venom, blood, and tissue fluid. The dark subcutaneous tissue has been damaged by the venom. The site of the bite is apparent, where the damaged subcutaneous tissue extends from the skin to the fascia. Notice that as the venom moves proximally it does so by going deep to the tissue planes around the fascia, not by the superficial lymphatics under the skin. Therefore, any recommendation to utilize a tourniquet to occlude superficial lymphatics is useless and serves only to increase the edema. The fasciotomy is longitudinal and extensive enough to restore circulation to the limb. The necessity for a fasciotomy may be determined by a flowmeter, measuring the tissue pressure of the closed space or by the clinical appearance of the extremity. This fasciotomy was performed on a 6-year-old patient with severe edema from the groin to the foot. The fang marks were 1 in. apart. No permanent tissue damage occurred in the patient, and the functional recovery was also complete.

The pharmacology of pit viper venoms is so similar that a single polyvalent antivenin is available which is effective for all of the rattlesnake and other pit viper bites in the Western Hemisphere. As the amount of antivenin required is determined by the severity of the symptoms, it is not necessary to identify that snake in order to select the proper antivenin, or to estimate the dosage (Arnold, 1979).

Finally, the time wasted looking for the snake can be better utilized in transporting the victim to a hospital.

Treatment of Bites

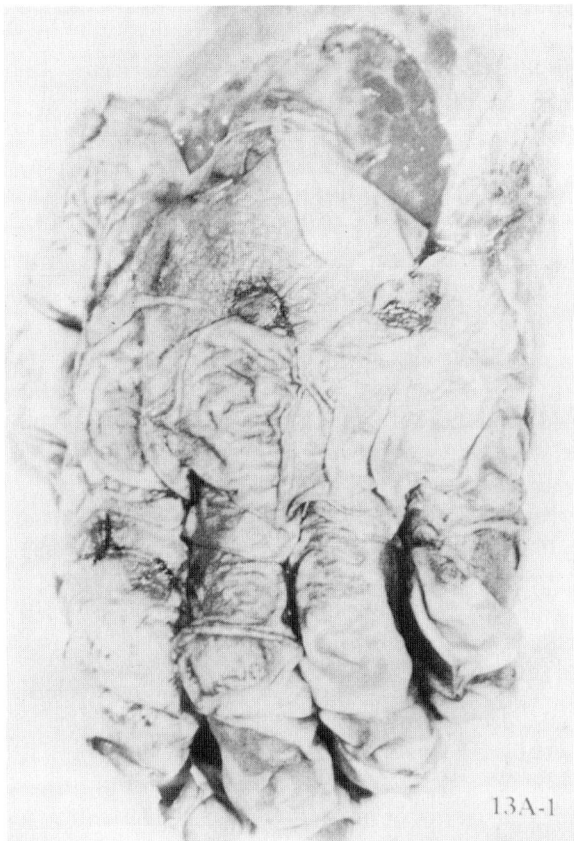

Figure 8 This is an example of the severe tissue damage resulting from the cryotherapy for the treatment of rattlesnake bite. This hand belonged to a 14-year-old patient bitten on the index finger. The hand was packed in ice. The tissue damage caused by this combination was so severe that this hand had to be amputated. (From McCullough and Gennaro, 1963.)

FACTORS AFFECTING MEDICAL TREATMENT

Tourniquets

Tourniquets have been used to treat rattlesnake bites for over 70 years, and their complications have been reported that long also. Tourniquets were recommended because they supposedly kept the venom isolated in the extremity. This was supposed to prevent systemic reactions and fatalities. Tourniquets also decrease tissue perfusion, depending on how tightly they are applied, and venom-damaged tissue tolerates ischemia poorly. This combination of poor circulation and keeping the venom concentrated increases tissue destruction and production of the biologically active products of tissue destruction. The blood trapped distal to a tourniquet also becomes acidotic, hyperkalemic, and a myocardial depressant, and a potentially lethal bolus is created (Bennett et al., 1961; Damus and Salzman, 1972; McCullough and Gennaro, 1968; Lockhart, 1970; Van Mierop, 1976; Snyder et al., 1972; Solomen et al., 1968). Shock, cardiac arrest, and fatalities have been reported when tourniquets were suddenly released (Laravuso, 1980; McCullough and Gennaro, 1963). Physicians with clinical experience in treating snakebites have noticed that venom which escapes the bite site is diluted enough to prevent tissue slough. Therefore, to aid in preventing tissue destruction, the proximal spread of venom should be enhanced, not retarded with a tourniquet.

Some first aid recommendations have attempted to compromise this recommendation by advising that the tourniquet should be applied loosely to occlude only the superficial lymphatics. A loosely applied tourniquet, however, will become an occlusive one when the edema is severe enough. As there has never been any evidence that tourniquets have reduced morbidity or mortality, there can be only one valid recommendation concerning tourniquets: Don't use them.

Incisions and Suction

Incision and suction is another recommendation that should be disregarded. Although studies in animals have demonstrated venom retrieval when incisions were quickly made at the site of envenomation and suction applied, this has never been shown to benefit human victims of rattlesnake bites (Gennaro and McCullough, 1961). Incisions should not be made by laymen administering first aid because of the risk of injuring vital structures. Mouth suction should never be considered because mouth bacteria can produce virulent infections (Arnold, 1979; Lockhart, 1970).

Some authors have recommended excision of the bite to mechanically remove the venom. The same hazards of incisions are present and increased in removing a plug of tissue (Lockhart, 1970); significant tissue slough and chronic osteomyelitis have been reported also. The reported results of treating rattlesnake bites by excision of the bite are so poor that this procedure should never be considered for the treatment of rattlesnake bites (Gennaro and McCullough, 1961; Huang et al., 1974).

Cryotherapy

Twenty years ago a handbook was published which advocated cryotherapy for snakebites, recommending immersion of the envenomated member in ice for 24-

72 hr. The simplicity of this approach and the availability of ice made this very attractive to physicians for emergency treatment. Many experienced people are reluctant to cut another person, tourniquets were usually not available, and yet, the general notion was that "something had to be done." Placing a snakebitten extremity in ice was easy, quick, and reassured the patient and practitioneer that everything possible was being done. This procedure was popularized by the lay press; consequently, cryotherapy was used extensively. It soon became apparent to the doctors taking care of these patients that the results were disastrous. Mortality among snakebite victims is minimal, and some of these victims are snake handlers who refuse medical treatment; others may have associated diseases which contribute to a fatal outcome. Cold, especially immersion in ice water, damages tissue and increases the number of tissue sloughs and amputations (Bennett et al., 1961; Lockhart, 1970; McCullough and Gennaro, 1963). The point is that the criteria for evaluating snakebite treatment is the avoidance of tissue slough, not death. Those unaware of this incorrectly assume that a living patient, albeit minus a hand or foot, is a successful result.

Unfortunately, by the time doctors treating these patients were able to evaluate cryotherapy and publish critical reports, cryotherapy had become so firmly entrenched in the popular press, and in some medical publications, that many patients still receive cryotherapy from enthusiastic but uninformed individuals.

There is now almost universal agreement among physicians experienced in treating snakebite that cryotherapy should never be considered (Arnold, 1979; McCullough and Gennaro, 1963; Gennaro and McCullough, 1961; Clark, 1971).

Alcohol

Alcohol has been a popular snakebite remedy for generations and is usually readily available. When our ancestors migrated west, they usually had a jug of alcohol for medicinal purposes. Doctors were scarce, snakes were plentiful, and the standard treatment for snakebite was to ply patients with whiskey until they passed out. When they woke up the next day, the snakebite was usually better. Thus, alcohol became the treatment of choice, but unfortunately it is ineffective. These patients may vomit and aspirate the vomitus if they are inebriated. Nothing should be given to these patients to drink or eat, especially alcohol.

Antivenin

Antivenin has been considered for first aid, and some people have suggested carrying antivenin in first aid kits, but it should not be used in these circumstances. It is too hazardous to use antivenin any place other than in a hospital equipped with resuscitative equipment and trained personnel.

Transportation

The second principle of snakebite first aid is to transport the patient to a hospital as soon as possible. The circumstances and severity of the bite may be so variable that common sense (or judgment) will be necessary. A reasonable course of action for mild envenomation may be inappropriate or impossible for a severe envenomation, especially if the bite occurs in a remote area.

The first thing to be done is to evaluate the bite. Remove shoes, socks, rings, bracelets, or anything else that could become a tourniquet if the swelling is severe. Then examine the size of the bite carefully, noting both the presence of fang marks, the progression of the swelling, and the severity of the pain.

Shock may be manifested by a rapid, thready pulse, cold clammy skin, nausea, and vomiting. Weakness, numbness and fainting also are symptoms indicative of severe envenomation. During this examination it is very important to reassure the patient. Hysteria is a frequent complication of snakebites and may manifest many bizarre symptoms. The physician's calm demeanor and purposeful unhurried movements assure the patient that he will be all right and are effective in preventing this.

If the patient is able to walk, he should do so, at least to the nearest vehicle for transportation to a hospital. If the patient is unable to walk, someone else should go for help if the patient is neither unconscious nor demonstrates any signs of paralysis such as drooping eyelids or the inability to swallow or talk. Never leave an unconscious patient because he may vomit and aspirate the vomitus and drown if the paralysis becomes severe. The patient may be unable to breathe and mouth to mouth resuscitation will be necessary until a respirator is available; this is an extremely serious situation, but total recovery is possible if he is ventilated. Therefore, always remain with the patient who demonstrates any paralysis. In the absence of coma or paralysis, make the patient as comfortable as possible, then seek help after the area is marked with some pieces of cloth tied to trees or bushes to facilitate finding the patient from the ground and air. While transporting the patient, splints are not only unnecessary, they may cause pain, or impair circulation if they become too tight. It is not necessary to place the envenomated extremity in any particular position, but permit the patient to be as comfortable as possible. Mouth to mouth resuscitation should be continued if necessary, while keeping the mouth cleaned of vomitus and secretions to prevent aspiration.

DEFINITIVE MEDICAL TREATMENT

Most physicians have had little or no experience in treating snakebite victims, and there are few controlled studies on snakebites. Many bites are trivial and the victim's recovery is incorrectly credited to the therapeutic measures the

Treatment of Bites

attending physician happened to be using at the time. The result of this *post hoc ergo propter hoc* reasoning has been a plethora of recommendations, some valid, some harmful, and some merely useless. Unfortunately, initial recommendations published in the literature are usually etched in the physician's mind, while the subsequent exposure of their hazards never is appreciated. It is important for the inexperienced physician treating a rattlesnake bite to understand that, while snakebites are rare, the clinical effects of envenomation are not; they are present in many other diseases and should be treated similarly. For example, the therapeutic measures effective for correction of hypovolemic shock will be just as effective whether the shock resulted from snakebite or any other cause of fluid loss. This concept is necessary in order for the physician with a limited experience with snakebite to have the confidence to rely on measures of proven therapeutic benefit instead of the latest fad.

Three things should be accomplished during the initial examination:

1. Reassure the patient.
2. Estimate the severity of the bite.
3. Plan a rational treatment.

Reassurance of the Patient

The importance of reassuring the patient is frequently overlooked. Snakebites are frightening experiences, and while the study of the pathology of fright is in its infancy, there can be little doubt of its harmful effect. Also, some of the evaluation of the severity of envenomation is subjective, and a patient who is frightened is difficult to evaluate.

Estimation of the Severity of the Bite

The severity of the envenomation can be reliably estimated by a careful examination. Toxicity is manifest by local and systemic signs and symptoms.

Rattlesnakes lose and replace fangs, so there may be one, two, or several fang puncture marks present, but at least one fang mark must be present. The proteolytic enzymes in venom cause significant tissue damage. The small blood vessels are vulnerable and their permeability is altered, resulting in leakage of plasma and blood into the tissues, producing ecchymosis, blistering, cyanosis, and edema. Edema is the most consistent local sign of envenomation; the severity and rapidity of its spread is proportional to the amount of the envenomation. Pain will be present and also proportional to envenomation and tissue destruction. In addition, the edema pressure in the limb may be great enough to compromise the circulation, resulting in a cold, cyanotic, pulseless limb.

The systemic manifestations of rattlesnake bites are shock, hemorrhage, neurotoxicity, and toxicity of the biologically active products of tissue destruction.

Shock results from a loss of fluid from the vascular compartment. The leakage of blood and plasma may be severe enough to cause hypovolemic shock. Hidden losses may also occur with bleeding into the retroperitoneal space, chest, or abdomen. Although venom contains components which theoretically could produce shock from other causes such as heart failure, both clinical and laboratory results implicate hypovolemia as the only significant cause of shock (Russell, 1971; Carlson, 1976).

Bleeding has been observed with pit viper bites for over 200 years. Several early attempts were made to classify snakes into those causing bleeding and those which do not. These efforts failed because the findings were so inconsistent. Several constituents of venom alter clotting. Some enzymes convert fibrinogen directly to fibrin without activating the clot-stabilizing factors, producing a soft, mushy clot which is quickly lysed, resulting in hypofibrinogenemia (Weiss et al., 1969; Hasiba et al., 1975). Bleeding in these patients is rarely severe, presumably because the platelets are unaffected and the platelets will form plugs and prevent bleeding even in the absence of fibrinogen. Thrombin-like procoagulants have also been demonstrated, as well as enzymes which will activate the clotting mechanism to produce thrombin (Van Mierop, 1976). Finally, envenomated tissue may release thromboplastic substances that initiate clotting and lysis, which has been described in other clinical conditions such as sepsis, trauma, and abruptio placentae and is known as diffuse intravascular clotting (DIC) (Lyons, 1971; Damus and Salzman, 1972; Van Mierop, 1976; Hasiba et al., 1975; Bradham, 1972). The prothrombin time (PTT) and thrombin times are abnormal, fibrinogen levels are low, and such large quantities of platelets are consumed that the resulting thrombocytopenia is the most consistent finding in acute DIC caused by snakebite. These various factors may be present individually or in combination, but severe bleeding accompanying snakebite resembles that caused by DIC. The resulting blood loss may be severe enough to require 20 or more units of blood, and bleeding into almost every organ has been described. The brain and kidneys are the most vulnerable organs to bleeding, and permanent damage of these organs has been described.

Neurotoxins

Neurotoxins are pharmacologically active components of venom which can produce failure of motor and sensory nerve impulse transmission. This may cause numbness, paresthesias, mydriasis, and in severe cases, paralysis. Neurotoxins have been demonstrated in the venom of some North American rattlesnakes, especially the Mojave rattlesnake (*Crotalus scutulatus scutulatus*)

(Rhoten and Gennaro, 1968). Although numbness and paresthesias have been reported with rattlesnake bites, these symptoms could be caused by hysteria as well as venom and are inconsequential. The tropical rattlesnake (*Crotalus durissus*) venom, on the other hand, is much more neurotoxic (da Silva et al., 1979). Paralysis severe enough to require a respirator has been reported from bites of this snake. Drooping eyelids, double vision, paralysis of the eyes and facial muscles, weakness, difficulty in talking and swallowing, and respiratory muscle weakness have also been described from bites of the tropical rattlesnake. The tropical rattlesnake, or cascabel, ranges from southern Mexico southward through Central America and eastern South America to Argentina. It is the only rattlesnake in all of this range except in Mexico.

Envenomation by *C. durissus* produces severe systemic symptoms with relatively minor local signs. This unusual pharmacologic phenomenon makes the tropical rattlesnake one of the most dangerous snakes in the world and it is the leading cause of snakebite in Brazil. The predominance of systemic symptoms, such as paralysis, distinguishes these bites from the North American rattlesnake bites which usually produce tissue destruction.

For this reason, anyone bitten by the cascabel should never be left alone, either in the field or the hospital, until their clinical condition has improved enough that respiratory arrest is no longer a consideration.

Planning a Rational Treatment

The envenomated limb should be examined and washed with soap and water if necessary. Remove rings, bracelets and other potential tourniquets. If fang marks are present, estimate the severity of the pain and determine the extent of the swelling. Many snakebites are dry; that is, no venom has been injected, and therefore, no symptoms are produced. Tetanus prophylaxis, broad spectrum antibiotics, and observation for 24 hr should be sufficient for these patients.

If significant envenomation is present, an intravenous infusion of Ringer's lactate solution should be started in the contralateral arm. The infusion rate should be enough to maintain tissue perfusion. Blood pressure, pulse, cerebration, respiration, and temperature should be recorded, and the circulation in the envenomated limb should be closely watched. Color, temperature, the severity of swelling, or a Doppler flow meter may be valuable for estimating perfusion.

The same reasons that tourniquets, incisions, suctions, cryotherapy, or excision of the bite are inappropriate first aid measures are also valid when the patient arrives at a hospital. An occlusive tourniquet should be removed as soon as possible, but only when adrenalin and resuscitative equipment are available, because sudden removal may cause shock and cardiac arrest. Remove all ice, splints, and bandages.

If there has already been enough tissue damage to jeopardize the viability of the limb, color photographs to document this are oftentimes invaluable and should be made at this time.

ASSOCIATED FACTORS OF MEDICAL TREATMENT

Pain

Rattlesnakes bites may be very painful. It is safe to use demerol or morphine for the relief of pain. Regional blocks are effective and have the advantage of controlling pain without affecting consciousness.

A complete blood and platelet count and a urinalysis should be obtained on every patient. These tests should be repeated every hour or so until the manifestations of snakebite are subsiding. Blood loss and thrombocytopenia indicates systemic reaction. Prothrombin time and fibrinogen determinations are valuable for estimating the severity of the coagulopathy and the effectiveness of the treatment. Blood, preferably fresh, should be available for these patients.

Bleeding may be sudden and severe, and if blood is not available, the patient's welfare may be jeopardized.

Urine volume and composition are important: volume as an indicator of perfusion, and composition for the early diagnosis of renal involvement. Anuria that does not respond to careful intravenous fluid management is a very serious sign of possible acute renal cortical necrosis and permanent renal damage.

The next consideration should be whether to use antivenin. The proper use of antivenin is imperative for the treatment of severe envenomations. Wyeth's Polyvalent Antivenin (Crotalidae) is the only commercially available antivenin in the United States that is effective for rattlesnake bites. Fortunately, it will neutralize the venom of all the known rattlesnakes in the world. It is a horse serum product with a significant allergenic potential. Many bites are so trivial that antivenin is unnecessary and should not be used, thereby avoiding any unnecessary allergic reactions. Zoo personnel and people who work in endemic areas are exposed to rattlesnakes so much that repeat bites are a possibility. If antivenin is used for mild reactions in these people, they may become so sensitized that it will be impossible to use the antivenin for the severe envenomations. Antivenin should be administered in the presence of all systemic reactions. Shock, renal failure, hemorrhage, coagulopathy, diffuse intravascular coagulation, or paralysis in the case of tropical rattlesnake bites indicate severe envenomation, and large amounts of antivenin will be needed to avert death or permanent brain or kidney damage (Russell, 1971).

Local reactions require some judgement and the realization that the manifestations of envenomation are dynamic instead of static. Rigid criteria based on the amount of edema are of limited value because the clinical appearance can change so rapidly. Most bites are seen within an hour; if the edema has progressed as much as 1 ft (30 cm) by then, 5-10 vials of antivenin should be sufficient. If the edema involves the entire extremity, 10-20 vials should be infused. Involvement of the trunk or systemic signs may require more than 20 vials. Children are more vulnerable to rattlesnake venom than adults, and comparable bites will require more antivenin for children.

Antivenin should be administered intravenously (McCullough and Gennaro, 1968; McCullough and Gennaro, 1963). Estimate the correct dose and add it to 500 or 1000 cc of lactated Ringer's, and then piggyback this via the previously started intravenous infusion and infuse at 200-300 cc/hr.

The advantages of the intravenous route are

1. The antivenin is injected directly into the blood stream and neutralization of the venom occurs quicker (McCullough and Gennaro, 1963).
2. It is easier to detect and combat allergic reactions. Permit a few drops of the antivenin to infuse, stop and observe for allergic reactions. If there are none, repeat this procedure with a few more cubic centimeters of antivenin. If there still is no reaction, permit the antivenin to infuse continuously. A syringe with 1 cc of 1:1000 adrenalin and resuscitative equipment should be readily available during the testing and administration of antivenin.

Antivenin allergy presents a serious dilemma. The demonstrated value of antivenin must be balanced against the danger of an anaphylactic reaction if antivenin allergy is present. In these cases, the *value of antivenin is greater than the hazard if the treating physician will remain with the patient* and observe closely for any allergic symptoms. A few drops of 1:1000 adrenalin in the intravenous infusion from time to time should prevent serious reactions. As many as eight vials of antivenin have been infused this way with no serious sequelae (Russell, 1978). Resuscitative equipment and trained personnel must be immediately available in these circumstances. A syringe loaded with adrenalin is also mandatory. In addition to Wyeth Laboratories, Inc., Marietta, Pennsylvania, antivenins effective against rattlesnake venoms are manufactured by

1. Instituto Butantan
   Caiza
   Postal 65, São Paulo, Brazil
2. Instituto Pinheiros Productos Therapeuticos
   Caixa
   Postal 951, São Paulo, Brazil

3. Instituto Nacional de Higiene
   Cxda M. Escobedo Number 20
   Mexico 17, D.F.

Cortisone

It is appropriate to discuss the use of cortisone because replacing antivenin with cortisone is probably the most serious error many physicians make in treating severe envenomations. Antivenin is imperative in severe envenomations, yet many physicians have been misled by recent articles in the medical literature advocating the use of cortisone instead (Glass, 1973). Cortisone has had extensive laboratory and clinical trials, and no benefit has ever been observed (Russell and Emery, 1961). Schottler conducted an extensive survey on several hundred mice and guinea pigs and concluded that cortisone offered no benefit in snakebite (Schottler, 1954). Clark later verified these results on dogs (Clark, 1971).

Finally, Cunningham et al. (1979) investigated corticosteroids in the following manner. There were 110 Sprague-Dawley rats divided into 4 groups. The 30 rats in group A were given 10.53 mg/kg (the approximate LD_{50}) of (*Crotalus adamanteus*) venom intramuscularly in the right hind leg. The 30 rats in group B were given the same dose of venom intramuscularly in the right hind leg and 5 min later, 30 mg/kg methylprednisolone sodium succinate was given intramuscularly in the left hind leg. The 30 rats in group C were given the same venom dose intramuscularly in the right hind leg and were given 150 mg/kg hydrocortisone sodium succinate in the left hind leg. Finally, the 20 rats in group D were given only 30 mg/kg methylprednisolone sodium succinate intramuscularly and were not given venom. There was a 37% mortality of the rats in group A. The mortality rate *jumped to 77%* in group C, the rats treated with hydrocortisone sodium succinate. The rats treated with methylprednisolone sodium succinate, group B, had the *highest mortality rate of all, 93%*. The rats in group D, which were not envenomated, had no deaths. It would seem that corticosteroids are harmful in treating rattlesnake bites. In summary:

1. Cortisone has not been demonstrated to have any beneficial effect in rattlesnake venom poisoning.
2. Cortisone has not been shown to have any effect on the biologically active substances resulting from tissue breakdown.
3. Cortisone has not been demonstrated to be of benefit in the treatment of systemic manifestations of rattlesnake envenomation. The fact that it has been used on patients who survive is no proof of efficacy (McCullough and Gennaro, 1963; Arnold, 1979).

4. Cortisone has been shown to interfere with immune responses. Laboratory evidence would indicate that it probably blocks the neutralization of venom with antivenin (McCullough and Gennaro, 1963; McCullough and Gennaro, 1970).

Although very few authors continue to recommend cortisone, their results either have not been reported or have been inferior to those of antivenin therapy (Glass, 1973; Arnold, 1976). Therefore, there is no evidence to support the use of cortisone for rattlesnake bites.

After the bite has been evaluated and treatment started, there is a lull in the action which presents an opportunity for the physician to examine the patient and the envenomation more thoroughly. If it appears that prolonged cooling, too tightly applied tourniquets, or severed nerves or tendons might jeopardize a satisfactory result, documentation of this with pictures and written description will be available. A complete history and physical examination should be done. Snakebites are stressful, and a history of conditions that are adversely affected by stress such as peptic ulcers and heart disease should be obtained. The physical examination should include a careful examination of the nervous and muscloskeletal systems so that previous infirmities can be carefully noted and recorded in order that later they are not considered a bad result of snakebite or treatment. The envenomated extremity should be examined very thoroughly for any signs of circulatory embarassment. Fingers and the fascial compartments of the leg are the most vulnerable to circulatory embarassment from edema pressure (Arnold, 1979). Relaxing incisions may rarely be necessary to avert permanent tissue damage in these cases. Fasciotomies should not be routine for every snakebite, but should only be performed when indicated. Fasciotomies should be longitudinal and long enough so the examining physician is certain that tissue perfusion is adequate. Debridement should not be necessary; although the damaged tissue may appear questionably viable, a delay in debridement is warranted because even severely damaged tissue may recover.

The patient should be reevaluated after an hour or two of treatment.
Blood pressure, pulse, and respiration should be determined every 30 min.
The urine should be examined every hour for volume and the presence of protein, blood, or hemoglobin. A blood count should also be obtained every hour to detect the loss of red cells or abnormal platelet consumption.
Subjective symptoms as well as the vital signs are important to evaluate the clinical course. The antivenin dosage may have to be adjusted by adding antivenin to the infusion.

Blood components such as platlets or fibrinogen will be unnecessary if fresh blood is transfused.

Heparin has been used in the management of snakebites. Although there has not been enough clinical experience to conclude that heparin is beneficial, the few cases reported indicate that it is polyanionic and may neutralize the polycationic components of venom (Weiss et al., 1973; Anonymous, 1973). The value of heparin, however, is that it blocks the formation of thrombin, thereby preventing clot formation and further consumption of the clotting factors (Bradham, 1972; Damus and Salzman, 1972). As the platelet count returns to normal, the platelet plugs are effective in controlling hemorrhage. Heparin also prevents cortical necrosis of the kidneys by preventing fibrin deposits in the small renal vessels. Heparinization is more effective before the coagulopathy is fulminant. When the platelet count drops below 25,000, 10,000 units of heparin should be infused initially and 5,000 units every 8 hr until the coagulopathy or thrombocytopenia improves. This may keep the bleeding diathesis from becoming severe and will protect the vital organs from fibrin deposits and circulatory embarrassment (Arnold, 1979). Pain should be relieved with morphine.

Kidney

Acute renal failure following snakebite is caused by the bleeding syndrome. Circulatory failure, shock, hemolysis, disseminated intravascular coagulation, and defibrination syndrome either singly or in combination may contribute to the development of acute tubular necrosis or cortical necrosis. Pathological findings described in the kidney include bilateral cortical necrosis (Arnold, 1979), proliferative glomerulonephritis (Hopkins, 1976), hemorrhagic glomerulonephritis, and interstitial nephritis (Chugh et al., 1975). Oliguria or anuria that does not respond to volume replacement or intravenous furosemide are presumptive evidence of renal involvement. If acute renal failure occurs, the electrolytes should be monitored, as well as the coagulopathy. Hyperkalemia, uremia, or acidosis indicate the need for dialysis. Peritoneal or hemodialysis are satisfactory until diuretic phase of recovery. This is usually 7-11 days after renal shutdown (da Silva et al., 1979). X-rays will reveal contracted dense kidneys with crenated wavy margins. Renal biopsy will confirm the diagnosis. Permanent dialysis or renal transplant will be necessary for these patients.

Acute respiratory failure has occurred after bites of the tropical rattlesnake (*Crotalus durissus terrificus*). The venom has enough neurotoxicity to cause weakness and palsy of respiratory muscles. A respirator with a cuffed endotracheal tube will be necessary to provide adequate oxygen and prevent aspiration of saliva. Ventilatory assistance may be necessary for 14 days. Serial arterial pO_2, pCO_2, and pH are valuable for determining the adequacy of ventilation (daSilva et al., 1979).

Tetanus Prophylaxis

Appropriate tetanus prophylaxis is necessary. This will vary, depending on the immunization history of the patient.

Antibiotics

Broad spectrum antibiotics are indicated because secondary infection has been described. A 7 day course of treatment should be sufficient.

Duration of Treatment

Intensive treatment with close observation should be continued until there is enough clinical improvement that the patient's life or limbs are no longer in jeopardy. If there is only a local reaction, improvement should be present after several hours of treatment. The patient should feel better with less pain and no further progression of the edema. If a fasciotomy has not been performed, a patient should be able to leave the hospital in 2-4 days. A fasciotomy should be repaired when all of the edema has subsided, which may be 3 or 4 weeks.

The systemic symptoms usually indicate such severe envenomation that intensive treatment and observation may be required for 7-14 days. Ventilatory assistance for 2 weeks had been reported (daSilva et al., 1979). Laboratory evidence of clotting abnormalities may persist for 2 weeks or more after the patient stops bleeding and is asymptomatic. If the platelet count and functions are within normal limits, hospitalization during this period should not be necessary. If debridement has been necessary and skin grafts are required, these should be done only after all of the systemic and local manifestations have returned to normal.

Some patients may be first seen at the hospital several days after they were bitten. Sometimes they did not seek medical attention and finally came to the hospital when their condition was critical. Others may have been treated, but their condition had deteriorated and they were transferred to a hospital with better facilities. Although snakebite was the original cause of their troubles, after a few days the envenomation is not a significant factor. Tissue deterioration, gangrene, and sepsis are common in these patients. Multiple organ failure secondary to sepsis may include kidneys, heart, liver, or lung. Debridement, amputation, and vital organ support may be necessary. It is necessary to treat the complications and not the snakebite in these patients, and antivenin is of little or no value if more than 1 day has elapsed since the bite.

SUMMARY

The severity of rattlesnake bites varies from inconsequential puncture wounds to those causing fatalities. There are also many different pathological effects of rattlesnake venom. In addition to tissue destruction, altered blood clotting, failure of nerve transmission, hemorrhage, shock, and brain and renal damage, occasionally an unexpected result, such as peritonitis, perforated ulcer, and even ruptured spleens, has been reported. It is impossible to treat each patient in a "cook book" manner, but each patient can be carefully examined and the observed pathology can be properly treated in an orderly manner. The pathological process is a dynamic one, and repeated examinations and close observation of the patient are necessary to detect either deterioration of the patient's condition or any new complications. Knowledge of the pathological effects of envenomation and the fundamentals of successful treatment will give the first aid practitioner or treating physician the confidence to rely on methods which are fundamentally sound and beneficial instead of the latest fad.

REFERENCES

Arnold, R. E. (1976). Treatment of snakebite. *JAMA, 236*:1843.
Arnold, R. E. (1979). Controversies and hazards in the treatment of pit viper bites. *South. Med. J.* 72:902–906.
Bennett, J. E., Brelsford, H. G., Lewis, S. R., and Blocker, T. G. (1961). Distal extremity necrosis after snake bite. *Plastic Reconstr. Surg.* 28:385–393.
Bradham, G. S. (1972). Treatment of poisonous snakebite. *J. S.C. Med. Assoc.* 68:449–454.
Carlson, R. W. (1976). Oncotic and hydrostatic forces in perfusion failure and volume loading. In *The Organ in Shock: A Scope Publication.* Upjohn Company, Kalamazoo, Mich. 50–56.
Chugh, K. S., Aikat, B. K., Sharma, B. K., Dash, S. C., Mathew, M. T., and Das, K. C. (1975). Acute Renal Failure Following Snakebite. *Am. J. Trop. Med. Hyg.* 24(4):692–697.
Clark, R. W. (1971). Cryotherapy and corticosteroids in the treatment of rattlesnake bite. *Mil. Med.* 136:42–44.
Cunningham, E. R., Sabback, M. S., Smith, R. M., and Fitts, C. T. (1979). Snakebite: Role of corticosteroids as immediate therapy in an animal model. *Am. Surg.* 45(12):757–759.
Damus, P. S., and Salzman, E. W. (1972). Disseminated intravascular coagulation. *Arch. Surg.* 104:262–265.

daSilva, O. A., Lopez, M., and Godoy, P. (1979). Intensive care unit treatment of acute renal failure following snake bite. *Am. J. Trop. Med. Hyg. 28*(2): 401-407.

Gennaro, J. F., and McCullough, N. C. (1961). Comments on the contemporary treatment of poisonous snakebite in North America. *Med. Rec. Ann. 46*: 224-225.

Glass, T. G. (1973). Early debridement in pit viper bite. *Surg. Gynecol. Obstet. 136*:774-776.

Hasiba, U., Rosenbach, L. M., Rockwell, D., and Lewis, J. H. (1975). DIC-like syndrome after envenomation by the snake *Crotalus horridus horridus*. *N. Engl. J. Med. 292*:505-506.

Hopkins, L. T. (1976). Snakes and snakebites: Characteristics, management and prognosis. *J. Kans. Med. Soc. 77*:193-196.

Huang, T. T., Lynch, J. B., Larson, D. L., and Lewis, S. R. (1974). The use of excisional therapy in the management of snakebite. *Ann. Surg. 179*:598-603.

Laravuso, R. B. (1980). Cortical blindness in a child after anesthesia. *JAMA, 243*:1187.

Lockhart, W. E. (1970). Pitfalls in rattlesnake bite. *Texas Med. J. 66*:42-42.

Lyons, W. J. (1971). Profound thrombocytopenia associated with *Crotalus ruber ruber* envenomation: A clinical case. *Toxicon 9*:237-240.

McCullough, N. C. and Gennaro, J. F. (1963). Evaluation of venomous snake bite in the southern United States from parallel clinical and laboratory investigations. *J. Fla. Med. Assoc. 49*:959-967.

McCullough, N. C., and Gennaro, J. F. (1968). Diagnosis, symptoms, treatment and sequelae of envenomation by *Crotalus adamanteus* and genus *Ancistrodon*. *J. Fla. Med. Assoc. 55*:327-329.

McCullough, N. C., and Gennaro, J. F. (1970). Treatment of venomous snakebite in the United States. *Clin. Toxicol. 3*:483-500.

Med. J. Aust. Comment, 1:322-323 (1973).

Rhoten, W. B., and Gennaro, J. F. (1968). Treatment of the bite of a Mojave rattlesnake. *J. Fla. Med. Assoc. 55*:338-340.

Russell, F. E. (1971). Venom poisoning. *Nat. Drug Ther. 5*:1-7.

Russell, F. E. (1978). Jaws that bite, things that sting. *Emerg. Med. 10*:24-59.

Russell, F. E., and Emery, J. A. (1961). Effect of corticosteroids on lethality of *Ancistrodon contortrix* venom. *Am. J. Med. Sci. 241*:501-511.

Schottler, W. H. A. (1954). Antihistamine, ACTH, cortisone hydrocortisone and anesthetics in snake bite. *Am. J. Trop. Med. 3*:1083-1091.

Snyder, C. C., Straight, R., and Glenn, J. (1972). The snakebitten hand. *Plast. Reconstr. Surg. 49*:275-282.

Solomen, K. A., Tarkkanen, L., Narvanen, S., and Gordin, R. (1968). Metabolic changes in the upper limb during tourniquet anesthesia. *Acta orthop Scand. 39*:20-32.

Van Mierop, L. H. S. (1976). Poisonous snakebite: A review. *J. Fla. Med. Assoc. 63*:191-210.

Weiss, H. J., Allan, S., Davidson, E., and Kochva, S. (1969). Afibrinogenemia in man following the bite of a rattlesnake (*Crotalus adamanteus*). *Am. J. Med.,* 47:625-634.
Weiss, H. J., Phillips, L. L., Hopewell, W. S., Phillips, G., Christy, N. P., and Nitt, J. F. (1973). Heparin therapy in a patient bitten by a saw scaled viper. *Am. J. Med.* 54:653-662.

7
Management of the Western Diamondback Rattlesnake Bite

THOMAS GRAHAM GLASS, JR.
The University of Texas Health Science Center, San Antonio, Texas

Introduction 340

Local Signs and Symptoms 341

Diagnosis of Rattlesnake Bite 341

Outline of Management of Rattlesnake Bite Victims (When Envenomation is Obvious) 342
Emergency Room • Operating Room • Operative and Postoperative Care • Specific Treatment Items

Evaluation of the Bite 345
The Snake (Identification) • Appearance at the Site of the Bite • Color of the Bite Site

Factors Associated with Medical Treatment 346
Corticosteroids • Antibiotics • Antivenin • Fibrinogen • Fresh Frozen Plasma • Packed Red Blood Cells • Platelets • Local Anesthetics • General Anesthetics • Laboratory Tests • Related Laboratory Studies • Errors

Statistics 357

Summary 360

---

*All the figures used in this chapter were taken from Glass, T. G. (1976). *Management of Poisonous Snakebite*. T. G. Glass, San Antonio, Texas.

INTRODUCTION

The Western diamondback rattlesnake is indigenous to south and west Texas. This snake is the largest and most dangerous poisonous snake that is native to the continental United States. In Texas there are up to seven recorded deaths each year from this snake's bite. Probably at least twice that many humans die each year in Texas, but only half of the deaths are recorded as being primarily due to the rattlesnake. An unknown number of victims of this snake's bite subsequently are crippled or have an extremity or part of an extremity amputated. The copperhead moccasin and the Western cottonmouth water moccasin are also abundant in Texas, but their bites are never as severe as those of the Western diamondback rattlesnake. No deaths have been recorded in Texas from bites of either of these two snakes, and seldom is there crippling of the human victim.

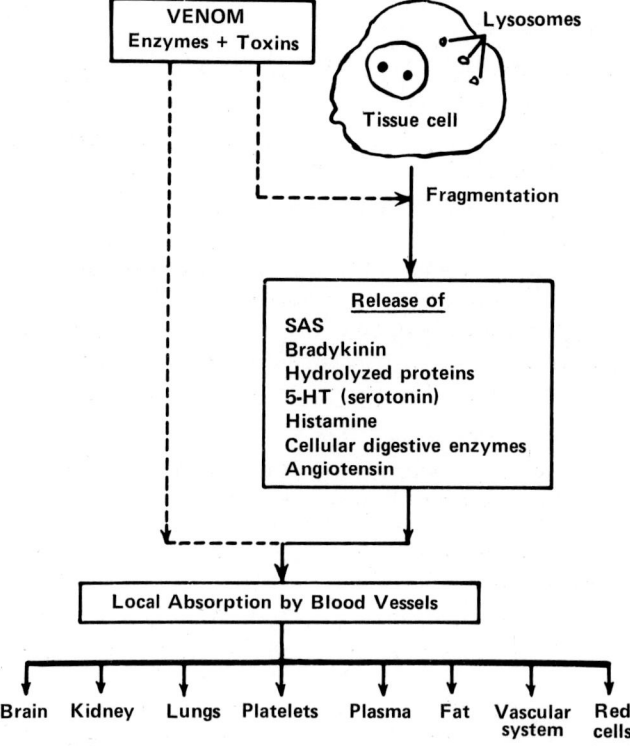

Figure 1   The venom and toxic products are distributed to organs remote from the bite site, causing many different clinical effects.

The deaths that I have seen (three) and those that I have investigated that resulted from the Western diamondback rattlesnake bite were due to the misdiagnosis of the bite, improper evaluation of the bite, prior alcoholic intoxication, or improper and/or delayed treatment. Two of these deaths occurred within an hour after the bite in victims that were toxic from alcohol ingestion prior to the bite. Both died from cardiac arrest with complete collapse of the myocardium.

Proper early care of the patient with Western diamondback rattlesnake envenomation will prevent the loss of life, tissue, limbs, and function. Adequate care of the snakebite victim requires that the bite be immediately evaluated and treated to prevent and/or alleviate the toxic symptoms, the local destruction of tissue at the bite site, and the defibrination of the plasma. This last problem often misleads the physician into thinking that no venom was injected into the tissues: A very serious mistake! Within 1 hr after a severe bite, skin blisters may develop at the fang puncture wounds (Figure 1).

LOCAL SIGNS AND SYMPTOMS

Envenomation causes a severe burning pain and tenderness at the bite. When venom is injected into the muscle there is severe tenderness over the muscle compartment. The peripheral pulses in almost all cases are unaffected even when muscle compartment damage is extensive. Waiting until peripheral pulses disappear or become very weak before surgical intervention will lead to disaster in these bites.

The symptoms of Western diamondback rattlesnake envenomation are as follows. Over 50% of all victims develop immediate nausea and vomiting. The pulse rate will rise rapidly or fall to 30 beats per minute in some cases. Hypotension of some degree is present in over 40% of the victims. Oliguria and anuria, convulsions, coma, and death follow if proper treatment is not instituted immediately in severe bites.

DIAGNOSIS OF RATTLESNAKE BITE

The correct early diagnosis (the diagnosis made when the patient arrives at the medical facility) completely answers the following questions. Was venom injected? How deep was the venom injected? Into what structures (fat, muscle, tendon sheath, or joint) was the venom injected, and approximately what quantity of venom was injected?

It is impossible for any physician to correctly answer these questions early after the bite by merely looking at the skin or trying to evaluate the symptoms and laboratory test results. To correctly answer these questions, a simple, but adequate, surgical incision must be made at the bite site, and the skin, fat, and underlying structures must be carefully examined. Surgical exploration, using local or general anesthesia, is simple, safe, and revealing. Delay in making the diagnosis of a Western diamondback rattlesnake bite leads to extensive loss of tissue and crippling even when antivenin is used.

OUTLINE OF MANAGEMENT OF RATTLESNAKE BITE VICTIMS (WHEN ENVENOMATION IS OBVIOUS)

Emergency Room

Start an intravenous infusion of 5% glucose with Ringer's lactate at 150 ml/hr in adults (use a slower rate in children). Keep an ice bag on the bite site and constricting band proximal and, if possible, distal to the bite site. Obtain legal permits as required. Obtain the following laboratory tests: the routine complete blood count, the urinalysis, the prothrombin time, the plasma fibrinogen level, and the platelet count. If there is some definite indication that packed red blood cells or fresh plasma will be required, ample blood should be obtained for typing and matching.

When venom has obviously been injected or there is the slightest suspicion that venom has been injected into muscle, joint, or tendon sheath, the victim is prepared for surgery with local or general anesthesia (Figures 2 and 3).

Operating Room

The cooling and bands are removed, the bite site area is prepared, and an incision is made between the fang marks or through the one fang mark. Preserve all of the skin unless it is very obviously necrotic! Inspect the fatty tissue and open the underlying structure and debride obviously necrotic fat and muscle. Irrigate joints and tendon sheaths and open tendon sheaths to inspect the associated muscle (venom migrates freely through the tendon sheaths and may cause extensive muscle damage at areas remote from the bite site). Open the fascia over involved muscle as extensively as necessary to allow complete debridement and to release the tension in the compartment. Place fine mesh Vaseline gauze over all exposed tissue, cover this with fluffs, and wrap the extremity loosely with a Kerlex or Kling roller bandage. Do not use elastic or other restricting types of bandages. Pressure dressings which prevent the free drainage of fluid from the tissues should be avoided.

Operative and Postoperative Care

Fluid administration, transfusions, correction of plasma fibrinogen levels, administration of platelet packs, cortisone, and antivenin treatment are used as indi-

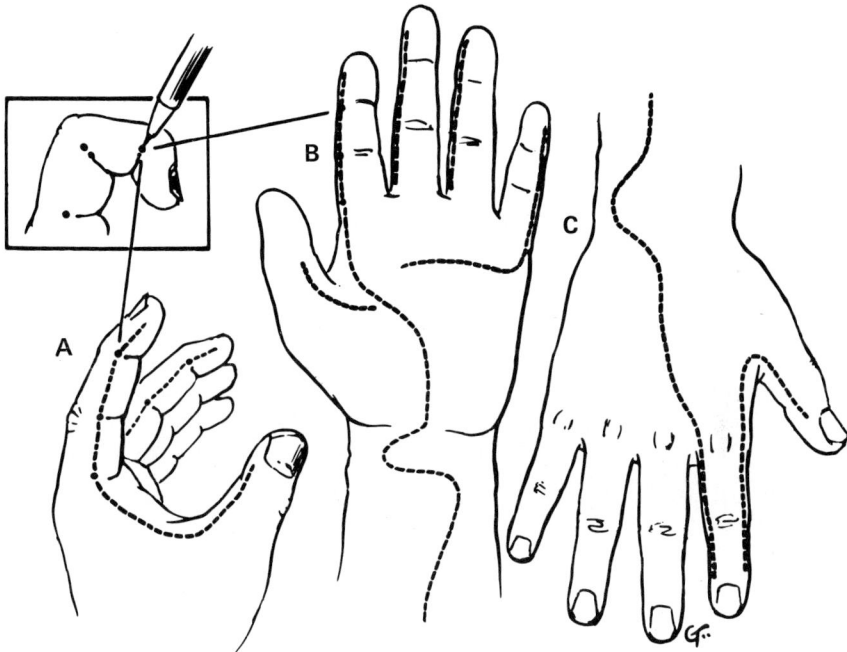

Figure 2 Incision path lines necessary for snakebite on the phalanges.

cated. The dressings are changed in the patient's room daily (or more often if necessary), and the skin (only) of the incision is closed loosely on the sixth or seventh postoperative day. The patient is then dismissed and is followed as an out-patient. When the edema is still extensive on the sixth or seventh day, a skin graft is applied to the area on the twelfth or thirteenth day (Figure 4).

Specific Treatment Items

Data

In 1966 I prepared a protocol for the management of pit viper bite victims which I have followed closely. In my private practice I have, since 1966, operated 160 patients with envenomation by the Western diamondback rattlesnake, and I have treated all of these patients following the same protocol. The use of the items and studies that I have discussed are based on this clinical experience plus the experience obtained from over 50 patients with envenomation by the Western diamondback rattlesnake treated by the same protocol by the residents and staff at the Bexar County Hospital system under the direction of the University of Texas Health Science Center at San Antonio, Texas.

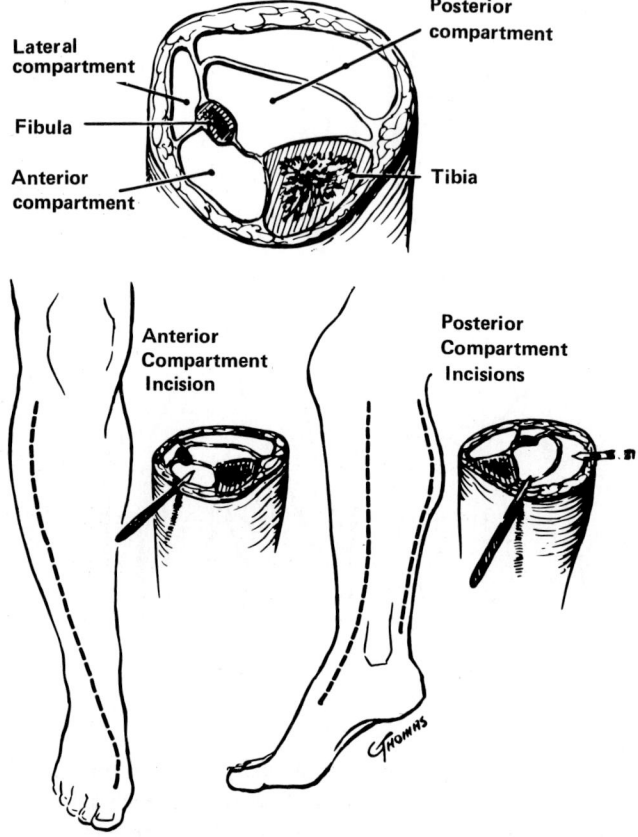

Figure 3 Incision path lines necessary for snakebite on the lower leg and foot.

Oral Intake

For the first 24 to 48 hr the oral fluid intake is restricted in case the patient needs another general anesthetic for further debridement and bleeding, or other causes. In almost all minor cases of envenomation the oral intake is started the day following the bite.

Intravenous Fluids

Intravenous (i.v.) fluids are started in the emergency area and are continued until the patient is taking oral fluids or until the antibiotics are discontinued. Since 1966 I have given only 5% G/RL. In some patients I have given over 11 liters of this fluid without causing any derangement of the electrolyte balance

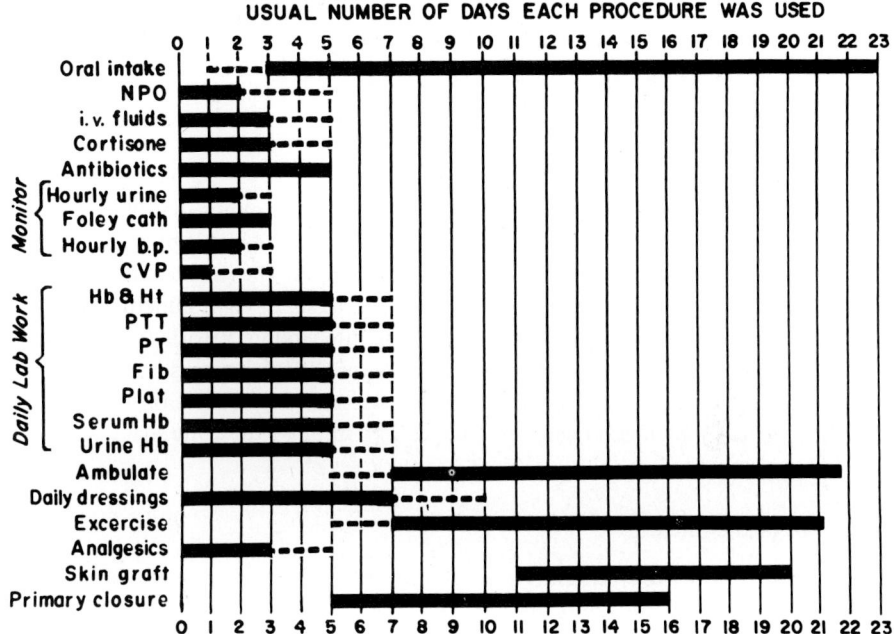

Figure 4 Management history showing 23 days of medical treatment.

or hypervolemia. The fluids are given at a rate such that the hourly urine output in the adult is at least 30 ml/hr, and proportionally slower rates are used for children. In those that require large volumes of fluids it is mandatory that some type of central venous pressure monitoring be present—preferably a subclavian catheter. This will hopefully prevent hypervolemia and pulmonary edema. Central venous pressures should be monitored hourly in the patients with severe envenomation.

EVALUATION OF THE BITE

The severity of the bite can be partially determined as soon as the victim arrives at the medical facility for care by the presence of toxic symptoms, by evidence of the envenomation at the bite site, and by blood coagulation studies.

The Snake (Identification)

The snake is usually brought to the medical facility by either the patient or someone present at the place where the bite occurred. If the snake is not brought in it has usually been identified. Every one of the rattlesnake bite victims I treated from 1966 to 1980 either brought in or identified the snake that bit them. A rather large number of the victims had handled the snake prior to the bite.

Appearance at the Site of the Bite

Bites by large snakes leave distinct and large puncture wounds that bleed freely. Fang puncture wounds may be almost invisible when the bite is by a small rattlesnake, and lacerations of the skin are not uncommon. The snake "strikes" from one to three or more times and leaves from one to six or eight puncture wounds. Most frequently there is only one visible puncture wound. Efforts to determine the type of snake by the shape or distribution of the teeth marks is a useless procedure. Three patients in this series of rattlesnakes bites were bitten three times before they could escape the snake. Two bites by the same snake were not uncommon.

Color of the Bite Site

When the rattlesnake venom is injected into the subcutaneous tissue the capillaries are opened and hemorrhage occurs. There is extravasation of fluid into the tissue and the hemorrhage and extravasation cause the skin to appear bluish in color, with swelling at the site. When all of the venom is injected into the underlying muscle, joint, or tendon sheath, there will be little or no swelling or blueness of the skin.

FACTORS ASSOCIATED WITH MEDICAL TREATMENT

Corticosteroids

Since 1962 I have administered very large doses of hyprocortisone sodium succinate (Solu-Cortef) to all victims of pit viper bites that have had toxic symptoms from the bite or that were to undergo a general anesthetic for debridement of the bite site during the first 48 hr after the bite. Either I administered a 1000-2000 mg, i.v. push as soon as toxic symptoms presented or arrangements were made to give the general anesthetic. After the debridement the cortisone was given in the same large doses every 6 hr for 72 hr. Then the cortisone was discontinued without the use of ACTH or without slowly decreasing the dosages of the cortisone. No patient that I treated with the cortisone had any untoward effect from the cortisone. It relieved the nausea and vomiting, decreased the heart rate in patients with tachycardia, increased the heart rate in patients with severe bradycardia, caused reestablishment of urine output (including one case in which anuria had been present for over 20 hr), and eliminated the toxic effects such as somnolence and drowsiness. There were two patients sent to me who had been treated only with antivenin for 18 and 40 hr, respectively, and had ruptured their esophagus from the continued vomiting and wretching that the antivenin did not relieve. The nausea and vomiting was relieved in every case I treated early with the cortisone, and none had a rupture of the esophagus. There was no evidence that the cortisone increased the infection rate, caused loss of tissue, slowed the healing process, or caused any other deleterious affect.

Antibiotics

At the time of the debridement, cultures of the tissue containing the venom at the bite site were taken from each patient for both aerobic and anaerobic organisms. In over 200 pit viper patients with envenomation, no organisms were grown on the culture medium from 1966 through 1979. Since the literature contains a number of articles describing the various pathogenic organisms grown from snakes mouths, I have routinely given antibiotics to all of those patients on whom I performed a surgical debridement. In adults I used either ampicillin or a tetracycline, and in children a penicillin. This was given intravenously and/or orally for 5 days after the bite.

Antivenin

Since 1966, 13% of the 160 Western diamondback rattlesnake bite victims were given antivenin. I have not used any "grading system" since 1966 because I consider them all to be useless in the proper care of the snakebite victim. I gave antivenin for the following reasons:

1. The parent of the child asked that antivenin be given.
2. The patient requested that antivenin be given.
3. Some other method of treatment was used (no antivenin in severe bites; too little antivenin when it was required in large doses; or incomplete or improper debridement of the bite site, with continued activity of the retained venom or failure to open involved muscle compartments which contained venom) by other physicians before I was called to see them because the treatment seemed to be failing.
4. The patient had obviously severe bites that appeared to be potentially lethal in nature (for example, a small child bitten by a 7 ft long Western diamondback rattlesnake).
5. The case was one in which there was some potentially dangerous legal situation that might develop as result of the treatment of the victim.
6. The patient had severe defibrination of the plasma in which bleeding was a problem.

Except in patients with severe defibrination of the plasma, I gave a starting dose of 50 ml of antivenin intravenously and another 50 ml if needed. The mixed antivenin was diluted with an equal amount of normal saline solution and given as fast as it would run in. When giant urticaria developed during the time of administration, the rate was slowed and 50 mg benadryl was given intravenously. It is important to perform a skin test prior to the administration of the antivenin, however, a positive skin test has not deterred me from giving the antivenin intravenously when the patients have needed it. It is my opinion that if the patient survives the skin test without an anaphylactic reaction, intravenous antivenin will not cause one. In all patients with a positive skin test prior to the administration of the antivenin, serum sickness subsequently developed.

In those patients that required treatment for the severe defibrination of the plasma, I gave from 10 to 40 ml antivenin. I feel that less than 10 ml of the antivenin will neutralize the effect of the enzymes in the venon which cause the defibrination. This small amount of antivenin will have less tendency to cause serum sickness (Figure 5).

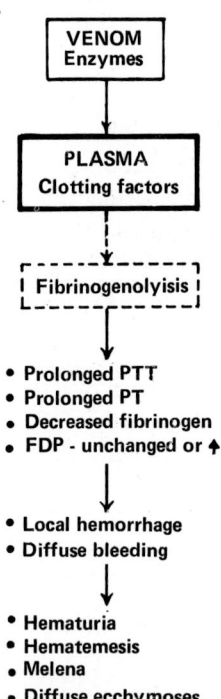

Figure 5 The effect of venom on the human blood coagulation system.

Fibrinogen

Prior to 3 years ago all patients that developed severe defibrination of the plasma were administered commercial fibrinogen in 2–8g doses. This amount seemed to be required to return the prothrombin activity to levels above 30%. The Federal Drug Administration removed this product from their approved list, and it is no longer available. Cryoprecipitate is available and can be used for replacement of the plasma fibrinogen; however, it is very expensive. In children it is useful because of its small fluid content. In children large volumes of fresh frozen plasma used to replace the fibrinogen deficits will cause hypervolemia.

Fresh Frozen Plasma

Presently, all patients except children are given fresh frozen plasma to restore the plasma fibrinogen to levels which will improve the prothrombin activity (Figures 6-11).

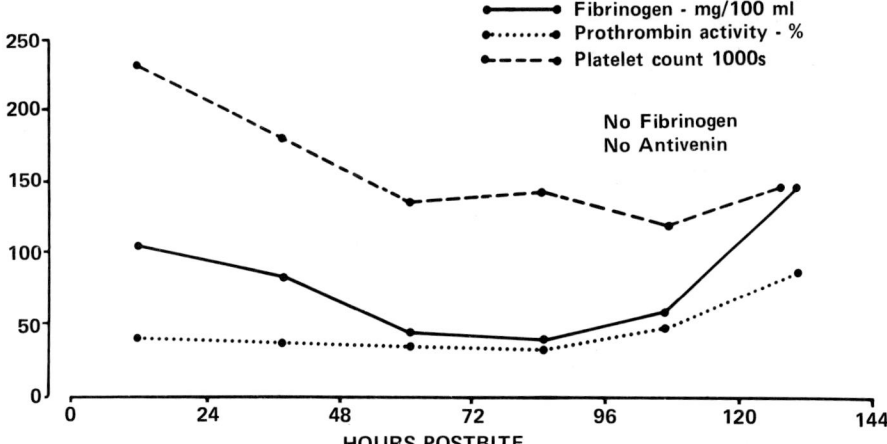

Figure 6   J. H., age 18, was bitten on the dorsum of the foot and ankle. The attending physician "excised" the bite and sutured the wound. On arrival in San Antonio 12 hr later, the leg was swollen to the knee and no pulses were palpable in the foot or ankle. A dermotomy and fasciotomy of the anterior compartment of the leg restored the circulation. Even though the prothrombin activity was below 50%, no bleeding occurred and the coagulation problem corrected itself after 5 days. The offending snake was 24 in. long.

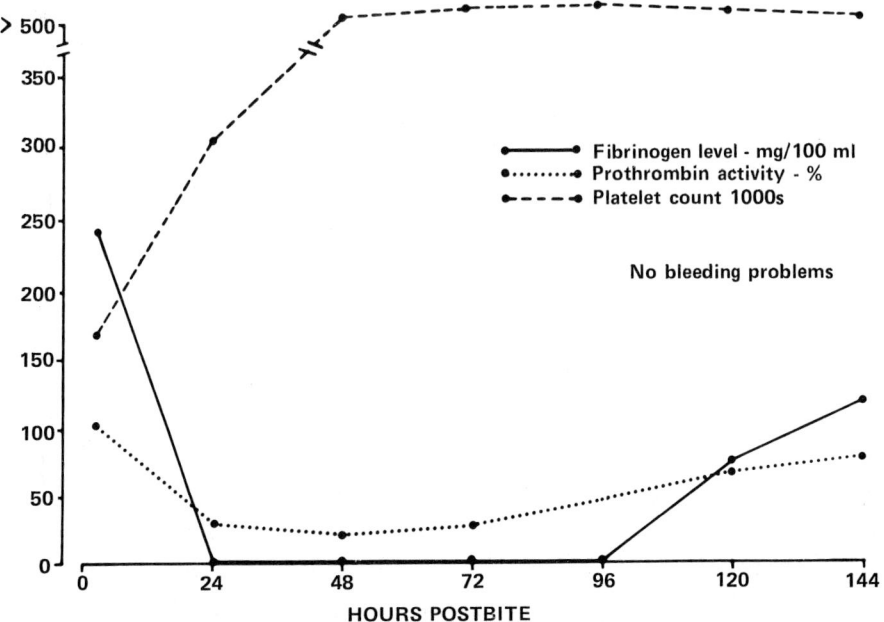

Figure 7 R. R., a 15-year-old male, was bitten on the anterior thigh by a large rattlesnake. Fasciotomy for a large intramuscular injection of venom was performed 2 hr after the bite. Defibrination was complete; however, his prothrombin activity remained above 20%. The platelet count rose to very high levels. No bleeding problems developed, and the defibrination was not treated.

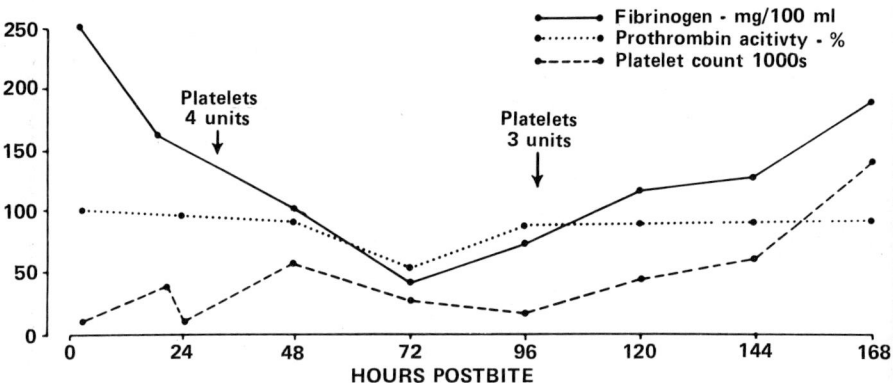

Figure 8 W. W., a 68-year-old female (white), was bitten on the hand by a large (5.5 ft long) rattlesnake. The plasma fibrinogen fell only to 50 mg% after 72 hr. The platelet count was severely depressed on admission to the hospital and was treated after 24 hr owing to bleeding from a dermotomy of the hand. Note the small change in the prothrombin activity during the course of treatment.

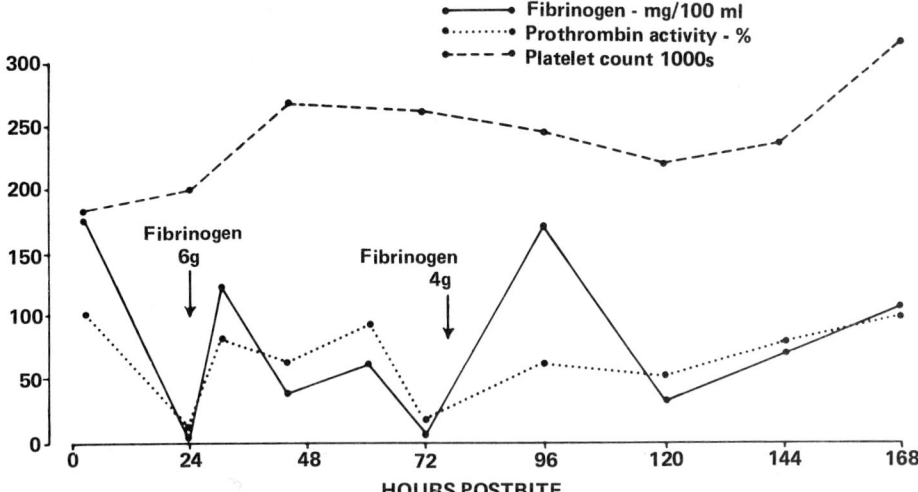

Figure 9 J. H., a 46-year-old male, was bitten on the hand by an 18 in. long rattlesnake. The local reaction was minimal except for swelling of the thumb. Defibrination occurred with severe depression of the prothrombin activity. Note the rise in platelet count. Note the response of the fibrinogen level and the prothrombin activity to treatment with commercial fibrinogen. No bleeding problems developed, but hepatitis developed 60 days after dismissal from the hospital.

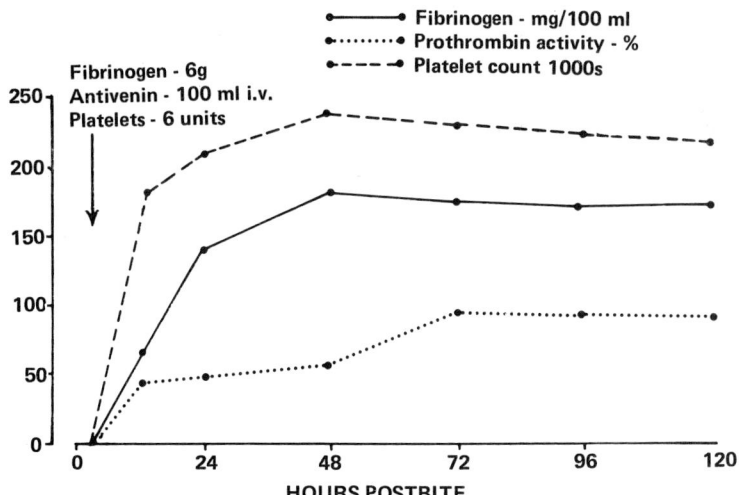

Figure 10 M. S., a 63-year-old female, was bitten on the right middle finger by an 18 in. long Western diamondback rattlesnake. Two and one-half hours later the platelet count was 2000 per cubic millimeter, PT = 0 activity, and the fibrinogen (plasma) was zero. Debridement was done after the coagulation problem was corrected. The incision was closed 7 days after the bite. There was no loss of function.

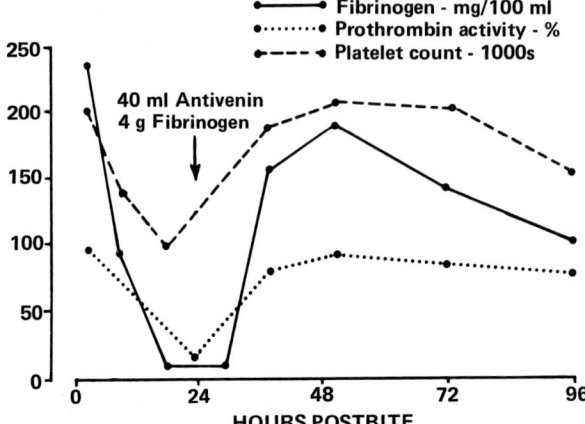

Figure 11 P. I., a 26-year-old-female, was bitten on the hand by a 20 in. long rattlesnake. The bite site was opened and affected muscle was removed at 2 hr after the bite. A seemingly better response in the prothrombin activity and fibrinogen levels occurred after the combined use of antivenin and fibrinogen. This patient developed severe unrelenting serum sickness.

Packed Red Blood Cells

Very seldom did patients develop anemia from the Western diamondback rattlesnake bite that required treatment. In those few patients that develop hemoglobin levels below 8 g per 100 ml, the transfusions are given with washed, packed red blood cells to bring the level of hemoglobin to at least 10 g%. No patients received whole blood transfusions.

Platelets

When bleeding was a problem and the platelet count was less than 50,000 platelets per cubic millimeter of blood, platelets were replaced with one platelet pack per 10,000 platelets below 50,000 per cubic millimeter. A platelet count of over 50,000 per cubic millimeter was usually necessary for proper coagulation of the blood. Quite often there will be platelet deficiency without abnormally low prothrombin activity. In these patients fresh frozen plasma is not needed.

Local Anesthetics

The only local anesthetic agent used in this series was 1% xylocaine in total volumes less than 20 ml. One elderly patient with a severe bite was given a 100 mg bolus dose of xylocaine for a cardiac arrhythmia and promptly had a grand mal convulsion. The xylocaine apparently caused enough added stimulation to the brain which was already irritable from the toxins in the venom to cause this convulsion. Otherwise no ill effects were noted in any patient.

General Anesthetics

Until 1966 it was felt by most "snakebite experts" that to administer a general anesthetic to an already toxic patient would result in death or some other major catastrophe. During the past 14 years, 124 of the total 160 patients I operated on for Western diamondback rattlesnake bite were administered a general anesthetic. Halothane, ketamine, nitrous oxide, sublimaze, pentothal, and fluothane were used without any misadventures. The first 10 or 12 patients that were administered general anesthetics when they were obviously toxic from the venom were watched very closely, and every vital sign was monitored with great care with every available type of equipment. Subsequently, only routine anesthetic care has been given, and there has been no great fear of anesthetizing these patients. Occasionally, an undetected hypovolemia was demonstrated when halothane was administered, but in those patients rapid fluid infusion always corrected the problem.

Laboratory Tests

A variety of tests has been done on the plasma and serum of the 160 victims of Western diamondback rattlesnake envenomation, but only a few of these have been of clinical benefit. From the outset I obtained the following tests at the time of admission to the emergency area, 8 hr after the bite and daily for 5 days:

1. Hematocrit and hemoglobin levels
2. Partial thromboplastin time
3. Prothrombin time
4. Platelet count
5. Fibrin split products
6. Plasma quantitative fibrinogen level
7. Plasma free hemoglobin
8. Urinary free hemoglobin

After about 10 years it was noted that the plasma free hemoglobin, urine free hemoglobin, and haptaglobin levels were not significantly changed in even the most severe rattlesnake bite victims to warrant their further analysis. The partial thromboplastin test has been obtained throughout the study, but I have found no clinical use for the study. The most readily available and reliable tests have been the prothrombin time (PT) and the platelet counts. The PT is done by different methods in some laboratories, and the results may on occasion be misleading; however, in spite of all of the laboratory tests, the real indication for treating any coagulation problem has been excessive bleeding at the site of surgery or elsewhere (gastrointestinal or urinary bleeding). There has been no evidence that there was a correlation between the severity of the changes in the coagulation studies and the severity of the envenomation.

Related Laboratory Studies

There have been no other laboratory studies done *routinely* on the victims of rattlesnake bite. Patients with hypertension, diabetes, leukemia, and other problems have been followed and treated as required by these diseases. The serum electrolyte studies have been followed in those patients that required large amounts of intravenous fluids, but no significant alterations have been seen. One patient developed a high output, salt-losing nephritis which required frequent serum sodium determinations. Arterial gas studies have not been useful in treating any of these patients.

Errors

The single most important error in the care of a victim of a severe Western diamondback rattlesnake bite has been the underestimation of the amount of envenomation. This has occurred because of the inability to properly assess the degree of envenomation by methods other than surgical exploration. The second most important error was when the physician assumed that muscle damage by the venom could be controlled by massive (300 ml or more) volumes of intravenous antivenin. In my experience the antivenin has failed to prevent loss of muscle tissue even when given in large amounts during the first hour after the bite occurred. When the circulation to a muscle compartment has been obstructed by hemorrhage and swelling in the compartment within the first hour after the bite, it is obvious that massive amounts of antivenin circulating in the vascular system will not enter the compartment. In these patients the circulation to the compartment must be restored by surgery to prevent loss of the entire compartment muscle. The third most important error was when the physician explored the wound surgically and did not thoroughly debride the involved muscle. This led to further destruction of the muscle. The fourth most important error was when the surgeon failed to explore the muscle, tendon, sheath, or joint deep to the bite site. The fifth most important error was when the physician gave a patient antivenin when none was needed. This sensitized the patient to the horse serum for no good reason. (Figures 12-15).

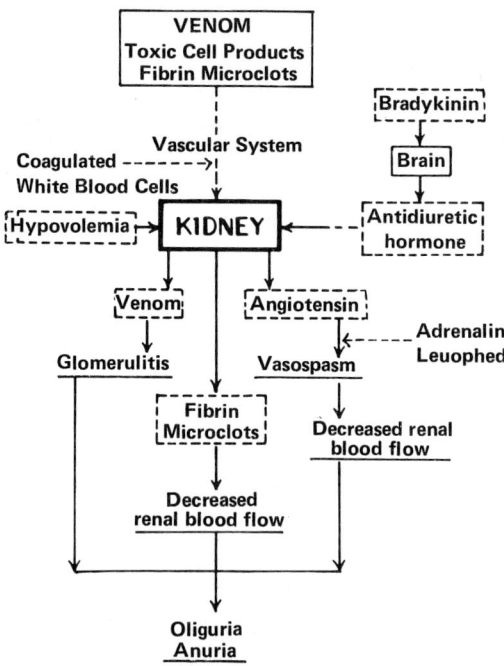

Figure 12 Venom and toxic cell products showing their effect on the kidneys.

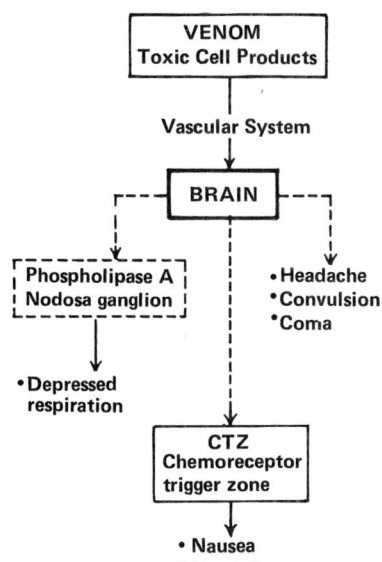

Figure 13 Venom and toxic cell products showing their effects on the central nervous system; specifically the brain.

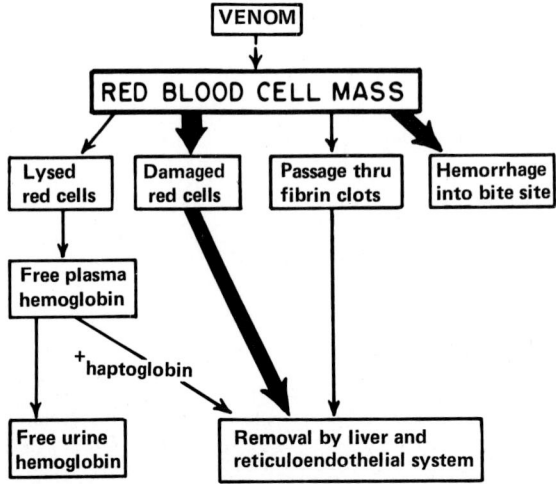

Figure 14 Venoms effect on the red blood cell mass.

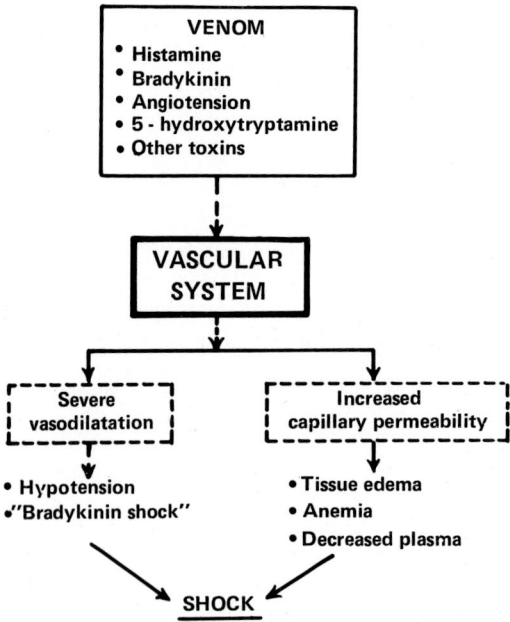

Figure 15 Increased capillary permeability and vasodilatation contribute to the early "shock." Later, the hypovolemia from the loss of fluids into the tissues is the major factor in the cause of the "shock."

Management of Western Diamondback Bite

STATISTICS

Cortisone: The same doses of cortisone were given to all age patients in 134 of the 160 rattlesnake bites.

Anesthetic: 124 of the 160 patients were administered general anesthesia.

Nausea and Vomiting: 55% of all victims presented with vomiting and/or nausea.

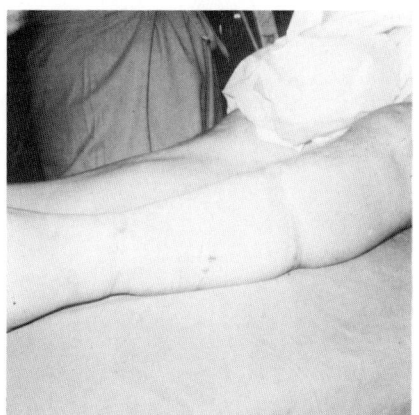
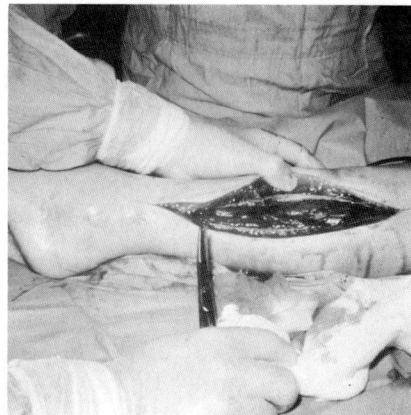
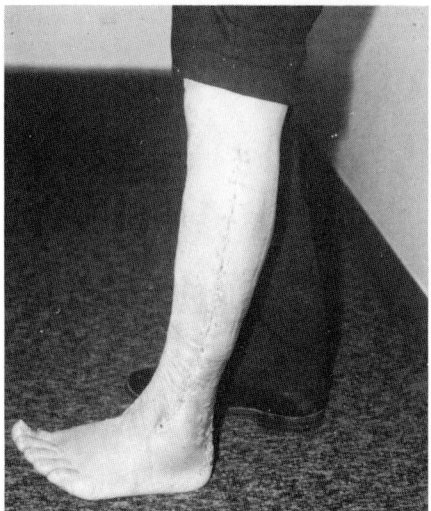

Figure 16 Early operative management of rattlesnake bite. A. Appearance 3 hr after the bite. B. At surgery. The swelling subsided and the wound was ready for closure. C. At 3 weeks after the bite there was full function with slight residual edema of the leg. No foot drop was present.

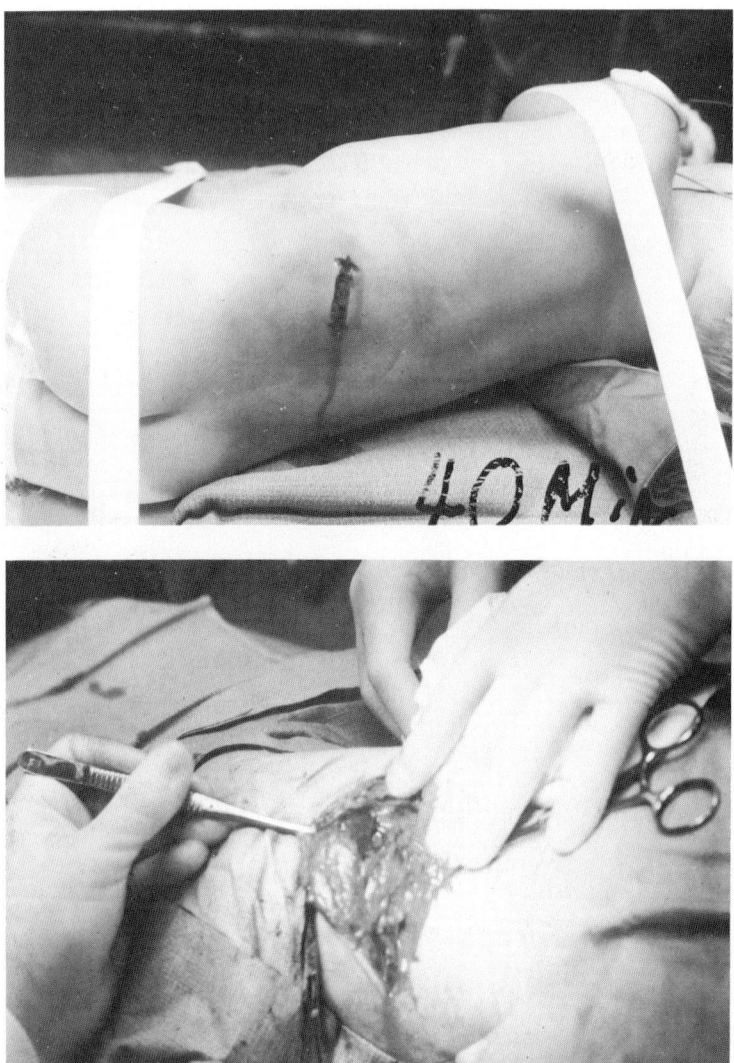

Figure 17 H. T., a 20-month-old male, was walking with his mother in southeast Bexar County around noon. A large 7 ft long rattlesnake struck H. T. in the left kidney area. The mother picked H. T. up and the snake bit him on the foot. Twenty minutes later H. T. was seen in a local emergency room, semicomatose. Fang marks measured 1 5/8 in. apart. A slash was made at the bite area. I saw H. T. 40 min after the bite. He was comatose, with no palpable pulse or respiration. There was 2000 mg Solu-Cortef given intravenously. The blood pressure

Management of Western Diamondback Bite

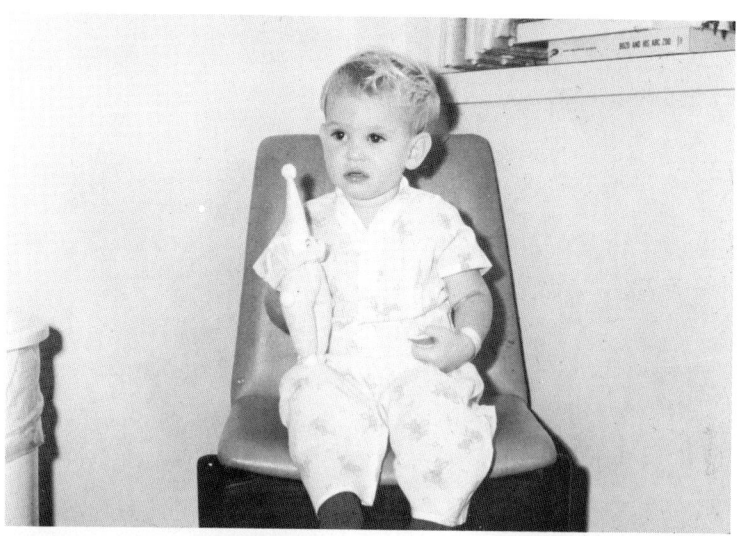

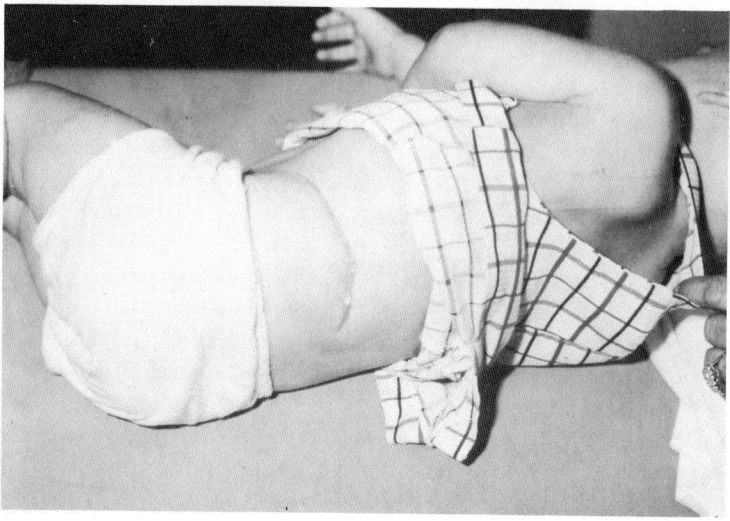

rose and H. T. responded. After a skin test was given, 100 ml antivenin was given rapidly, intravenously. Surgery was performed immediately. A. A slash was made through the bite area. B. H. T. in surgery; the wound widely excised to the kidney. C. H. T. responded well postoperatively and was treated for defibrination of the plasma and thrombocytopenia. He was up and about at 48 hr and was dismissed about 10 days after the bite. D. H. T. at 6 weeks after the bite.

Site of Bite:
  Finger, 35
  Hand, 19
  Arm, 3
  Body, 1
  Thigh, 3
  Knee, 1
  Leg (anterior), 19
  Leg (posterior), 29
  Ankle, 10
  Foot, 40

  Total, 160

See Figures 16 and 17.

SUMMARY

Envenomation by the Western diamondback rattlesnake causes a very complex combination of events to occur. The management of these events requires that a complex variety of decisions be made and treatments instituted. It is foolhardy to think that any one treatment method (such as the use of antivenin alone) will prevent loss of life, amputations, or crippling. The combined use of cortisone for the prevention or alleviation of the toxic symptoms, antivenin where indicated, and the surgical and mechanical removal of the necrotic tissue and venom has proved safe and effective and has prevented amputations and crippling when instituted within the first 6 hr after the bite. The administration of adequate amounts of cortisone has alleviated the toxic symptoms and prevented the side effects of these symptoms (for example, ruptured esophagus). The surgical exploration of the bite site and debridement have allowed for the immediate evaluation of the degree of envenomation by visual inspection, the removal of necrotic tissue and venom, the release of excessive intracompartment pressures with restoration of circulation to the muscle, and the irrigation of joints injected with venom which otherwise would have become ankylosed.

Appendix: Common Names of Rattlesnakes

Crotalus adamanteus Eastern diamondback rattlesnake
Crotalus atrox Western diamondback rattlesnake
Crotalus basiliscus basiliscus Mexican west coast rattlesnake
Crotalus basiliscus oaxacus Oaxacan rattlesnake
Crotalus catalinensis Santa Catalina Island rattlesnake
Crotalus cerastes cerastes Mojave Desert sidewinder
Crotalus cerastes cercobombus Sonoran Desert sidewinder
Crotalus cerastes laterorepens Colorado Desert sidewinder
Crotalus durissus durissus Central American rattlesnake
Crotalus durissus culminatus Northwestern neotropical rattlesnake
Crotalus durissus terrificus South American rattlesnake
Crotalus durissus totonacus Totonacan rattlesnake
Crotalus durissus tzabcan Yucatan neotropical rattlesnake
Crotalus enyo enyo lower California rattlesnake
Crotalus enyo cerralvensis Cerralvo Island rattlesnake
Crotalus enyo furvus Rosario rattlesnake
Crotalus exsul Cedros Island diamond rattlesnake
Crotalus horridus horridus timber rattlesnake
Crotalus horridus atricaudatus canebrake rattlesnake
Crotalus intermedius intermedius Totalcan small-headed rattlesnake
Crotalus intermedius gloydi Oaxacan small-headed rattlesnake
Crotalus intermedius omiltemanus Omilteman small-headed rattlesnake
Crotalus lannomi Autlan rattlesnake
Crotalus lepidus lepidus mottled rock rattlesnake
Crotalus lepidus klauberi banded rock rattlesnake
Crotalus lepidus maculosus Durango rock rattlesnake
Crotalus lepidus morulus Tamaulipan rock rattlesnake
Crotalus mitchellii mitchellii San Lucan speckled rattlesnake
Crotalus mitchellii angelensis Angel de la Guarda Island speckled rattlesnake
Crotalus mitchellii muertensis El Muerto Island speckled rattlesnake

Crotalus mitchellii pyrrhus Southwestern speckled rattlesnake
Crotalus mitchellii stephensi Panamint rattlesnake
Crotalus molossus molossus Northern blacktail rattlesnake
Crotalus molossus estebanensis San Esteban Island rattlesnake
Crotalus molossus nigrescens Mexican blacktail rattlesnake
Crotalus polystictus Mexican lance-headed rattlesnake
Crotalus pricei pricei Western twin-spotted rattlesnake
Crotalus pricei miquihuanus Eastern twin-spotted rattlesnake
Crotalus pusillus Tancitaran dusky rattlesnake
Crotalus ruber ruber red diamond rattlesnake
Crotalus ruber lorenzoensis San Lorenzo Island diamond rattlesnake
Crotalus ruber lucasensis San Lucan diamond rattlesnake
Crotalus scutulatus scutulatus Mojave rattlesnake
Crotalus scutulatus salvini Huamantlan rattlesnake
Crotalus stejnegeri long-tailed rattlesnake
Crotalus tigris tiger rattlesnake
Crotalus tortugensis Tortuga Island diamond rattlesnake
Crotalus transversus cross-banded mountain rattlesnake
Crotalus triseriatus triseriatus central plateau dusky rattlesnake
Crotalus triseriatus aquilus Queretaran dusky rattlesnake
Crotalus triseriatus armstrongi Armstrong's dusky rattlesnake
Crotalus unicolor Aruba Island rattlesnake
Crotalus vegrandis Uracoan rattlesnake
Crotalus viridis viridis prairie rattlesnake
Crotalus viridis abyssus Grand Canyon rattlesnake
Crotalus viridis caliginis Coronado Island rattlesnake
Crotalus viridis cerberus Arizona black rattlesnake
Crotalus viridis concolor midget faded rattlesnake
Crotalus viridis helleri southern Pacific rattlesnake
Crotalus viridis lutosus Great Basin rattlesnake
Crotalus viridis nuntius Hopi rattlesnake
Crotalus viridis oreganus northern Pacific rattlesnake
Crotalus willardi willardi Arizona ridgenose rattlesnake
Crotalus willardi amabilis Del Nido ridgenose rattlesnake
Crotalus willardi meridionalis Southern ridgenose rattlesnake
Crotalus willardi obscurus New Mexican ridgenose rattlesnake
Crotalus willardi silus West Chihuahua ridgenose rattlesnake
Sistrurus catenatus catenatus Eastern massasauga

Appendix

Sistrurus catenatus edwardsii   desert massasauga
Sistrurus catenatus tergeminus   Western massasauga
Sistrurus miliarius miliarius   Carolina pygmy rattlesnake
Sistrurus miliarius barbouri   Eastern pygmy rattlesnake
Sistrurus miliarius streckeri   Western pygmy rattlesnake
Sistrurus ravus ravus   Mexican pygmy rattlesnake
Sistrurus ravus brunneus   Oaxacan pygmy rattlesnake
Sistrurus ravus exiguus   Guerreran pygmy rattlesnake

Author Index

Italic numbers give page on which the complete reference is listed.

A

Abalos, J. W., 62, 83, 84, *113*
Abdelbaki, Y. Z., 200, *208*
Abe, T., 139, 145, *153*
Abel, J. H., 76, *112*, 192, 193, *203*
Abels, G. H., 199, *204*
Aikat, B. K., 334, *336*
Alagón, A. C., 85, *117*, 135, *160*, 251, 280, *308, 310*
Aleksiev, B. V., 144, *154*
Alema, S., 139, *153*
Allan, S., 193, *209*, 288, *311*, 328, *338*
Allen, R., 64, 65, 66, 74, 75, 77, 108, *111*
Almagro, D., 194, 195, *204*, 290, 291, *302*
Alper, C. A., 294, *299*
Alvarez, J. M., 282, *305*
Amaral, A. D., 63, 72, 74, 77, 81, 87, 91, 93, 95, 96, 100, 103, 108, *111*
Amorim, M. F., 199, 200, *203*
Andreasen, T. J., 145, *153*
Andrews, C. E., 193, *203*
Andrews, T. F., 293, *304*
Angeletti, P. U., 285, *312*

Angeletti, R. H., 285, *299*
Anton, A. H., 279, *299*
Apsalon, V. R., 145, *159*
Armstrong, B. L., 4, 7, 8, 9, 95, 100, 101, *112*
Arnold, R. E., 316, 318, 322, 324, 325, 332, 333, 334, *336*
Ashley, B. D., 85, *116*

B

Babilli, S., 299, *311*
Bailey, G. S., 180, *203*, 273, *300*
Bajwa, S. S., 289, 290, 291, *300, 308*
Baker, J. O., 265, *311*
Ballester, A., 268, *310*
Banèrjee, S., 194, *204*, 287, *302*
Bardawill, G. J., 74, *114*
Barrabin, H., 150, *154*
Barrio, A., 150, *154, 161*, 282, *309*
Barthel, W., 129, 131, *159*, 298, 299, *306*
Basu, A. S., 251, *300*
Baumgartner, R., 289, 290, *300, 308*
Bdolah, A., 59, 60, 63, *113, 117*
Beamer, P. D., 164, 179, 190, 196, 197, 199, *203*
Bechtel, H. B., 8, *112*

Beck, C., 287, *307*
Belluomini, H. E., 62, 69, 83, 84, *112*
Benkovic, S. J., 261, *301*
Bennet, E. L., 228, *243*
Bennett, J. E., 324, 325, *336*
Beraldo, W. T., 151, *160*, 296, *309*
Berman, J. M., 123, *161*
Bernick, J. J., 267, *300*
Bertke, E. M., 251, *300*
Bhargava, N., 129, 131, 132, 136, *154*, *161*, 191, *203*
Bieber, A. L., 99, *112*, 126, 127, 133, 146, *154*, *155*, 189, 202, *203*, *204*, 212, 231, 232, 233, 234, 235, 237, 240, *242*, *243*, *245*, 251, 257, *300*, *301*
Bier, O. G., 257, *304*
Billings, T. A., 75, 78, 85, 89, 99, 104, 105, 108, *116*, 236, 237, *245*
Birdsey, V., 294, *300*
Bishop, R. C., 194, 195, *204*, 290, 291, *302*
Bjarnason, J. B., 132, 136, *160*, 169, 171, 175, 176, 177, 178, 188, 190, *203*, *207*, 253, 263, 264, 276, 277, 278, 283, 292, 299, *300*, *307*
Blanchard, R. J., 191, *209*
Blinov, N. O., 144, *154*
Blocker, T. G., 324, 325, *336*
Böhm, G. M., 132, *154*
Bohr, V. C., 201, *208*
Bolaños, R., 62, 83, 84, 85, *112*
Bomar, H., 287, *307*
Bon, C., 138, 143, 144, 145, *154*, 225, 229, *242*
Bonilla, C. A., 64, 66, 70, 75, 76, 104, *112*, *113*, 126, 129, 148, *154*, *159*, 192, 193, 194, *203*, 279, 280, 289, *300*, *301*
Bonta, I. L., 129, 131, 132, 136, *154*, *161*, 191, *203*
Bourillet, F., 147, 148, *156*, 187, 188, *204*
Bowman, W. C., 123, 128, 151, *154*

Boys, F. E., 164, 179, 190, 196, 197, 199, *203*
Bradham, G. S., 328, 334, *336*
Brazil, O. V., 85, *112*, 122, 127, 130, 132, 133, 136, 137, 143, 144, 145, 147, 149, 151, 152, *154*, *155*, *160*, 187, 188, 198, 200, *203*, *204*, *206*, *207*, 223, 225, 229, *242*, *244*, 273, 282, *301*, *303*, *304*, *308*
Breithaupt, H., 127, 133, 136, 144, 145, *155*, *157*, 198, *204*, 215, 216, 218, 219, 220, 223, 224, 225, 226, 227, *242*, *243*, *244*, *245*, 257, *301*
Brelsford, H. G., 324, 325, *336*
Broad, A. J., 75, *112*
Broman, T., 150, *155*
Brown, J. H., 191, *204*
Bruce, S., 127, *159*, 296, *305*
Bryant, F. R., 261, *301*
Buess, F. W., 192, 196, 197, *208*, 270, *309*
Burchfield, P. M., 85, *117*, 135, *160*, 251, *308*
Burgers, P. M. J., 261, *301*

C

Calissano, P., 287, *305*
Call, A., 68, *115*
Callahan, G., 123, 128, 135, *161*
Callahan, W. P., III, 60, *113*
Cameron, D. L., 180, 181, 185, 187, 188, 190, *204*, *207*, 253, 271, 272, 273, 277, 278, 280, *301*, *307*
Campbell, J. A., 4, 6, 7, 8, 9, 54, 56, *112*
Carlson, R. W., 105, *118*, 122, 123, 124, 128, 129, 131, 132, 135, *155*, *161*, 190, 191, 192, 193, 196, *204*, *207*, *208*, 232, *246*, 291, *309*, 328, *336*
Carmichael, E. B., 164, 169, 190, 196, *205*
Carrillo, M., 265, *305*
Castilonia, R. R., *243*

Author Index

Catanese, J., 266, 283, *305*
Cate, R. L., 146, *155,* 189, *204,* 233, 234, 235, 237, 240, *243,* 257, *301*
Cavanagh, J. G., 150, *158*
Chan, T. K., 276, *305*
Chang, C. C., 136, 137, 142, 143, 144, 145, 147, 148, *155,* 188, *204,* 225, 229, *243,* 248, 273, *301*
Changeux, J.-P., 138, 143, 144, *154,* 225, 229, *242*
Chao, P. Y., 191, *206*
Chaves, F., 271, *303*
Chen, I., 137, *155*
Chen, S. S., 269, 270, *301*
Chen, Y. M., 144, 146, *159*
Cheng-Raude, D., 150, 151, *157*
Cheymol, J., 147, 148, *156,* 187, 188, *204*
Chignard, M., 130, *162, 311*
Chou, C. H., 235, 236, *246*
Christy, N. P., 334, *338*
Chua, A., *311*
Chugh, K. S., 334, *336*
Chute, A. L., 298, *301*
Cini, C., 270, *301*
Ciuchta, H. P., 75, 78, *119,* 192, 198, 201, *209*
Clark, R. W., 325, 332, *336*
Clark, W. C., 293, *302*
Clauss, L. M., 270, *312*
Claypool, W., 293, *309*
Clement, J. F., 190, 192, 193, 195, *204*
Cliff, F. S., 5, 9, 53, 55, 86, 92, *112*
Cohen, S., 285, *302*
Collins, J. T., 87, *117*
Colman, R. W., 293, *309*
Conant, R., 9, 51, 56, *112*
Condrea, E., 151, *156*
Connor, J. D., 150, *160*
Copley, A. L., 194, *204,* 287, *302*
Coulter, A. R., 75, *112*
Crawford, I. L., 150, *160*
Criley, B. R., 75, 79, 80, 85, 89, 97, *112*

Cull-Candy, S. G., 137, 142, *156*
Cunningham, E. R., 195, *208,* 332, *336*
Curti, B., 268, 269, *302*

D

Daly, J. G., 236, 237, *245*
Damerau, B., 127, 128, *156*
Damus, P. S., 132, *156,* 194, *204, 206,* 289, *306,* 324, 328, 334, *336*
Daniels, M. P., 140, *156*
Danzig, L. E., 199, *204*
Dao Hai, N., 127, *161*
DaRoza, D., 268, *305*
Das, K. C., 334, *336*
Dash, S. C., 334, *336*
Da Silva, M. H., 257, *304*
da Silva, O. A., 329, 334, 335, *337*
Davey, M. G., 193, *204*
David, M. M., 74, *114*
Davidson, E., 193, *209,* 288, *311,* 328, *338*
Davidson, T. M., 132, *156,* 190, 191, 194, *204, 205, 206*
De, S. S., 216, *243*
de Haas, G. H., 256, *311*
Deichmann, W. B., 74, 75, *112*
DeKok, A., 269, *302*
DeLucca, F. L., 58, 59, 83, *112, 113, 119*
DeMarco, C., 270, *301*
Dempster, D. W., 146, *156*
Denson, K. W. E., 194, 195, *204,* 287, 288, 290, 291, *302, 306*
de Oliveira, V. A., 122, 127, 130, 132, 133, *154*
Deutsch, H. F., 71, *114,* 126, *156,* 272, 296, *302, 303*
de Vos, C. J., 129, 131, *154,* 191, *203*
Dias Fontana, M., 147, *160*
Dickerson, G. D., 191, *204*
Didisheim, P., 194, *205,* 287, 291, *302*

Diniz, C. R., 126, *156*, 296, *302*
Diniz Filho, I., 282, *305*
Ditada, I. E., 62, 83, 84, *113*
Dixon, J. R., 4, 5, 54, *118*
Doerge, D. R., 145, *153*
Dolapchiev, L. B., 259, 260, 261, *302, 311*
Dörfel, H., 215, 216, *243*
Doublas, V., 293, *306*
Doucet, M. E., 62, 83, 84, *113*
Dowdall, M. J., 143, *156*
Dragstedt, C. A., 298, *302*
Duerre, J. A., 269, 270, *301*
Duncan, C. J., 189, *205*
Duran-Reynals, F., 268, *302*

E

Eagle, H, 194, *205*, 287, *302*
Eaker, D., 137, 143, 147, *155, 156, 159*, 225, 229, *243*
Eckstein, F., 261, *301*
Edwards, S. R., 87, *117*
Efrati, P., 164, *205*
Egmond, M. R., 256, *311*
Elliott, W. B., 135, *160*, 233, 234, 236, 239, *245*, 258, *303*
Elzinga, M., 180, *205*, 271, 276, 278, *303*
Emery, J. A., 75, 78, 81, 89, 92, 94, 97, 99, 105, *113, 117, 118*, 332, *337*
Engle, R. J., 191, *204*
Essex, H. E., 122, 123, 124, 127, 136, 151, *156*, 164, 196, 197, 199, *208*, 298, *308*
Eterovic, V. A., 228, *243*
Eventov, R., 75, 76, *118*, 270, *309*
Excell, B. J., 136, 137, 143, *154, 155*, 225, 229, *242*
Eyer, S. W., 298, *302*
Ezekiel, E., 293, *308*

F

Faber, V., 295, *306*
Fabiano, R. J., 188, *205*, 276, 277, 278, *302*
Fahlman, J., 225, 229, *243*
Faith, M. R., 70, *112*
Farber, J. L., 189, *208*
Fariña, R., 122, 127, 130, 132, 133, *154*
Farr, E., 252, 279, *310*
Farrell, J. J., 74, 75, *112*
Favilli, G., 268, *302*
Fein, A., 63, *113*
Feldberg, W., 298, 299, *302, 303*
Ferreira, S. H., 127, *156*, 298, *303*
Fidler, H. K., 164, 169, 190, 196, *205*
Fiero, M. K., 64, 66, 70, 75, 76, 104, *112, 113*, 192, *203*, 279, *300*
Filho, A. P., 273, *303*
Fisher, F. G., 215, 216, *243*
Fitts, C. T., 195, *208*, 332, *336*
Fletcher, J. E., 151, *156*, 258, *303*
Fletcher, T., 289, *300*
Flexner, S., 164, 166, *205*, 271, *303*
Fohlman, J., 137, 142, 143, *155, 156*
Fontana, M. D., 137, 143, 144, 145, *155*, 188, *207*, 273, *303*
Foote, R., 252, *303*
Foppoli, C., 270, *301*
Foscarini, L. G., 282, *305*
Fox, J. W., 180, *205*, 271, 276, 278, *303*
Fraenkel-Conrat, H., 138, 143, 144, *154, 158*, 201, *208*, 214, 215, 216, 219, 221, 222, 223, 225, 226, 227, 229, *242, 243, 245, 246*, 257, 280, *304*
Fraenkel-Conrat, J., 216, 222, *243*
Franceschi, J. P., 85, *112*
Frank, L. P., 75, 76, *112*, 279, *300*

Friederich, C., 75, 78, 85, 89, 104, 108, *113*, 164, *205*, 223, *243*, 253, 254, *303*
Fujiwara, M., 152, *160*

G

Gans, C., 58, 59, 60, *113*, *116*
Garcia, C. A., 123, *161*
Garcia, V. E., 78, *117*, 293, *307*
Garfin, S. R., 190, 191, *205*
Geiger, R., 267, *303*
Gennaro, J. F., 60, 75, 78, *113*, *114*, 279, *299*, 323, 324, 325, 329, 331, 332, 333, *337*
George, I. D., 67, 84, *113*
Geren, C. R., 89, *117*, 251, 252, 279, *306*, *310*
Gerencser, G. A., 149, *156*, 189, *205*
Gewurz, H., 294, *300*
Ghosh, B. N., 216, *243*
Giltner, J. B., 193, *209*
Gingrich, W. C., 75, 78, 79, 80, 85, 97, 104, 105, *113*
Ginn, F. L., 188, *205*
Gitlin, T. S., 70, 74, 78, 79, 80, 81, 84, 85, 89, 91, 92, 93, 94, 95, 96, 97, 98, 99, 100, 103, 104, 105, 106, 107, 108, *113*, *114*
Glasgow, R. D., 164, 169, 190, 196, *205*
Glass, T. G., 292, *303*, 332, 333, *337*
Glenn, J. L., 59, 62, 63, 65, 68, 71, 72, 74, 75, 77, 78, 85, 98, 99, 100, 103, 104, 105, *114*, 135, *157*, 231, 232, 238, *244*, *246*, 293, *310*, 324, *337*
Glissmeyer, H. R., 63, *114*
Gloyd, H. K., 3, 6, 7, 8, 9, 50, 51, 53, 54, 55, 56, 79, 83, *114*
Glusa, E., 129, 131, *159*, 298, 299, *306*
Godoy, P., 282, *309*, 329, 334, 335, *337*

Goetz, J. C., 85, *116*
Gonçalves, J. M., 71, *114*, 147, 148, *156*, 187, 188, *204*, *205*, 272, 273, 298, *303*, *304*
Goncalves, R. P., 58, 59, *119*
Gopalakrishnakone, P., 126, 127, 133, 135, 137, 140, 141, 142, 146, *156*, *157*, 188, 189, 197, 198, 202, *205*, 212, 225, 238, *244*, 248, *303*
Gordin, R., 324, *337*
Gornall, A. G., 74, *114*
Gotgilf, I. M., 145, *159*
Gralen, N., 214, *244*
Grishin, E. V., 144, *154*
Gulland, J. M., 258, *303*
Gundersen, C. B., 143, *157*
Gustavsson, D., 137, 142, *156*
Gutierrez, J. M., 271, *303*
Guzman, F., 191, *204*, *206*

H

Habermann, E., 133, 136, 144, 150, 151, *157*, 179, 201, *206*, 215, 216, 218, 219, 220, 223, 224, 225, 226, *243*, *244*, *245*, 257, 296, *303*
Hachimori, Y., 236, 239, 240, *245*
Haddad, A., 58, 59, 83, *113*, *119*
Hadler, W. A., 200, *206*, 223, *244*, 282, *304*
Hafner, E. W., 269, *304*
Hall, H. P., 75, *114*
Hall, K., 287, *307*
Halmagyi, D. F. J., 122, 123, 127, 128, 136, *157*, *159*, 192, 197, 198, *206*, 296, *305*
Hampe, O. G., 273, *304*
Hanahan, D. J., 236, 239, 240, *245*, *246*, 255, *312*
Hanashiro, M. A., 257, *304*
Hanley, M. R., 143, 144, *157*, 225, 228, 229, *243*, *244*, 258, *304*
Harris, H. S., Jr., 4, 5, 7, 9, 54, 55, 56,

[Harris, H. S., Jr.] 86, 92, *114, 115, 118*
Harris, J. B., 146, 152, *157, 158*
Hasiba, U., 193, 194, *206,* 328, *337*
Hawgood, B. J., 126, 127, 133, 135, 136, 137, 139, 142, 143, 144, 145, 146, *156, 157, 158,* 188, 197, 198, 202, *205,* 212, 225, 228, 229, 238, *244,* 248, *303*
Haws, W. C., 78, *117,* 293, *307*
Hayashi, K., 152, *160*
Hayes, M. B., 269, *304*
Heilbronn, E., 139, *158*
Heinrikson, R. L., 256, 257, *304*
Heluany, N. F., 137, 143, 144, 145, *155*
Hendon, R. A., 85, 99, *115,* 138, *158,* 215, 216, 217, 219, 220, 223, 224, 225, 226, 227, 229, 230, 231, 233, *244, 245,* 257, 280, *304*
Henrickson, R. L., 145, *158*
Henriques, S. B., 265, *305*
Heppel, L. A., 258, *304*
Herbert, M. S., 228, *243*
Hertel, R., 130, 149, *160*
Heuser, J. E., 137, 139, 142, *158, 161*
Heyrend, F. L., 68, *115*
Higashi, H., 152, *158*
Hilmoe, R. J., 258, *304*
Ho, C. L., 147, *159*
Högberg, B., 298, *304*
Hoffman, A., 129, 131, *159,* 298, 299, *306*
Hoge, R. A., 4, 9, 52, 71, 82, 83, *115*
Hohenadel, J. C., 75, 78, 79, 80, 85, 97, 104, 105, *113*
Holbrooke, S. E., 126, 127, 133, 137, 146, *157,* 188, 197, 198, 202, *205,* 212, 238, *244,* 248, *303*
Homma, M., 89, *118,* 164, 169, 180, 190, 193, *206, 209,* 271, *304, 311*
Hong, B-S., 89, *118,* 164, *209, 311*

Hopewell, W. S., 334, *338*
Hopkins, L. T., 318, 334, *337*
Hoppe, J., 64, 66, 89, *115*
Horner, G. J., 122, 123, 127, 128, 136, *157, 159,* 192, 197, 198, *206,* 296, *305*
Horst, J., 219, 227, *245*
Houssay, B. A., 69, 83, 84, *115*
Howell, D. E., 276, *305*
Hsiang, M., 215, 216, 219, 221, *243*
Huang, K.-S., 257, *304*
Huang, M.-C., 129, 148, *159*
Huang, T. T., 324, *337*
Hunneman, D. H., 261, *301*
Hunt, R. D., 189, 190, 191, *208*

I

Imaizumi, M. T., 59, *112*
Ishay, J., 258, *303*
Ishizaki, H., 247, *311*
Iwanaga, S., 255, 265, *307*
Iwanaga, W., 264, *310*

J

Jackson, E. M., 258, *303*
Jacobs, J. M., 150, *158*
James, G. P., *311*
Janszky, B., 290, *304*
Jenden, D. J., 143, *157*
Jeng, T. W., 138, 143, 144, *154, 158,* 215, 216, 219, 221, 223, 225, 226, 229, *242, 243, 245,* 257, 280, *304*
Jennings, B. M., III, 78, *117,* 293, *307*
Jimenez-Porras, J. M., 77, *115,* 194, *206,* 258, *304*
Johnson, B. D., 64, 66, 89, *115,* 251, *304*
Johnson, C. M., 62, *115*

Johnson, M. A., 146, *157, 158*
Jones, T. C., 189, 190, 191, *208*
Julia, J.Z., 280, *310*
Just, M., 179, 201, *206*

K

Kainer, R. A., 169, 171, 172, 173, 179, 180, 185, 190, 193, *207, 208,* 271, 276, 291, 299, *307, 310*
Kalas, J. P., 255, *312*
Kamenskaya, M., 139, 142, *158*
Kane, A. B., 189, *208*
Karlsson, E., 144, 146, *159*
Keegan, H. L., 293, *304*
Keim, P. S., 145, *158,* 256, 257, *304*
Kelen, E. M. A., 194, 195, *207,* 282, *309*
Kellaway, C. H., 298, *303*
Kelly, R. B., 137, 139, *158, 161*
Kennicott, T., 231, *245*
Keplinger, M. L., 74, 75, *112*
Kilmon, J. A., 293, *304*
Kim, S., 269, *307*
Kinamon, S., 60, *117*
Kirakossian, H., 290, *308*
Klauber, L. M., 3, 4, 5, 6, 7, 8, 9, 50, 51, 52, 53, 54, 55, 56, 61, 62, 63, 64, 65, 66, 67, 68, 73, 74, 75, 76, 77, 79, 81, 82, 84, 86, 87, 88, 91, 92, 93, 95, 96, 98, 100, 101, 103, 106, 107, 108, *115, 116*
Kocholaty, W. F., 75, 78, 85, 89, 99, 104, 105, 108, *116,* 236, 237, *245*
Kochva, E., 58, 59, 60, 63, 69, 83, *113, 116, 118*
Kochva, S., 193, *209,* 288, *311,* 328, *338*
Kortmann, H., 267, *303*
Koumanov, K. S., 261, *302*
Kowalski, D., 261, *308*
Krauss, A., 265, *307*

Kress, L. F., 262, 263, 264, 265, 266, 268, 283, 285, *305, 310*
Krueger, E. T., 145, *158,* 256, 257, *304*
Kunze, H., 299, *311*
Kurecki, T., 263, 264, 265, 266, *305*

L

LaGrange, R. G., 193, *206*
Lancini, A. R., 82, 85, 86, 102, *118,* 253, 255, 268, *309*
Lang, J., 257, *301*
Laravuso, R. B., 324, *337*
Larson, D. L., 324, *337*
Laseter, Y., 293, *308*
Laskowski, M., 261, 263, 264, 265, 268, 285, *305, 308, 310*
Lassignal, N. L., 142, *158*
Laure, C. J., 147, *155, 160,* 187, 188, 204, *207,* 272, 273, 278, *301, 303, 305*
Leander, S., 143, *156*
Ledford, E. B., 75, 78, 85, 89, 99, 104, 105, 108, *116,* 236, 237, *245*
Lee, C. K., 276, *305*
Lee, C. Y., 122, 129, 137, 138, 144, 146, 147, 148, 152, *155, 159, 160,* 192, *206,* 250, *305*
Lee, J. D., 136, 137, 142, 143, 145, *155,* 180, *203,* 225, 229, *243,* 248, 273, *300, 301*
Lee, S. Y., 192, *206*
Lefort, J., 130, *162, 311*
Lege, L., 127, 128, *156*
Leopold, R. S., 78, *113*
Levi, A., 194, *204,* 287, *302*
Levi-Montalcini, R., 285, 287, *305, 312*
Lewis, J. H., 193, 194, *205, 206,* 287, 291, *302,* 328, *337*
Lewis, S. R., 324, 325, *336, 337*

Li, C. H., 214, *245*
Lim, R. K. S., 191, *204, 206*
Lindberg-Broman, A. M., 150, *155*
Lipp, J., 201, *209*
Lobo de Araujo, A., 130, 149, *160*
Lockhart, W. E., 324, 325, *337*
Lomba, M. G., 133, *159*
Lombard, E. A., 122, 124, 132, *162*, 192, 196, *209*
Londorfer, J., 294, *300*
Long, T. E., 92, 94, *118*, 135, *161*
López, M., 282, *305, 309*, 329, 334, 335, *337*
Lorincz, A. F., 60, *113*
Luft, E., 299, *311*
Lüllman-Rauch, R., 137, 142, 143, 152, *156, 159*
Luscher, E. F., 193, *204*
Lynch, J. B., 324, *337*
Lyons, W. J., 193, *206*, 292, *305*, 328, *337*

M

McCranie, J. R., 4, 82, *116*
McCreary, T., 193, *206*
McCullough, N. C., 323, 324, 325, 331, 332, 333, *337*
MacDonald, W. E., 74, 75, *112*
MacDonell, C. A., 146, *157, 158*
MacFarlane, R. G., 287, 288, 291, *306*
MacFarlane, R. M., 150, *158*
McGiff, J. C., 127, *159*
Macht, D. I., 78, 81, 89, 92, 94, 95, 99, 104, 105, *116*
MacMahon, J. A., 252, *303*
McNamee, M. G., 145, *153*
Madinaveitia, J., 268, *305*
Maeda, N., 129, 148, *159*, 273, *305*
Magazanik, L. G., 139, 145, *159*
Magener, M., 259, *305*
Maier, E., 64, 65, 66, 74, 75, 77, 108, *111*

Malik, K. V., 127, *159*
Man, D. P., 295, *305, 306*
Man, M. P., 295, *306*
Mandelbaum, F. R., 265, *305*
Mannherz, H. G., 259, *305*
Manthei, J. H., 75, 78, *119*, 192, 198, *209*
Margolis, J., 127, *159*, 296, *305*
Markland, F. S., 132, *156*, 194, 195, *204, 206, 207*, 289, 290, 291, *300, 306, 308*
Markowitz, J., 122, 123, 124, 127, 136, 151, *156*
Markwardt, F., 129, 131, *159*, 298, 299, *306*
Marlas, G., 130, 144, *159, 162, 311*
Marra, U. D., 282, *305*
Marsh, N. A., 126, 127, 133, 137, 146, *157*, 188, 192, 197, 198, 202, *205, 206*, 212, 238, *244*, 248, *303*
Martiarena, J. L., 150, *154*
Martin, R., 202, *206*
Martori, R. A., 62, 83, 84, *113*
Maslin, T. P., 4, 6, 7, 50, 97, *117*
Mason, E., 256, *310*
Massey, V., 268, 269, *302*
Mathew, M. T., 334, *336*
Matsuo, Y., 265, *307*
Mays, C. E., 8, *117*
Mead, F. B., 298, *302*
Mebs, D., 126, *160*, 296, *306*
Meister, A., 268, 269, *306, 312*
Mello, R. F., 199, 200, *203*
Merriam, W. M., 78, *113*
Meyer, U., 299, *311*
Michaels, S., 122, 123, 131, *155*, 192, 196, *204*
Michl, H., 267, *306*
Miledi, R., 139, 145, *153*
Miller, D. G., 191, *206*
Miller, D. M., 251, 252, *312*
Minta, J. O., 295, *305, 306*
Minton, M. R., 81, 99, *116*

Author Index

Minton, S. A., 64, 65, 66, 69, 70, 71, 75, 78, 79, 81, 87, 88, 89, 90, 91, 95, 97, 99, 105, 107, 108, 109, *112, 116,* 135, *160,* 164, 195, *206*
Miroshnikov, A. I., 145, *159*
Mitchell, S. W., 61, *117*
Miura, A., 152, *160*
Molzer, H., 267, *306*
Moran, J. B., 89, *117,* 251, *306*
Mubarak, S. J., 190, 191, *205*
Muramatsu, I., 152, *160*
Murdock, D. S., 267, 268, *311*
Murphy, J. B., 4, 95, 100, 101, *112*

N

Nagasawa, S., 264, *310*
Nahas, L., 194, 195, *207,* 287, 288, 291, *306*
Nair, B. C., 135, *160,* 233, 234, 236, 239, *245*
Nair, C., 135, *160,* 233, 234, 236, 239, *245*
Nakazone, A. K., 257, *306*
Nalech, P., 224, *244*
Narvanen, S., 324, *337*
Nelson, A. W., 76, *112,* 192, 193, *203*
Neumann, N. P., 216, *245*
Newton, M. W., 143, *157*
Nichol, A. A., 293, *306*
Nickerson, M. A., 8, *117*
Nielsen, H., 295, *306*
Nikai, T., 265, *311*
Nitt, J. F., 334, *338*
Noguchi, H., 164, 166, *205,* 271, *303*
Nonaka, S., 247, *311*
Novak, J., 279, *301*
Nudel, F., 282, *309*

O

Oberg, S. G., 139, *158*

Ochsner, D. C., 251, 252, *312*
O'Connor, R., 251, *300*
Odell, G. V., 187, *207,* 276, *305, 307*
Ohashi, M., 179, *207*
Ohlsson, J. T., 270, *312*
Ohsaka, A., 179, 190, 201, *206, 207,* 265, *310*
Oldendorf, W. H., 149, *160*
Oldigs, H.-D., 127, 128, *156*
Omori-Satoh, T., 257, *301*
Oron, U., 59, 60, *117*
Osborne, A. H., 190, 191, *208,* 232, *246*
Oshima, G., 126, *160,* 194, *207,* 255, 265, 268, 287, 296, *307*
Ostrowski, W., 260, *302*
Ownby, C. L., 132, 136, *160,* 169, 171, 172, 173, 175, 176, 177, 178, 179, 180, 181, 185, 187, 188, 190, 193, *207,* 271, 276, 277, 278, 291, 292, 299, *307*

P

Page, D. S., 269, *307*
Paik, W. K., 269, *307*
Painter, R. H., 295, *306*
Paradies, H. H., 227, *245*
Pardridge, W. M., 150, *160*
Parker, R., 251, *300*
Paroski, E. A., 262, *305*
Parra, N. D., 71, 82, 83, 84, 85, 86, *117,* 252, 255, *309*
Passey, R. B., 255, *311*
Pattabhiraman, T. R., 99, 105, *117,* 129, 148, *159, 161,* 196, *208,* 233, 235, *243, 245,* 273, *305*
Pearce, R.M., 164, 166, 199, *207*
Peck, L., 293, *306*
Pellegrini-Filho, A., 147, *160,* 188, *207*
Pérez, J. C., 78, *117,* 293, *307*
Pérez-Polo, J., 287, *307*
Perry, J. F., 191, *209*

Pfleiderer, G., 263, 265, *307, 312*
Phillips, G., 334, *338*
Phillips, L. L., 334, *338*
Phillips, S. J., 192, *207*
Philpot, V. B., 293, *308*
Pichyangkul, S., 293, *307*
Pietrusko, R. G., 190, 192, 193, 195, *204*
Pirkle, H. C., 195, *207*, 289, 290, *300, 306, 308*
Pisani, G. R., 87, *117*
Polley, E. H., 201, *209*
Pool, W. R., 240, *245*
Possani, L. D., 84, *117*, 135, *160*, 251, 280, *308, 310*
Pough, F. H., 262, *310*
Prado-Franceschi, J., 130, 133, 147, 149, 151, 152, *155, 160, 162*, 187, *204*, 273, 282, *301, 308, 311*
Prescott, B., 235, 236, *246*
Prescott, J. M., 265, *311*
Primosigh, J., 216, *246*
Pritchard, A. E., 261, *308*
Procópio, N. M. M., 282, *305*
Purdon, A. D., 255, 256, *308, 310*
Puri, V. K., 123, 128, 135, *161*

R

Radcliffe, G. W., 4, 6, 7, 50, 97, *117*
Radomski, J. L., 74, 75, *112*
Rammel, O. J., 126, *154*, 279, 280, *300*
Rand, M. J., 123, 128, 151, *154*
Rapuano, B. E., 151, *156*
Raudonat, H. W., 253, *308*
Rawitch, A. B., 269, *302*
Raynald, A. C., 123, *161*
Reid, H. A., 64, 65, 66, 70, 72, 76, 78, 79, 88, *117*, 195, *207*, 287, *308*
Reif, L., 164, *205*

Remington, J. W., 122, 124, 132, *162*, 192, 196, *209*
Rhoten, W. B., 329, *337*
Roch-Areviller, M., 147, 148, *156*, 187, 188, *204*
Rocha e Silva, M., 151, *160*, 296, 298, *308, 309*
Rockwell, D., 193, 194, *206*, 328, *337*
Rodgers, D. W., 191, *204, 206*
Rodgers, R. W., 191, *206*
Rodriguez, O. G., 71, 82, 83, 84, 85, 86, 102, *117, 118*, 253, 255, 258, 259, 268, *309*
Rogers, R., 64, 66, 89, *115*
Romano, S. A., 9, 52, *115*
Ronchi, M., 150, *162*
Rosenbach, L. M., 193, 194, *206*, 328, *337*
Rosenberg, P., 151, *156*, 202, *206*, 258, *303*
Rosenfeld, G., 151, *160*, 194, 195, *207*, 282, 296, *309*
Rothschild, A. M., 59, 83, *113*, 126, 127, 128, 129, 131, *160*, 298, 299, *309*
Rothschild, Z., 126, 127, 128, 131, *160*, 299, *309*
Rovers, A. A., 123, *161*
Roy, D., *245*
Rübsamen, K., 215, 216, 218, 219, 220, 223, 224, 225, 226, *243, 244, 245*, 257, *303*
Ruiz, C. E., 131, 132, *161*, 193, 196, *207*, 291, *309*
Russell, F. E., 60, 75, 76, 78, 81, 89, 92, 94, 97, 99, 104, 105, *113, 117, 118*, 122, 123, 124, 128, 129, 131, 135, 148, *155, 159, 161*, 164, 190, 191, 192, 193, 194, 195, 196, 197, 201, *204, 206, 207, 208*, 231, 232, 233, 235, *243, 245, 246*, 270, 273, 290, 291, 296, *300, 302, 305, 309*, 328, 330, 331, 332, *337*

Author Index

Ruzic, N., 60, *118*
Ryan, M., 151, *156*

S

Sabback, M. S., 195, *208,* 332, *336*
Saenger, W., 261, *301*
Salzman, E. W., 324, 328, 334, *336*
Sanroman, A., 123, *161*
Santana de Sa, S., 126, 127, 133, 136, 137, 139, 143, 145, 146, *155, 157, 158,* 188, 197, 198, 202, *205,* 212, 225, 229, 238, *242, 244,* 248, *303*
Sathyanarayana, K., 261, *301*
Sato, T., 264, *310*
Sato-Ohmori, T., 126, *160,* 194, *207,* 268, 287, 296, *307*
Sawin, H. L., 4, *118*
Scannone, H. R., 71, 82, 83, 84, 85, 86, 102, *117, 118,* 253, 255, 258, 259, 268, *309*
Schaeffer, R. C., 105, *118,* 122, 123, 124, 128, 129, 131, 132, 135, *155, 161,* 192, 193, 196, *204, 207, 208,* 291, *309*
Schanne, F. A. X., 189, *208*
Schenberg, S., 272, *309*
Schmaier, A. H., 293, *309*
Schmidt, M. E., 200, *208*
Schoettler, W. H. A., 78, 79, 80, 84, 85, 89, 107, *118,* 332, *337*
Schöttler, W. A., 223, *246*
Seki, C., 150, *161,* 282, *309*
Shaham, N., 59, *118*
Shanley, J. D., 132, *156,* 194, *204,* 206
Sharma, B. K., 334, *336*
Shelburne, J. D., 188, *205*
Shih, T. Y., 191, *206*
Siefert, M. W., 64, 66, 70, 104, *113*

Silva, O. A., 282, *309*
Simmons, R. S., 4, 7, 9, 55, 56, 86, 92, *114, 115*
Simpson, J. W., 267, *300, 309, 310*
Singer, B., 216, *243*
Sjödin, T., 143, *156*
Slavnova, T. I., 139, 145, *159*
Sloan, A. J., 86, 92, *118*
Slotta, K. H., 201, *208,* 211, 213, 214, 216, 222, 224, *246,* 268, *310*
Smith, H. A., 189, 190, 191, *208*
Smith, H. M., 4, *118,* 164, 179, 190, 196, 197, 199, *203*
Smith, J., 225, 228, 229, *244*
Smith, J. W., 136, 137, 139, 142, 143, 144, 145, 146, *158*
Smith, R. B., 4, *118*
Smith, R. G., 293, *308*
Smith, R. M., 332, *336*
Snyder, C. C., 59, 62, 63, 65, 74, 77, 78, *114,* 238, *246,* 293, *310,* 324, *337*
Solie, T. N., 268, *310*
Solomen, K. A., 324, *337*
Sørensen, H., 295, *306*
Sosa, B. P., 85, *117,* 135, *160,* 251, 280, *308, 310*
Soule, M., 86, 92, *118*
Spiekerman, A. M., 265, *311*
Spiro, L., 255, *308*
Stadleman, R. E., 63, 73, 74, 76, *118*
Stahnke, H. L., 164, *208*
Stanhke, H. L., 64, 66, 89, *115*
Starzecki, B., 127, *159,* 192, 197, 198, *206,* 296, *305*
Starzecki, G., 122, 123, 128, 136, *157*
Stermitz, J., 247, *311*
Stjernholm, R. L., 293, *308*
Stocker, K., 290, *310*
Straight, R. C., 59, 62, 63, 65, 68, 71, 72, 74, 75, 77, 78, 85, 98, 99, 100, 103, 104, 105, *114,* 135, *157,* 231, 232, 238, *244, 246,* 293, *310,* 324, *337*

Strassberg, J., 192, 196, 197, *208*
Stringer, J. M., 180, 185, *208,* 271, *310*
Strong, P. N., 137, 139, *158, 161*
Su, M. J., 137, 144, *155,* 225, 229, *243*
Sulkowski, E., 285, *310*
Sullivan, J. A., 252, 279, *310*
Sumyk, G., 263, *307*
Sutherland, S. K., 75, *112*
Suzuki, K., 179, *207*
Suzuki, T., 126, *160,* 194, *207,* 264, 265, 268, 287, 296, *307, 310*
Svedberg, T., 214, *244*
Svehag, S. E., 295, *306*
Syed, M., 150, *162*

T

Takahashi, T., 265, *310*
Tamiya, N., 129, 148, *159,* 273, *305*
Tanner, W. W., 4, 5, 54, 90, *118*
Tarabini-Castellani, G., 268, *310*
Tarkkanen, L., 324, *337*
Taube, H. N., 164, 196, 197, 199, *208*
Tavares, D. G., 130, 149, *160*
Taylor, A. C., 267, *310*
Tchorbanov, B. P., 144, *154*
Terrill, J. B., 89, *118, 311*
Theakston, R. D. G., 64, 65, 66, 70, 72, 76, 78, 79, 88, *117, 118,* 135, *157,* 195, *207,* 287, *308*
Theodor, I., 195, *207,* 289, 290, *300, 308*
Thesleff, S., 137, 139, 142, 152, *156, 158, 159*
Thomas, G. J., Jr., 235, 236, *246*
Thomas, R. G., 262, *310*
Tijs, T., 129, 131, *154,* 191, *203*
Tinker, D. O., 255, 256, *308, 310*
Tomes, C. S., 61, *118*

Toom, P. M., 255, 267, 268, *310, 311*
Trevino, C. H., 4, *119*
Trevino, G. S., 164, 169, 179, *208*
Trump, B. F., 188, *205*
Tsai, M.-D., 259, *310*
Tseng, K. H., 147, 148, *155,* 188, *204,* 273, *301*
Tu, A. T., 75, 78, 85, 89, 99, 104, 108, *112, 113, 115, 118,* 122, 126, 127, 132, 133, 135, 136, 137, 146, 149, *154, 156, 157, 160, 161,* 164, 169, 171, 172, 173, 175, 176, 177, 178, 179, 180, 181, 185, 187, 188, 189, 190, 192, 193, 195, 197, 198, 200, 201, 202, *203, 204, 205, 206, 207, 208, 209,* 212, 222, 223, 229, 230, 231, 232, 233, 235, 236, 238, *242, 243, 244, 245, 246,* 247, 248, 250, 251, 253, 254, 255, 257, 260, 263, 264, 265, 267, 268, 270, 271, 272, 273, 276, 277, 278, 280, 282, 283, 291, 292, 294, 299, *300, 301, 302, 303, 304, 307, 310, 311*
Tu, T., 99, *112,* 126, 127, 133, *154,* 202, *203,* 212, 231, 233, 235, *242,* 251, *300*

U

Ubuka, T., 269, *311*
Üvnas, B., 298, *304*

V

Valeri, V., 58, 59, 83, *113, 119*
Vane, J. R., 127, *156*
Van Etten, R. L., 269, *307*
Van Mierop, L. H. S., 324, 328, *337*
Vargaftig, B. B., 127, 128, 129, 130, 131, 132, 136, *154, 161, 162,* 191, *203, 311*

Author Index

Vassileva, R. A., 261, *302, 311*
Verheij, H. M., 256, *311*
Vick, J. A., 75, 76, 78, 81, 85, 108, *118, 119,* 124, 128, 132, 135, 136, *162,* 190, 191, 192, 198, 201, *209,* 293, *312*
Vidal, J. C., 150, *154, 161,* 282, *309*
Villalaz, A. R. L., 102, *119*
Vogel, Z., 140, *156*
Vogt, W., 127, 128, *156, 162,* 299, *311*
von Wedel, R. J., 139, *158*
Voris, H. K., 293, *302*
Vozari-Hampe, M. M., 273, *304*

W

Wagner, F. W., 265, *311*
Wagner, G. M., 139, *158*
Wainschel, J., 190, 191, *208,* 232, *246*
Waisbich, E., 85, *112,* 133, 151, 152, *155*
Walgate, J. H., 269, 270, *301*
Walsh, P., 133, 144, *157,* 216, 218, 219, *243*
Walsmann, P., 129, *159*
Wang, S. K., 191, *206*
Ward, B. C., 276, *305*
Warshawsky, H., 58, 59, *119*
Wasi, S., 295, *306*
Wasylewska, E., 260, *302*
Wasylewski, Z., 260, *302*
Waters, E. T., 298, *301*
Watt, D. D., 251, *300*
Weaver, T. J., 64, 66, 70, 104, *113*
Weber, M., 260, *302*
Wei, J., 256, *310*
Weil, M. H., 105, *118,* 122, 123, 124, 128, 129, 131, 132, 135, *155, 161,* 192, 193, 196, *204, 207, 208,* 291, *309*
Weiss, H. J., 193, *209,* 288, *311,* 328, 334, *338*
Wellner, D., 268, 269, *304, 306, 312*

Wells, M. A., 236, 239, 240, *245, 246,* 255, *312*
Welsh, J. H., 279, *312*
Werner, R. M., 293, *312*
Whaler, B., 192, *206*
Whigham, H., 99, 105, *117, 118,* 122, 123, 124, 131, *155, 161,* 192, 196, *204, 208,* 235, *245*
Wilson, L. D., 4, 82, *116*
Witham, A. C., 122, 124, 132, *162,* 192, 196, *209*
Wohlman, A., 150, *162*
Wolff, N. O'C., 70, 75, 78, 79, 80, 81, 85, 89, 91, 92, 94, 95, 97, 99, 100, 104, 105, 106, 107, 108, *113, 114*
Woo, M. Y., 270, *309*
Woods, W. M., 187, *207,* 276, *307*
Wurzel, H., 193, *206*

Y

Yaeger, R. G., 293, *308*
Yang, C.-C., 151, *156,* 252, *312*
Yao, K., 269, *311*
Yoshida, L., 122, 127, 130, 132, 133, *154*
Young, E. E., 189, *208*
Young, M., 287, *312*
Young, R. A., 251, 252, *312*

Z

Zanini, A., 285, *312*
Zar, M. A., 152, *157*
Zarafonetis, C. J., 255, *312*
Zeigler, A., 152, *162*
Zeller, E. A., 270, *312*
Zemlicka, J., 261, *312*
Zertuche, J. J., 4, *119*
Zirinis, P., 131, *154*
Zmudka, M., 268, 269, *302*
Zwilling, V. R., 263, *312*
Zwisler, V. O., 251, *312*

Subject Index

A

Abdomen, 382
Acetylcholine, 212, 223, 229
Acetylcholine esterase, 251
Acidic subunit, 229, 231
Acidosis, 324, 334
Action potential, 145, 147
Acute tubular necrosis, 334
Adrenalin, 329, 331
Agkistrodon controtrix, 63
A. halys blomhoffii, 265
A. piscivorus, 191, 202
A. piscivorus leucostoma, 265
A. rhodostoma, 289
Alcohol, 325
Allergic reactions, 331
Alpha-bungarotoxins, 228
Alveolar necrosis, 197
Amine-liberating factor, 129, 131
Amino acid composition, 215, 216, 220, 234, 264
Amino acid sequences, 221, 257, 276, 290
Amputation, 318, 335
Anaphylactic reaction, 331
Angel de la Guarda Island speckled rattlesnake, 5, 26, 92, 362
Angiotensin, 355
Angiotensin converting enzyme inhibitor, 251, 297
Anoxia, 197
Antibiotics, 329, 334, 347, 355
Anticomplement action, 247, 293–295
Antivenin (antiserum), 131, 133, 135, 321, 322, 325, 330, 331, 332, 347
Antivenin allergy, 331
Antivenom activity, 247, 293
Ants, 214
Anuria, 330, 334
Aphids, 214
Arachnids, 214
Arginine esterase, 251, 267
Arginine ester hydrolase, 252, 267
Arizona black rattlesnake, 6, 40, 102, 362
Arizona ridgenose rattlesnake, 7, 43, 106, 363
Armstrong's dusky rattlesnake, 6, 37, 101, 362
Arterial blood pressure, 122–124
Arterial lesions, 169
Aruba Island rattlesnake, 6, 37, 101, 362
ATPase, 262
Autacoids, 126–131, 149, 152
Autlan rattlesnake, 5, 23, 90, 362
Autonomic nervous system, 150, 151, 152
Autopharmacologic action, 247, 295–299

B

Banded rock rattlesnake, 5, 24, 90, 362
Basic subunit, 229, 230
Beta-bungarotoxin, 137, 138, 139, 142, 143, 144, 146, 152, 228
Bifunctional cross-linking agent, 229
Blacktail rattlesnake, 94
Bleached rattlesnake, 92
Bleeding, 328, 330
Blistering, 327
Blood, 328, 330
Blood coagulation, 193, 195, 247, 252, 287–291
Blood loss, 330
Blood pressure, 329
Blood volume, 123
Bothrops jararaca, 265
Bradykinin, 126, 127, 128, 151, 190, 191, 295–296, 355
Bradykinin-potentiating factor, 127, 297
Bradykinin release, 299
Bradykinin-releasing enzyme, 251, 252
Bradykinin shock, 356
Brain, 136, 149–151, 328, 330, 336
Bronchoconstriction, 130
Bungarus multicinctus, 257

C

Calcium, 254
Canebrake rattlesnake, 5, 21, 361
Capillary, 166, 167, 170, 171, 172, 173, 175, 176, 177, 178, 179, 200, 201, 202
Cardiac arrest, 324, 329
Cardiac dysrhythmia, 124, 125, 126, 136
Cardiac output, 122, 192
Cardiovascular effect, 247, 279–280
Cardiovascular system, 122–132, 191, 192, 198
Carolina pygmy rattlesnake, 7, 47, 107, 363
Cascabel, 329
Cedros Island diamond rattlesnake, 5, 20, 87, 361
Central America rattlesnake, 5, 16, 81, 361
Central origin, 136
Central plateau dusky rattlesnake, 6, 36, 101, 362
Cerebral intraventricular injection, 150, 151
Cerebration, 329
Cerebrospinal fluid, 133
Cerralvo Island rattlesnake, 5, 19, 86, 361
"Chaperone" concept, 229, 230
Chemical modification, 216, 222
Chest, 328
Circular dichroism, 228
Circulatory failure, 334
Clot-stabilizing factors, 328
Clotting, 328
Coagulopathy, 330, 334
Cobra venom factor, 293
Collagenase, 252, 267
Colloidal fluid replacement, 123, 126
Colorado Desert sidewinder, 5, 15, 81, 361
Colubridae, 266
Coma, 326
Composition of rattlesnake venoms, 247
Contractile response, 145, 147, 148, 149
Contracture, 145, 147
Convulsions, 149–151, 355
Convulxin, 130, 132, 149
Copper, 254
Copperhead moccasin, 340
Coral snakes, 321
Coronado Island rattlesnake, 6, 39, 102, 362

Subject Index

Cortical necrosis, 334
Corticosteroids (cortisone), 332-334, 346
Cortisone, 332
Cross-banded mountain rattlesnake, 6, 35, 101, 362
Cross-linked crotoxin, 230
Crotactin, 218
Crotalidae, 4, 58, 59, 60, 231
Crotalus, 3, 4, 5, 6, 7, 8, 9, 10, 11
Crotalus adamanteus, 5, 12, 60, 61, 63, 64, 65, 66, 73-75, 76, 78, 79, 89, 96, 97, 109, 110, 124, 126, 127, 128, 129, 131, 132, 136, 148, 164, 165, 166, 169, 180, 191, 192, 194, 198, 199, 201, 202, 224, 226, 255, 256, 257, 258, 259, 260, 261, 263, 264, 265, 266, 267, 268, 269, 270, 279, 285, 286, 287, 288, 289, 291, 293, 298, 361
  distribution of, 50
  length of, 73
  photograph of, 12
  skull of, 60
  venom toxicity of, 76, 110
  venom yield of, 73, 74, 110
C. atrox, 4, 5, 12, 50, 60, 61, 62, 63, 64, 65, 66, 67, 69, 70, 72, 75-79, 89, 96, 97, 100, 109, 110, 114, 115, 116, 117, 118, 122, 123, 124, 126, 127, 128, 129, 131, 132, 135, 136, 149, 164, 166, 169, 174, 179, 180, 190, 192, 194, 195, 197, 198, 200, 201, 232, 253, 255, 256, 258, 263, 264, 265, 267, 268, 276, 277, 278, 281, 283, 285, 287, 291, 292, 293, 297, 298, 361
  distribution of, 50
  length of, 75
  photograph of, 12
  venom toxicity of, 78, 79, 110
  venom yield of, 75, 76, 77, 78, 110

C. basiliscus, 5, 13, 79-80, 83, 126, 135, 164, 212, 267, 291, 361
  distribution of, 51
  length of, 79
  photograph of, 12, 13
  venom toxicity of, 79, 80
  venom yield of, 79
C. basiliscus oaxacus, 5, 13, 51, 79, 80, 83, 361
C. catalinensis, 5, 7, 14, 55, 57, 80-81, 361
  distribution of, 55
  length of, 80
  photograph of, 14
  venom toxicity of, 80
  venom yield of, 80
C. cerastes, 5, 14, 55, 68, 80, 81, 82, 110, 115, 192, 361
  distribution of, 55
  length of, 80
  photograph of, 14, 15
  venom toxicity of, 81, 110
  venom yield of, 81, 110
C. cerastes cercobombus, 5, 15, 81, 82, 110, 115, 361
C. cerastes laterorepens, 5, 15, 81, 82, 110, 115, 361
C. durissus, 4, 57, 69, 81, 82, 83, 101, 102, 109, 164, 329
  distribution of, 51, 52, 82
  group, 83
  length of, 81, 82
  photograph of, 16, 17, 18
  venom toxicity of, 83, 85, 86, 110, 111
  venom yield of, 83, 84, 110
C. durissus atricaudatus, 267
C. durissus cascavella, 82, 130
C. durissus collilineatus, 82
C. durissus culminatus, 5, 16, 51, 57, 79, 81, 83, 86, 101, 102, 361
C. durissus cumanensis, 71, 82, 253, 255, 258

C. durissus dryinus, 82
C. durissus durissus, 5, 16, 51, 57, 63, 69, 79, 81, 83, 84, 85, 86, 109, 110, 114, 116, 180, 194, 195, 212, 213, 272, 278, 290, 291, 361
C. durissus marajoensis, 82
C. durissus ruruima, 82
C. durissus terrificus, 4, 5, 17, 52, 57, 59, 69, 70, 71, 79, 82, 83, 84, 85, 86, 89, 100, 101, 102, 109, 110, 113, 119, 122, 123, 126, 127, 128, 129, 130, 131, 132, 133, 134, 135, 148, 149, 151, 164, 180, 187, 188, 189, 195, 197, 198, 201, 202, 211, 212, 213, 214, 222, 223, 231, 248, 251, 253, 261, 267, 268, 269, 270, 272, 280, 282, 285, 287, 290, 291, 293, 297, 298, 334, 361
C. durissus terrificus var. *crotaminicus,* 147, 187
C. durissus terrificus var. *cumanensis,* 100
C. durissus totonacus, 5, 10, 17, 51, 57, 79, 82, 83, 84, 85, 86, 89, 101, 102, 109, 110, 117, 164, 169, 180, 361
C. durissus tzabcan, 5, 18, 51, 57, 79, 82, 83, 86, 89, 101, 102, 109, 361
C. enyo, 86
   distribution of, 55
   length of, 86
   photograph of, 18, 19
   venom toxicity of, 86
   venom yield of, 86
C. enyo cerralvensis, 5, 19, 55, 86, 115, 361
C. enyo enyo, 5, 18, 55, 86, 115, 361
C. enyo furvus, 5, 19, 55, 86, 115, 361

C. exsul, 5, 20, 50, 87, 361
   distribution of, 50
   length of, 87
   photograph of, 20
   venom toxicity of, 87
   venom yield of, 87
C. horridus, 83, 87, 88, 196, 201, 258, 285, 290, 291, 298
   distribution of, 51
   length of, 87
   photograph of, 20, 21
   status of, 87
   venom toxicity of, 88, 89, 110, 111
   venom yield of, 87, 88, 110
C. horridus atricaudatus, 5, 21, 51, 65, 66, 70, 71, 83, 87, 88, 89, 109, 110, 115, 134, 135, 148, 149, 164, 180, 192, 279, 361
C. horridus horridus, 5, 20, 51, 65, 66, 71, 83, 84, 88, 89, 110, 112, 116, 122, 123, 124, 126, 128, 129, 131, 132, 134, 135, 136, 148, 149, 151, 164, 180, 192, 193, 194, 267, 279, 289, 293, 361
C. intermedius, 57, 90, 108
   distribution of, 54
   length of, 90
   photograph of, 21, 22
   venom toxicity of, 90
   venom yield of, 90
C. intermedius glyodi, 5, 21, 22, 54, 57, 90, 107, 108, 362
C. intermedius intermedius, 5, 21, 54, 57, 90, 107, 108, 361
C. intermedius omiltemanus, 5, 21, 22, 57, 90, 107, 108, 362
C. lannomi, 5, 9, 23, 54, 90, 362
   distribution of, 54
   length of, 90
   photograph of, 23
   venom toxicity of, 90
   venom yield of, 90

Subject Index

C. lepidus, 57, 90, 91
  distribution of, 54
  length of, 90
  photograph of, 23, 24, 25
  venom toxicity of, 91, 111
  venom yield of, 90, 91
C. lepidus klauberi, 5, 24, 54, 57,
    61, 71, 90, 91, 109, 118, 362
C. lepidus lepidus, 5, 23, 54, 57, 90,
    91, 118, 362
C. lepidus maculosus, 4, 5, 23, 24,
    54, 57, 71, 90, 91, 118, 362
C. lepidus morulus, 5, 25, 54, 57, 90,
    91, 118, 362
C. mitchellii, 57, 92
  crown scales of, 11
  distribution of, 53
  length of, 92
  photograph of, 25, 26, 27
  venom toxicity of, 92, 93, 94, 110,
    111
  venom yield of, 92, 93, 110
C. mitchellii angelensis, 5, 26, 53,
    57, 92, 94, 100, 362
C. mitchellii mitchellii, 5, 25, 53, 57,
    92, 93, 94, 100, 109, 110, 115,
    362
C. mitchellii muertensis, 5, 26, 53,
    57, 92, 94, 100, 362
C. mitchellii pyrrhus, 6, 11, 27, 53,
    92, 93, 94, 100, 109, 110, 115,
    362
C. mitchellii stephensi, 6, 27, 53,
    92, 93, 94, 100, 109, 115, 362
C. molossus, 83, 94
  distribution of, 51
  length of, 94
  photograph of, 28, 29
  venom toxicity of, 95
  venom yield of, 95
C. molossus estebanensis, 6, 28, 51,
    57, 94, 95, 109, 362
C. molossus molossus, 6, 28, 51, 94,
    95, 109, 362

C. molossus nigrescens, 6, 29, 51, 94,
    95, 109, 362
C. polystictus, 6, 29, 51, 57, 61, 94,
    95, 109, 362
  distribution of, 55
  length of, 95
  photograph of, 29
  venom toxicity of, 95
  venom yield of, 95
C. pricei, 57, 61, 95, 108
  distribution of, 54
  length of, 95
  photograph of, 30
  venom toxicity of, 95
  venom yield of, 95
C. pricei miquihuanus, 6, 30, 54, 57,
    95, 107, 108, 362
C. pricei pricei, 6, 30, 54, 57, 95,
    107, 108, 362
C. pusillus, 6, 31, 55, 57, 95, 362
  distribution of, 55
  length of, 95
  photograph of, 31
  venom toxicity of, 95
  venom yield of, 95
C. ruber, 87, 109
  distribution of, 50
  length of, 96
  photograph of, 31, 32
  venom toxicity of, 97
  venom yield of, 67, 96, 97
C. ruber lorenzoensis, 4, 6, 7, 32, 50,
    96, 97, 109, 117, 362
C. ruber lucasensis, 6, 32, 50, 62, 63,
    65, 96, 97, 110, 362
C. ruber ruber, 6, 31, 61, 65, 66, 67,
    89, 96, 97, 109, 110, 117, 193,
    292, 362
C. scutulatus, 164, 188, 197, 198,
    202, 212, 222, 231, 232, 248,
    251, 276, 282
  crown scales of, 11
  distribution of, 55
  length of, 97

[*Crotalus scutulatus*]
  photograph of, 33
  venom toxicity of, 72, 98, 99, 100,
    110, 111
  venom yield of, 98, 110
C. scutulatus salvini, 6, 33, 55, 57,
  97, 98, 99, 100, 109, 112, 135,
  362
C. scutulatus scutulatus, 6, 11, 33,
  55, 68, 70, 71, 72, 97, 98, 99,
  100, 109, 110, 112, 114, 115,
  117, 124, 125, 128, 133, 134,
  148, 149, 194, 195, 327, 362
C. stejnegeri, 6, 34, 55, 57, 100, 362
  distribution of, 55
  length of, 100
  photograph of, 34
  venom toxicity of, 100
  venom yield of, 100
C. terrificus basiliscus, 290
C. terrificus terrificus, 193, 200, 213
C. tigris, 6, 34, 53, 57, 100, 107,
  108, 115, 362
  distribution of, 53
  length of, 100
  photograph of, 34
  venom toxicity of, 100
  venom yield of, 100
C. tortugensis, 6, 35, 50, 57, 100,
  362
  distribution of, 50
  length of, 100
  photograph of, 35
  venom toxicity of, 101
  venom yield of, 101
C. transversus, 6, 35, 54, 57, 101,
  362
  distribution of, 54
  length of, 101
  photograph of, 35
  venom toxicity of, 101
  venom yield of, 101
C. triseriatus, 57, 95, 101, 108
  distribution of, 54

[*Crotalus triseriatus*]
  length of, 101
  photograph of, 36, 37
  venom toxicity of, 101
  venom yield of, 101
C. triseriatus aquilus, 6, 36, 54, 57,
  95, 101, 107, 108, 362
C. triseriatus armstrongi, 4, 6, 37,
  54, 57, 95, 101, 107, 108, 362
C. triseriatus triseriatus, 6, 36, 54,
  57, 95, 101, 107, 108, 362
C. unicolor, 6, 37, 52, 83, 101, 107,
  362
  distribution of, 52
  length of, 101
  photograph of, 37
  venom toxicity of, 101
  venom yield of, 101
C. vegrandis, 6, 38, 52, 83, 86, 100,
  102, 109, 118, 119, 212, 253,
  255, 362
  distribution of, 52
  length of, 102
  photograph of, 38
  venom toxicity of, 102
  venom yield of, 102
C. viridis, 102
  distribution of, 53
  length of, 102
  photograph of, 9, 38, 39, 40, 41,
    42
  venom toxicity of, 104, 105
  venom yield of, 102, 103, 105
C. viridis abyssus, 6, 39, 53, 102,
  103, 104, 105, 362
C. viridis caliginis, 6, 39, 53, 102,
  103, 104, 362
C. viridis cerberus, 6, 40, 53, 102,
  103, 104, 105, 362
C. viridis concolor, 6, 40, 53, 68,
  72, 102, 103, 104, 105, 108,
  109, 110, 114, 212, 293, 362
C. viridis helleri, 6, 41, 53, 102, 103,
  104, 105, 109, 110, 116, 117,

Subject Index

[*Crotalus viridis helleri*]
    118, 122, 123, 126, 128, 129, 131, 132, 148, 192, 193, 194, 195, 196, 212, 273, 290, 291, 298, 362
C. viridis lutosus, 6, 9, 41, 102, 103, 105, 110, 362
C. viridis nuntius, 6, 42, 53, 102, 103, 105, 362
C. viridis oreganus, 6, 42, 53, 102, 103, 105, 109, 110, 363
C. viridis organus helleri, 103
C. viridis viridis, 6, 38, 53, 64, 65, 66, 67, 70, 102, 103, 104, 109, 110, 113, 164, 180, 185, 186, 187, 188, 198, 253, 273, 277, 278, 282, 285, 297, 362
C. willardi, 57, 106
  distribution of, 55
  length of, 106
  photograph of, 43, 44, 45
  venom toxicity of, 106
  venom yield of, 106
C. willardi amabilis, 7, 43, 55, 57, 106, 114, 363
C. willardi meridionalis, 7, 44, 55, 57, 106, 114, 363
C. willardi obscurus, 4, 7, 44, 55, 57, 106, 114, 363
C. willardi silus, 7, 45, 55, 57, 106, 114, 363
C. willardi willardi, 7, 43, 55, 57, 106, 114, 363
Crotamine, 128, 129, 147, 148, 149, 150, 151, 187, 188, 272, 273
Crotamine-like polypeptides, 129, 148, 149
Crotapotin, 218, 223
Crotoxin, 127, 129, 131–133, 135–146, 150, 152, 188, 189, 197, 198, 200, 201, 202, 211, 212, 214, 216, 219, 222, 223, 224, 225, 226, 227, 228, 229, 231, 252, 257

Crotoxin A, 218, 219, 221, 223, 224, 226, 227, 230, 231
Crotoxin B, 218, 219, 221, 223, 224, 226, 227, 230, 231
Crotoxin complex, 217
Crotoxin subunits, 215, 223, 225
Cryotherapy, 323, 324, 325
Crystallization, 214
Curare, 222
Cyanosis, 327
Cytotoxicity, 281
Cytotoxins, 193, 252

D

Debridement, 333, 335
Defecation, 151, 152
Defibrination, 334
Definitive medical treatment, 326–330
Del Nido ridgenose rattlesnake, 7, 43, 106, 363
Demerol, 330
Depolarizing action, 223
Desert massasauga, 7, 46, 106, 363
Detoxification, 222
Development, growth, and aging, 63–68
Diagnosis of rattlesnake bite, 341
Dialysis, 334
Diffuse introvascular clotting (DIC), 328, 330
Dimethyl suberimidate, 229
Direct hemolytic factor (DHF), 280
Direct lytic factor (DLF), 280
Disseminated intravascular coagulation, 334
Dissociable crotoxin complex, 229
Distribution, 4, 8, 9, 50–56, 57, 71, 72
Diuretic, 334
Doppler flow meter, 329
Drooping eyelids, 326, 329

Durango rock rattlesnake, 5, 24, 90, 362
Duration of treatment, 335
Dusky rattlesnake, 101

E

Eastern diamondback rattlesnake, 5, 12, 73–75, 165, 288, 361
Eastern massasauga, 7, 45, 106, 363
Eastern pygmy rattlesnake, 7, 47, 107, 363
Eastern twin-spotted rattlesnake, 6, 30, 95, 363
Eccymosis, 327
Edema, 163, 164, 166, 177, 189, 191, 192, 197, 200, 202, 317, 327, 331
Elapidae, 251, 256, 268, 321
Elastase, 252, 267
Electrocardiogram, 124, 126
Electroencephalogram, 136
Electron micrograph, 171, 172, 175, 176, 177, 178, 181, 182, 183, 184, 185, 272, 273, 274, 275, 292
Electroplaques, 229
El Muerto Island speckled rattlesnake, 5, 26, 92, 362
Emboli, 122, 149
Emergency room, 342
Emergency treatment, 325
Endopeptidase, 252, 263
Endothelium, 166, 167, 168, 169, 170, 171, 172, 173, 175, 176, 178, 179, 194, 199, 200
Envenomated limb, 329
Environmental and seasonal effects on venom yield, 68–69
Enzymes, absence in venoms, 270
Erythrocyte ghosts (RBC), 226
Erythrocyte membranes (RBC), 226
Erythrocytes, 229

Evaluation of bite, 345
Excision, 324
Exoenzyme, 226
Exonuclease, 259, 260
Exopeptidase, 252, 266

F

Factors affecting medical treatment, 323–326
Factors affecting venom lethal toxicity, 69–72
Factors affecting venom yield, 61–69
Factor X activating enzyme, 291
Fainting, 326
Fangs, 58, 59, 60, 61, 73, 81, 95, 100, 115, 249
Fasciotomy, 322, 335
Fibrin, 194, 195, 201, 328
Fibrinogen, 328, 330, 334
Fibrinogenolytic action, 290
Fibrinolytic action, 290
First aid, 319, 326
Flaccid paralysis, 133
Fluorescence, 228
Fluorodinitrobenzene, 216, 217
Fright, 327
Furosemide, 334

G

Gangrene, 335
General anesthetics, 353
Geographic and individual variation in lethal toxicity, 71, 72
Glomerulonephritis, 324, 355
Glycoprotein, 219
Grand Canyon rattlesnake, 6, 39, 102, 362
Great Basin rattlesnake, 6, 41, 102, 362
Guerreran pygmy rattlesnake, 7, 49, 108, 363

Gyroxin, 149, 150

H

Head (crown) scales, 8, 10, 11
Heart failure, 328
Heart/myocardium, 192, 193, 196
Hemodyalisis, 334
Hemolysis, 181, 195, 199, 224, 225, 226, 247, 280, 284, 334
Hemolytic activity, direct, 225
  indirect, 225
Hemorrhage, 164, 166, 167, 168, 169, 181, 195, 197, 202, 282, 328, 330, 336
Hemorrhage, local, 163, 164, 166, 169, 171, 174, 175, 177, 179, 181, 185, 189, 190, 192, 202, 276–279, 328
Hemorrhage per diapedesis, 166, 167
Hemorrhage per rhexis, 164, 168, 169, 175, 179
Hemorrhage, systemic, 195, 196, 197, 199, 201
Hemorrhagic glomerulonephritis, 334
Hemorrhagic toxin *a* (HTa), 171, 175, 176, 177, 179, 263
Hemorrhagic toxin *b* (HTb), 171, 177, 178, 179, 188, 189
Hemorrhagic toxin *e* (HTe), 171, 175, 177, 179
Hemorrhagic toxins, 132, 166, 167, 170, 171, 172, 173, 174, 175, 176, 177, 178, 179, 200, 201, 202, 251, 252, 264, 278, 284, 285
Hemostasis, 131–132
Heparin, 334
Histamine, 128, 129, 130, 131, 149, 151, 152, 190, 191, 283, 355
Histamine release, 298–299
Hopi rattlesnake, 6, 42, 102, 362

Huamantlan rattlesnake, 6, 33, 97, 362
Hyaluronidase, 252
Hydrolytic enzymes, 252
Hydrophiidae, 251, 268
5-Hydroxytryptamine, 356
Hyperkalemia, 324, 334
Hypocortisone sodium succinate (Solu-Cortef), 346
Hypofibrinogenemia, 193, 328
Hypolemic shock, 327, 328
Hypoproteinemia, 192
Hypotension, 122, 123, 124, 125, 126, 127, 128, 129, 130, 192, 202
Hypoventilation, 123, 133, 136
Hypovolemia, 355
Hysteria, 326

I

Ice, 325
Immune, 333
Immunological studies, 211
Immunology, 248
Incision and suction, 324
Incisions, 324, 343, 344
Indirect hemolytic factor, 224, 229, 284
Indomethacin, 127
Inflammation/inflammatory reaction, 189, 190, 200
Inflammatory edema, 191
Inhibitors of prostaglandin synthetase, 125, 127, 128, 130
Inorganic constituents, 253
Interstitial nephritis, 324
Intraperitoneal LD_{50}, 135
Intraperitoneally, 228
Intravascular coagulation, 324
Intravenous LD_{50}, 147, 149
Iron, 254
Ischemia, 192, 324

Isoelectric focusing, 219
Isoelectric point, 214, 219, 221
Isoelectric precipitation, 214

J

Jaws, 249, 250
Jellyfish, 214

K

Kidney, 196, 199, 328, 330, 334–335
Kininogenases, 126, 151, 264

L

Laboratory tests, 353
L-amino acid oxidase, 252, 268–270
LD_{50}, 58, 70, 71, 72, 75, 76, 78, 79, 80, 82, 85, 88, 89, 91, 94, 95, 97, 99, 100, 102, 104, 105, 107, 108, 109, 110, 111, 180, 285, 332
Lethal activity, 226
Lethal effects (death), 198, 199, 203
Lethal toxicity, 57–111
Lethal toxins, 279
Light micrograph, 173, 174
Liver, 335
Local anesthetics, 352–353
Local effects, 163, 164–191
Local tissue damage, 164, 190, 247, 271, 317–323
Long-tailed rattlesnake, 6, 34, 100, 362
Lower California rattlesnake, 5, 18, 86, 361
Lungs, 196, 197, 335
Lymphatics, 322
Lysis, 328

Lysophosphatidylcholine, 224, 226
Lysophospholipase, 258

M

Magnesium, 254
Malayan pit viper, 288
Management of rattlesnake bite, 342
Manganese, 254
Massasauga, 106
Mast cells, 128, 131
Medical treatment, 330
Metabolic acidosis, 123
Metals, 253
Methyl maleic anhydride, 218
Methylprednisolone sodium succinate, 332
Mexican blacktail rattlesnake, 6, 29, 94, 362
Mexican lance-headed rattlesnake, 6, 29, 95, 362
Mexican pygmy rattlesnake, 7, 48, 108, 363
Mexican west coast rattlesnake, 5, 13, 361
Midget faded rattlesnake, 6, 40, 102 362
Miniature endplate potentials, 137, 138, 139, 143
Minimum lethal dose (MLD), 73, 85, 88, 94, 99
Mitochondria, 137, 138, 139, 140, 141, 146
Mojave Desert sidewinder, 5, 14, 81, 361
Mojave rattlesnake, 6, 33, 97, 222, 231, 248, 328, 362
Mojave toxin, 126, 127, 133, 137, 146, 188, 189, 198, 202, 211, 212, 252
Molecular weight, 214, 217, 219, 263
Morphine, 330

Subject Index

Morphologic changes, 164
Motor coordination, 149-151
Motor nerve terminals, 136-146
Motor paralysis, 133-136
Mottled rock rattlesnake, 5, 23, 90, 362
Muscle excitability, 145
Muscle/muscle cells, 173, 175, 177, 178, 179, 180, 181, 182, 183, 184, 185, 186, 187
Mydriasis, 328
Myocardial depression, 324
Myocardial failure, 192
Myocardial hemorrhages, 192
Myofibrils, 180, 187
Myonecrosis, 146, 163, 164, 166, 179, 180, 185, 186, 187, 188, 189, 202, 271, 282
Myotoxicity, 145, 146
Myotoxin *a*, 129, 180, 183, 184, 185, 186, 187, 188, 189, 271, 273, 274, 275, 276, 278, 280, 285
Myotoxins, 251, 252, 284

N

NAD nucleosidase, 252
Naja naja atra, 281
Nausea, 326, 355
Necrosis, 169, 177, 179, 180, 188, 189, 197, 199
Neotropical rattlesnake, 4, 81, 222, 248
Nephrotoxins, 201, 252
Nerve Growth Factor (NGF), 247, 252, 285-287
Nervous system, 197, 201, 202
Neuroelectrophysiology, 214
Neuromuscular junction, 197, 202
Neuromuscular transmission blockade, 133, 134, 136-145, 152, 212, 222, 223

Neurotoxicity, 211, 228, 229, 230, 328, 334
Neurotoxins, 211, 212, 213, 224, 226, 227, 251, 252, 321, 328
New Mexican ridgenose rattlesnake, 7, 44, 106, 363
Nondepolarizing type, 223
Nonmetals, 255
Nonneurotoxic phospholipase A_2, 255
Northern blacktail rattlesnake, 6, 28, 94, 362
Northern Pacific rattlesnake, 6, 42, 102, 363
Northwestern neotropical rattlesnake, 5, 16, 81, 361
5'-Nucleotidase, 252, 258, 259
Numbness, 326, 328, 329
Nystagmus, 149, 150

O

Oaxacan pygmy rattlesnake, 7, 49, 108, 363
Oaxacan rattlesnake, 5, 13, 361
Oaxacan small-headed rattlesnake, 5, 22, 90, 363
Oliguria, 334
Omilteman small-headed rattlesnake, 5, 22, 90, 363
Ontogenic effects on venom lethal toxicity, 70, 71
Operating room, 342
Operative and postoperative care, 342
Osteomyelitis, 324
Overall toxic action, 282

P

Packed red blood cells, 352
Pain, 163, 189, 191, 327, 330-332

Panamint rattlesnake, 6, 27, 92, 362
Paralysis, 321, 326, 328, 329, 330
Paresthesias, 328, 329
Pathologic changes, 163, 164, 169, 187, 197
Pelamis platurus, 281, 291
Per diapedesis, 164, 167
Perforated ulcer,
Peripheral origin, 133–135
Peritonitis, 336
Per rhexis, 164, 168
Pharmacokinetic, 144
Phosphatidylcholine, 224, 226
Phosphodiesterase, 252, 259, 260, 261, 262
Phospholipase A_2, 127, 128, 130, 135, 136, 137, 138, 139, 143, 144, 145, 146, 151, 152, 188, 189, 193, 197, 211, 212, 218, 223, 225, 226, 229, 251, 252, 255, 256, 257, 258, 280, 284, 298, 299, 355
Phospholipid, 226
Phospholipolytic, 224, 226
Phosphomonoesterase, 252, 262, 270
Pinheiros, 331
Plasma, 328
Plasma volume, 123
Platelet aggregation, 169, 173, 175, 176, 177, 247, 291–293, 299
Platelet count, 330, 334, 349–352
Platelets, 191, 193, 195, 328, 350, 352
Polyvalent antivenin, 321
Postsynaptic neurotoxins, 143, 144, 145, 212, 225, 228, 251
Potassium, 254
Potentiation of bradykinin action, 296–298
Potentiation of the twitch amplitude, 147, 148
Prairie rattlesnake, 6, 38, 102, 362
Presynaptic, 136–144, 152, 225, 231, 251

Presynaptic neurotoxic activity, 212, 229
Presynaptic neurotoxins, 137, 138, 139, 142, 143, 146, 152, 154, 229, 251
Presynaptic vesicles, 212
Proliferative glomerulonephritis, 334
Prostaglandins, 127, 128, 130
Proteinase inhibitor, 266
Proteolytic enzymes, 251, 252, 262, 263, 264, 265, 267, 283, 287, 327
Prothrombin activator, 288
Prothrombin activity, 349–352
Prothrombin time (PTT), 328, 330
Pufferfish, 214
Pulmonary arterial pressure, 122
Pulmonary hypertension, 122
Pulse, 329
Pulseless, 327
Pygmy rattlesnake, 107

Q

Queretaran dusky rattlesnake, 6, 36, 101, 362

R

Rational treatment, 329
Rattle, 7, 8
Rattlesnake distribution, 1, 4, 8, 9, 50, 51, 53, 54, 55, 56, 57, 71, 72
Rattlesnake photographs, 12–49
Receptor, 223, 231
Recombination ratio, 227
Reconstituted complex, 226
Red cell mass, 123
Red diamond rattlesnake, 6, 31, 96, 362
Reflex stimulation, 132, 133
Regeneration, 146

Subject Index

Regional blocks, 330
Release of neurotransmitter, 136–139, 142, 143, 148, 152
Release of neurotransmitter, evoked, 136, 137, 138, 139, 142, 143, 148, 152
   spontaneous, 137, 138, 139, 142, 143, 148
Renal biopsy, 334
Renal circulation, 123
Renal cortical necrosis, 330
Renal damage, 330, 336
Renal failure, 199, 330, 334
Renal involvement, 330
Respirator, 329, 334
Respiratory system, 132–136, 196, 202
Resting membrane potential, 145, 146, 147
Retroperitoneal shock, 328
Ridgenose rattlesnake, 106
Ringer's lactate, 329
Rock rattlesnake, 90
Rosario rattlesnake, 5, 19, 86, 361
Ruptured spleen, 336

S

Salivation, 151, 152
San Esteban Island rattlesnake, 6, 28, 94, 362
San Lorenzo Island diamond rattlesnake, 6, 32, 96, 362
San Lucan diamond rattlesnake, 6, 32, 96, 362
San Lucan speckled rattlesnake, 5, 25, 92, 362
Santa Catalina Island rattlesnake, 5, 14, 80–81, 361
Sarcolemma, 179, 180, 185, 187, 188, 202
Sarcoplasmic reticulum (SR), 146, [Sarcoplasmic reticulum (SR)] 179, 180, 181, 182, 183, 184, 186, 187, 189, 192
Scorpions, 214
Sepsis, 328
Serotonin, 129, 130, 131, 132, 149, 151, 190, 191, 299
Severity of bite, 327
Sexual dimorphism, 68
Shock, 122, 123, 124, 126, 128, 129, 131, 132, 133, 202, 324, 326, 328, 329, 330, 334, 336, 356
Sidewinder, 81
Sistrurus, 3, 4, 5, 6, 7, 8, 9, 10, 11
Sistrurus catenatus, 106, 126, 128, 131, 135
   crown scales of, 10
   distribution of, 56
   length of, 106
   photograph of, 45, 46
   venom toxicity of, 107, 110, 111
   venom yield of, 106, 107
S. catenatus catenatus, 7, 45, 56, 106, 107, 108, 109, 110, 363
S. catenatus edwardsii, 7, 46, 56, 106, 107, 363
S. catenatus tergeminus, 7, 10, 46, 56, 106, 107, 108, 363
S. miliarius, 107, 108, 126, 135
   distribution of, 56
   length of, 107
   photograph of, 47, 48
   venom toxicity of, 108, 109, 110
   venom yield of, 106, 107
S. miliarius barbouri, 7, 47, 56, 107, 108, 109, 110, 164, 195, 267, 279, 291, 363
S. miliarius miliarius, 7, 47, 56, 107, 108, 363
S. miliarius streckeri, 7, 48, 56, 107, 108, 363
S. ravus, 57, 108

[*Sistrurus ravus*]
  distribution of, 56
  length of, 108, 109
  photograph of, 48, 49
  venom toxicity of, 109
  venom yield of, 109
S. ravus brunneus, 4, 7, 8, 48, 56, 57, 108, 109, 112, 363
S. ravus exiguus, 4, 7, 48, 56, 57, 108, 109, 112, 363
S. ravus ravus, 7, 8, 48, 56, 57, 108, 109, 363
Skeletal muscle, 145–149
Skeletal muscle cells, 185, 187, 189
Skull, 60
Small-headed rattlesnake, 90
Sodium, 254
Sodium permeability, 147
Sodium transport, 149
Solu-Cortef (Hypocortisone sodium succinate), 346
Sonoran Desert sidewinder, 5, 15, 81, 361
South American rattlesnake, 5, 17, 82, 361
Southern Pacific rattlesnake, 6, 41, 102, 362
Southern ridgenose rattlesnake, 7, 44, 106, 363
Southwestern speckled rattlesnake, 6, 27, 92, 362
Spastic paralysis, 147, 151
Species specificity, 222
Splanchnic visceral volume, 122
Splints, 326
Spontaneous contractile activity, 147, 148, 149
Spreading factor, 268
Subunit, 212, 217, 218, 221, 227, 230
Subunit interaction, 211, 228
Subunit neurotoxin, 212, 231, 248
Suction, 324
Swelling, 282

Synaptic transmission blocking activities, 228, 229
Synaptic vesicles, 137, 138, 139, 140, 141, 142
Synergistic interaction, 227
Systemic arterial resistance, 122, 124, 126
Systemic effects, 163, 191, 335
Systemic hemorrhage, 195

T

Tachyphylaxis, 125, 127, 150
Taipoxin, 137, 138, 139, 142, 143, 144, 146, 152
Tamaulipan rock rattlesnake, 5, 25, 90, 362
Tancitaran dusky rattlesnake, 6, 31, 95, 362
Tetanus, 335
Tetanus prophylaxia, 329, 335
Thrombin, 194, 328
Thrombin-like enzymes, 194, 195, 287–291
Thrombin-like procoagulants, 328
Thrombocytopenia, 130, 131, 132, 193, 328, 330
Tiger rattlesnake, 6, 34, 100, 362
Timber rattlesnake, 5, 20, 361
Tissue damaging toxins, 278
Tortuga Island diamond rattlesnake, 6, 35, 100, 362
Totalcan small-headed rattlesnake, 5, 21, 90, 361
Totonacan rattlesnake, 5, 17, 82, 361
Tourniquets, 323, 326, 329
Toxic action, lethal, 247, 270
Toxic action, overall, 247
Toxins with myotoxic and hemorrhagic activities, 277
Trauma, 328
Trimeresurus flavoviridis, 179, 265

Subject Index

Tropical rattlesnake, 329
Twin-spotted rattlesnake, 95

U

Unconsciousness, 326
Uracoan rattlesnake, 6, 38, 102, 362
Uremia, 334
Urinalysis, 330
Urinary system, 199, 202
Urination, 151, 152
Urine volume, 330

V

Vacuolation, 187
Vascular permeability, 123, 128, 129, 131, 132
Vasculature of skeletal muscle, 122
Vasodilation, 356
Vasospasm, 355
Venom apparatus, 58-61
Venom collection, 62
Venom extraction frequency, 62-63
Venom glands, 58-60, 62, 63, 65
Venom yields, 57-111
Venom yields in neonates and juveniles, 63-68
Ventilated, 326
Ventilatory assistance, 335
Vesicles, 229
Vipera palaestinae, 59, 63, 69
Vipera russellii, 291
Viperidae, 7, 58, 60, 256, 266, 268

Viriditoxin, 188, 189, 277, 278, 285
Visceral smooth muscle, 151, 152
Volvatoxin, 227
Vomiting, 151, 152, 326, 355

W

Wasps, 214
Weakness, 326, 329
West Chihuahua ridgenose rattlesnake, 7, 45, 106, 363
Western cottonmouth water moccasin, 340
Western diamondback rattlesnake, 5, 12, 75-79, 340, 341, 360, 361
Western massasauga, 46, 73, 106, 363
Western pygmy rattlesnake, 7, 48, 107, 363
Western rattlesnake, 102
Western twin-spotted rattlesnake, 6, 30, 95, 363
Wyeth Laboratories, 331
Wyeth's polyvalent antivenin, 330

Y

Yield data, 64
Yucatan neotropical rattlesnake, 5, 18, 82, 361

Z

Zinc, 253, 254